Benchmark Papers
in Animal Behavior

Series Editor: Martin W. Schein
West Virginia University

PUBLISHED VOLUMES

HORMONES AND SEXUAL BEHAVIOR
  Carol Sue Carter
TERRITORY
  Allen W. Stokes
SOCIAL HIERARCHY AND DOMINANCE
  Martin W. Schein
EXTERNAL CONSTRUCTION BY ANIMALS
  Nicholas E. Collias and Elsie C. Collias

VOLUMES IN PREPARATION

PSYCHOPHYSIOLOGY
  Stephen W. Porges and M. G. H. Coles
IMPRINTING
  E. H. Hess
SOUND RECEPTION IN FISH
  William N. Tavolga
PLAY
  Dietland Müller-Schwarze
PARENTAL BEHAVIOR IN BIRDS
  Rae Silver
VERTEBRATE SOCIAL ORGANIZATION
  Edwin M. Banks

Benchmark Papers
in Animal Behavior / 4

A BENCHMARK® Books Series

EXTERNAL
CONSTRUCTION
BY ANIMALS

Edited by

NICHOLAS E. COLLIAS
University of California, Los Angeles
and **ELSIE C. COLLIAS**
Los Angeles County Museum of
Natural History

Dowden, Hutchinson & Ross, Inc.

STROUDSBURG, PENNSYLVANIA

Distributed by

HALSTED
PRESS

A division of
John Wiley & Sons, Inc.

Copyright © 1976 by **Dowden, Hutchinson & Ross, Inc.**
Benchmark Papers in Animal Behavior, Volume 4
Library of Congress Catalog Card Number: 75-34185
ISBN: 0-470-16543-X

78 77 76 1 2 3 4 5
Manufactured in the United States of America.

LIBRARY OF CONGRESS CATALOGING IN PUBLICATION DATA

Main entry under title:
External construction by animals
   (Benchmark papers in animal behavior/4)

   Includes bibliographical references and indexes.
   1. Animals, Habitations of.   I. Collias, Nicholas Elias, 1914–   II. Collias,
Elsie C.
QL756.E9   591.5'6   75-34185
ISBN 0-470-16543-X

Exclusive Distributor: **Halsted Press**
A Division of John Wiley & Sons, Inc.

ACKNOWLEDGMENTS AND PERMISSIONS

ACKNOWLEDGMENTS

THE AMERICAN ORNITHOLOGISTS UNION—*The Auk*
    The Development of Nest-Building Behavior in a Weaverbird
    An Experimental Study of the Mechanisms of Nest Building in a Weaverbird

PERMISSIONS

The following papers have been reprinted with the permission of the authors and copyright holders.

AMERICAN ASSOCIATION FOR THE ADVANCEMENT OF SCIENCE—*Science*
    Avian Incubation

THE AMERICAN MUSEUM OF NATURAL HISTORY—*Natural History*
    Tents and Tactics of Caterpillars

AMERICAN SOCIETY OF ZOOLOGISTS—*American Zoologist*
    Apicotermes Nests
    Evolution of Caddisworm Cases and Nets
    The Evolution of Nests and Nest-Building in Birds
    The Evolution of Spider Webs
    Evolution of the Nests of Bees
    Maturing and Coordination of Web-Building Activity
    Synthesis of Silk, Mechanism, and Location

BALLIERE AND TINDALL—*Animal Behavior Monographs*
    The Behaviour of Free-Living Chimpanzees in the Gombe Stream Reserve

DR. FRANK A. BEACH—*Sex and Behavior*
    Interaction Between Internal and External Environments in the Regulation of the Reproductive Cycle of the Ring Dove

COLUMBIA UNIVERSITY PRESS—*Ants: Their Structure, Development and Behavior*
    Ant-Nests

COMMONWEALTH SCIENTIFIC AND INDUSTRIAL RESEARCH ORGANIZATION, AUSTRALIA—*Commonwealth Scientific and Industrial Research Organization Wildlife Research*
    Temperature Regulation in the Nesting Mounds of the Mallee-Fowl, *Leipoa ocellata* Gould

DODD, MEAD & COMPANY and HUGHES MASSIE LTD.—*The Mason-Bees*
    Some Reflections upon Insect Psychology

Acknowledgments and Permissions

DUKE UNIVERSITY PRESS FOR THE ECOLOGICAL SOCIETY OF AMERICA—*Ecological Monographs*
Termite Nests—A Study of the Phylogeny of Behavior

THE IBIS JOURNAL OF THE BRITISH ORNITHOLOGISTS UNION—*Ibis*
The Thermal Significance of the Nest of the Sociable Weaver *Philetairus Socius:* Winter Observations

THE UNIVERSITY OF CHICAGO PRESS—*Physiological Zoölogy*
The Oxygen Consumption and Bioenergetics of Harvest Mice

VILTREVY, SWEDISH WILDLIFE—*Viltrevy, Swedish Wildlife*
Observations and Experiments on the Ethology of the European Beaver (*Castor fiber L.*)

SERIES EDITOR'S PREFACE

It was not too many years ago that virtually all research publications dealing with animal behavior could be housed within the covers of a very few hard-bound volumes easily accessible to the few workers in the field. Times have changed! Today's students of animal behavior have all they can do to keep abreast of developments within their own area of special interest, let alone in the field as a whole.

Even fewer years ago those who taught animal behavior courses could easily choose a suitable textbook from among the very few that were available; all "covered" the field, according to the bias of the author. Students working on a special project used *the* text and *the* journal as reference sources, and for the most part successfully covered their assigned topics. Times have changed! Today's teachers of animal behavior are confronted with a bewildering array of books to choose among, some purported to be all-encompassing, others confessing to strictly delimited coverage, and still others being simply collections of recent and profound writings.

In response to the problem of the steadily increasing and over-whelming volume of information in the area, the Benchmark Papers in Animal Behavior was launched as a series of single-topic volumes designed to be some things to some people. Each volume contains a collection of what an expert considers to be *the* significant research papers in a given topic area. Each volume serves several purposes. To teachers, a Benchmark volume serves as a supplement to other written materials assigned to students; it permits in-depth consideration of a particular topic while confronting students (often for the first time) with original research papers of outstanding quality. To researchers, a Benchmark volume serves to save countless hours digging through the various journals to find *the* basic articles in their area of interest; often the journals are not easily available. To students, a Benchmark volume provides a readily accessible set of original papers on the topic, a set that forms the core of the more extensive bibliography that they are likely to compile; it also permits them to see at first hand what an "expert" thinks is important in the area, and to react accordingly. To librarians, a Benchmark volume represents a collection of important papers from many diverse sources, thus making readily available materials that might otherwise not be economically possible to obtain or physically possible to keep in stock.

The choice of topics to be covered in this series is no small matter. Each of us could come up with a long list of possible topics and then search for potential volume editors. Alternatively, we could draw up long lists of recognized and prominent scholars and try to persuade them to do a volume on a topic of their choice. For the most part, I have followed a mix of both approaches: match a distinguished researcher with a desired topic, and the results should be outstanding. And so it is with the present volume.

The Colliases, Nicholas and Elsie, were the most natural first choice for the present topic, but I would have been pleased to have them do any topic they chose. Their films on weaverbirds, on nest construction, and on jungle fowl are in common use in our classroom and serve as testimony to the meticulous care and the breadth with which they approach a subject. They have devoted many years (and tens of thousands of miles) to the study of the behavioral ecology of weaverbirds and have published extensively on the topic as well as on many other topics in the field of behavioral ecology. I am especially pleased that they have agreed to put together this book; the match of scholars to subject is perfect.

MARTIN W. SCHEIN

PREFACE

The study of external constructions built by animals is important to the understanding of behavioral biology (ethology) and ecology. The building of a home, or habitation, is an attempt by the animal to modify those aspects of its environment that are of real and often of central significance to it. In addition to providing insights into habitat control, external construction represents an unequivocal, species-specific, and relatively permanent record of an important aspect of the animal's behavior. Such construction can therefore be used to help analyze behavioral mechanisms and the development of behavioral capacity. One can often compare different species by the structures they build as readily as by their morphology or anatomy.

The study of external construction by animals brings into focus the significant habitat relations of animals. Ecologists, who are concerned with the interrelationships of organisms with their environment, should by the very meaning of the word pay particular attention to the homes or habitations of animals. The prefix of the term "ecology" was derived from an ancient Greek word "oikos," meaning "home," and ecology literally means "study of the home."

Animal architecture, like human architecture, is the art or practice of designing and building structures, and especially habitable ones. However, in the case of animal architecture, it is natural selection that has been the designer, and the plans and capacity for construction have themselves been built into each species in the course of its evolution. The term "architect" further implies a "master builder," and we have in this volume attempted to select some of the most spectacular and complex examples of construction by animals for special consideration. The term "external construction" is broader than the term "architecture" and implies, for example, the making of tools, whereas architecture is more restricted in its implications to habitations.

The term "external construction" can be used in a broad, or in a narrow, sense. In a broad sense it may be used to include the physiologically secreted outer parts of an animal, such as the shells of mollusks, or even the exoskeleton of an insect. In a more narrow sense, which is the way we use the term in this book, external construction refers specifically to the relatively permanent structures built by an animal by means of special movements, or the process of building itself. Often special building organs are involved, such as jaws, limbs, or spinnerets. There is no sharp line between structures built as a result of physiological secretions or excretions and those built by manipulatory movements of the animals, but one can readily recognize a preponderance of one or the other method in particular instances.

The articles selected for presentation are grouped into four sections, covering, respectively, historical and philosophical background, the evolution of external construction by animals, habitat control and behavioral energetics, and the mechanisms and ontogeny of building behavior.

Throughout the editing of this book we have been greatly encouraged by the friendly and generous cooperation and aid given by the authors of the articles, as well as by the series editor and by the publishers. This volume is dedicated to Alfred E. Emerson, who first inspired our interest in the evolution of nests and other external constructions in animals.

NICHOLAS E. COLLIAS
ELSIE C. COLLIAS

CONTENTS

Acknowledgements and Permissions   v
Series Editor's Preface   vii
Preface   ix
Contents by Author   xv

Introduction   1

PART I: HISTORICAL AND PHILOSOPHICAL BACKGROUND

Editors' Comments on Papers 1 Through 3   8

1 **DARWIN, C.:** Instinct   21
       On the Origin of Species, John Murray, London, 1859, pp. 207–212, 224–238,
                            242–244

2 **FABRE, J. H.:** Some Reflections upon Insect Psychology   43
       The Mason-Bees, A. T. de Mattos, trans., Dodd, Mead & Co., 1914, pp.
                            158–179

3 **WHEELER, W. M.:** Ant-Nests   64
       Ants: Their Structure, Development, and Behavior, Columbia University
                            Press, 1910, pp. 192–199, 216–224

PART II: EVOLUTION OF EXTERNAL CONSTRUCTION BY ANIMALS

Editors' Comments on Papers 4 Through 10   82

4 **EMERSON, A. E.:** Termite Nests—A Study of the Phylogeny of
       Behavior   95
       Ecol. Monogr., **8**(2), 249–284 (1938)

5 **COLLIAS, N. E.:** The Evolution of Nests and Nest-Building in Birds   134
       Symposium on the Evolution of External Construction by Animals (1964)
       Am. Zool., **4**(2), 175–190 (1964)

6 **KASTON, B. J.:** The Evolution of Spider Webs   152
       Symposium on the Evolution of External Construction by Animals (1964)
       Am. Zool., **4**(2), 191–207 (1964)

Contents

7   ROSS, H. H.: Evolution of Caddisworm Cases and Nets   173
     Symposium on the Evolution of External Construction by Animals (1964)
                   Am. Zool., **4**(2), 209–220 (1964)

8   SCHMIDT, R. S.: *Apicotermes* Nests   185
     Symposium on the Evolution of External Construction by Animals (1964)
                   Am. Zool., **4**(2), 221–225 (1964)

9   MICHENER, C. D.: Evolution of the Nests of Bees   191
     Symposium on the Evolution of External Construction by Animals (1964)
                   Am. Zool., **4**(2), 227–239 (1964)

10   LAWICK-GOODALL, J. VAN: The Behaviour of Free-Living
                   Chimpanzees in the Gombe Stream Reserve   204
     Anim. Behav. Monogr., **1,** Pt. 3, 165, 194–210, 298–300 (1968)

PART III. HABITAT CONTROL AND BEHAVIORAL ENERGETICS

Editors' Comments on Papers 11 Through 15   226

11   WELLINGTON, W. G.: Tents and Tactics of Caterpillars   235
                   Nat. Hist., **83**(1), 65–72 (1974)

12   PEARSON, O. P.: The Oxygen Consumption and Bioenergetics of
                   Harvest Mice   243
                   Physiol. Zool., **33**(2), 152–160 (1960)

13   FRITH, H. J.: Temperature Regulation in the Nesting Mounds of the
                   Mallee-Fowl, *Leipoa ocellata Gould*   252
                   C.S.I.R.O. Wildl. Res., **1**(2), 79–95 (1956)

14   WHITE, F. N., and J. L. KINNEY: Avian Incubation   271
                   Science, **186**(4159), 107–115 (1974)

15   WHITE, F. N., G. A. BARTHOLOMEW, and T. R. HOWELL: The Thermal
                   Significance of the Nest of the Sociable Weaver
                   Philetairus socius: Winter Observations   280
                   Ibis, **117**(2), 171–179 (1975)

PART IV: MECHANISMS AND ONTOGENY OF BUILDING BEHAVIOR

Editors' Comments on Papers 16 Through 21   292

16   PETERS, H. M.: Maturing and Coordination of Web-Building Activity   304
                   Am. Zool., **9**(1), 223–227 (1969)

17   PEAKALL, D. B.: Synthesis of Silk, Mechanism, and Location   309
                   Am. Zool., **9**(1), 71–79 (1969)

18   LEHRMAN, D. S.: Interaction Between Internal and External
                   Environments in the Regulation of the Reproductive
                   Cycle of the Ring Dove   318
     Sex and Behavior, F. A. Beach, ed., John Wiley & Sons, Inc., 1965, pp.
                   355–369, 378–380

19   COLLIAS, N. E., and E. C. COLLIAS: An Experimental Study of the
Mechanisms of Nest Building in a Weaverbird   335
Auk, **79**(4), 568–595 (1962)

20   COLLIAS, E. C., and N. E. COLLIAS: The Development of Nest-Building
Behavior in a Weaverbird   363
Auk, **81**(1), 42–52 (1964)

21   WILSSON, L.:   Observations and Experiments on the Ethology of the
European Beaver *(Castor fiber L.)*   374
Viltrevy, Swedish Wildl., **8**(3), 160–165, 168, 182–203, 254–260 (1971)

Author Citation Index   405
Subject Index   410
About the Editors   414

CONTENTS BY AUTHOR

Bartholomew, G. A., 280
Collias, E. C., 335, 363
Collias, N. E., 134, 335, 363
Darwin, C., 21
Emerson, A. E., 95
Fabre, J. H., 43
Frith, H. J., 252
Howell, T. R., 280
Kaston, B. J., 152
Kinney, J. L., 271
Lawick-Goodall, J. van, 204

Lehrman, D. S., 318
Michener, C. D., 191
Peakall, D. B., 309
Pearson, O. P., 243
Peters, H. M., 304
Ross, H. H., 173
Schmidt, R. S., 185
Wellington, W. G., 235
Wheeler, W. M., 64
White, F. N., 271, 280
Wilsson, L., 374

EXTERNAL CONSTRUCTION BY ANIMALS

INTRODUCTION

Although people from time immemorial have observed and wondered at the more specialized habitations built by animals, the study of animal construction as a scientific discipline dates essentially from the time of Charles Darwin, who provided the basic, objective, and empirical philosophy for all modern biology. A large part of the chapter "Instinct" in his *Origin of Species* (1859) is focused on the nests of animals, particularly of birds and social insects. The philosophical controversy between Darwin's ideas of the evolution of species versus their origin by special creation dominated much of the latter half of the nineteenth century.

Nevertheless, because the first task of science is the accurate description of natural phenomena, we can say that a science of ecology or ethology, including habitat relations and modifications by animals, began whenever and wherever there was accurate and objective observation, recording, and adequate communication of the relevant phenomena. Leaving out the pre-Darwinians, we can see in Part I that Darwin, J. Henri Fabre, and W. M. Wheeler provide a succession of early and influential exemplars of such methodology.

Part II, on the evolution of external construction by animals, attempts to give some idea of the present status of this aspect of our subject. It is the students of highly social insects who have led the way. In 1938, Alfred E. Emerson showed in some detail how termite nests could be studied as examples of the phylogeny of behavior (Paper 4 of this volume). Interest in the architecture of the more social insects is long-standing and continues to the present day. The first comprehensive treatise on insect societies, since the two books by Wheeler published in the first third of this century, appeared in 1971, by E. O. Wilson. Some important recent reviews of the nests of different groups of social insects include studies of termites (Howse, 1970), ants (Sudd, 1967), wasps (Evans and Eberhard, 1970; Spradbery, 1973), and social bees (Michener, 1974).

Other widely studied animals that build specialized constructs are spiders, silkworms, caddisworms, birds, and beaver. Most animals, however, have attracted little attention as regards the study of external construction; the literature is widely scattered. In 1964 the first general symposium on the evolution of specialized external construction by animals was published in *American Zoologist*. In 1969 a symposium on web-building spiders was published in the same journal. There is as yet no really comprehensive and thoroughly documented modern treatment of external constructions built by all animal groups. Two popular and very readable books on the structures built by animals have recently appeared, one by a zoo architect (Hancocks, 1973) and one by a biologist (von Frisch, 1974). The former has a useful bibliography, the latter has no documentation or references. Both books contain many excellent pictures of animal architecture.

The following brief classification of external construction by animals should help provide an overview of this important class of behavioral phenomena. In a broad sense, external construction is found throughout the animal kingdom from protozoa to man. Examples are given for each special category of external construction; intermediate examples are of common occurrence. Various types of classification could be given, depending on the purpose of the classification. When classified by function the examples illustrate the universal tendency of animals toward some form of habitat control and stability, that is, toward individual or social homeostasis. Emerson (1956) has given many specific examples of social homeostasis for termites and their nests. Many of the examples used in the following classification are elaborated in other sections of this volume.

A Classification of External Construction by Animals

I. Classified by building process and materials used
   A. Primarily a physiological process with little or no manipulation

1. Excretion: fecal components of soil, e.g., earthworm castings
2. Secretions: mucous tubes; bags and nets of many marine organisms; lime shells of mollusks; limestone; coral reefs
 B. Primarily a behavioral process, involving manipulation
1. Building materials secreted or excreted by builder: silkworm cocoon; spider web; wax honeycomb; fecal pellets of *Apicotermes* and *Macrotermes* nests; saliva nest of edible-nest swiftlet
2. Building largely by manipulation of external environment
Excavated burrows: widespread among animals
Construction of materials gathered from environment: nests of social insects, of birds, etc.

II. Classified by function (habitat control)
 A. Control of physical environment
1. Soil: turned over by earthworms and ants
2. Temperature: tent caterpillars; mallee fowl mound; sociable weaver nest
3. Humidity: *Nasutitermes exitosus* nest
4. Drought: lungfish casts
5. Rain-shedding devices: chevrons over nest of *Procubitermes*
6. Floods and water level: beaver dams
7. Wind: burrows of many terrestrial animals of open country
8. Ventilation: pores of *Apicotermes nests;* bubble nests of water spider (*Argyroneta*) and fighting fish (*Betta*)
9. Resting place: sleeping nests of mice and apes
10. Dispersal: ballooning threads of young spiders
 B. Control of biotic environment
1. Food collection: storehouses of rodents and social insects; mucous bag of annelid *Chaetopterus* for collecting organic detritus
2. Agriculture: fungus gardens of *Macrotermes* and attine ants
3. Capture of prey: snares and webs of spiders
4. Defense from predators: caddisworm cases; masking crabs
5. Protection from disease: mold-inhibitor lining of cell in ground-nesting bees

3

6. Securing a mate: silk case—gift of male dance fly to female; bowerbirds
7. Raising of young: egg cases or nests of many animals
C. Multiple functions of a given external construction: any combination of preceding categories; typical of burrows and nests

Modern research on the building activities of animals, like scientific research in general, has been greatly aided and expanded, especially during the second half of this century, by funding from various foundations and similar organizations. In the United States support from the National Science Foundation, established in 1950, has been particularly important for research on behavioral ecology of the sort that we are concerned with in this volume.

The probable future course of the study of external construction by animals is not easy to delineate. It is already evident, however, that there will be more emphasis on the quantitative aspects of the subject, including measurements directed toward elucidating homeostatic mechanisms made possible by animal construction. The measurement of the amount of work required to build, balanced against the benefits conferred by the structure built, is an area of behavioral energetics that has attracted recent and continuing attention. Some examples of temperature regulation and energetics of external constructions are elaborated in Part III.

Descriptive and developmental aspects of external construction by animals are considered in Part IV, the last section of this volume. Subsidiary problems include selection of a site and of materials for building, factors determining form and structure, and termination of building. Use of modern techniques, such as time–motion film analysis, makes possible increased precision and more accurate description. Experimentation is important, especially in relating external and internal factors in construction work.

Better understanding of the function, mechanisms of construction, and ontogeny of building abilities in turn should make possible renewed perspective and progress in our understanding of evolutionary trends and of the mechanisms of evolution of animal architecture and other external constructs.

The selection of articles reproduced in this volume is based primarily on the attempt to illustrate the significance, history, present status, and possible future course of the subject. We have selected articles that seem to us to have been, or promise to be, of service as guidelines to the study of external construction by animals. As our emphasis is on the more highly specialized structures built by animals, the reader can get a better picture of overall evolution by examining graded series from simple to the more complex constructions. Furthermore, the most complex constructions provide highly organized patterns well suited to the analysis of functional and energetic relations, or to the analysis of mechanisms of building and the ontogeny of building behavior.

For the sake of continuity and integration we provide a single commentary on all the papers together for each of the four parts of the volume, rather than introduce each article and author separately.

Almost all the authors are naturalists; they represent seven different countries. Many of them have traveled widely in the course of their investigations, and among them they have done research on all continents. Almost all have worked for one or more decades on a particular group of animals or subject relevant to this book.

Aside from Darwin's general chapter on instinct, there are 10 articles on invertebrates (insects, 7; spiders, 3) and 10 on vertebrates (birds, 7; mammals, 3). Because of their elaborate building activities, spiders, social insects, and birds have been of greatest interest to most investigators.

In addition to the articles reproduced in whole or in part, we give brief descriptions of a few other outstanding studies to help round out the story in each commentary. We also mention in every part a few of the more recent studies both to help indicate current trends and future possibilities and as a key to some of the more important current literature.

Many important and desirable articles and some active areas of inquiry have had to be omitted for reasons of space, most notably the large amount of work done on silkworms (e.g., Van der Kloot and Williams, 1953) and on marine animals. The book by the MacGinities (1949), *Natural History of Marine Animals,* is a classic, with a good deal of information, much of it original, on structures built by animals in the sea. The present survey is restricted to terrestrial and freshwater animals.

REFERENCES

Darwin, C. 1859. *On the Origin of Species by Means of Natural Selection, or the Preservation of favored Races in the Struggle for Life.* John Murray, London, 502 pp.

Emerson, A. E. 1956. Regenerative behavior and social homeostasis of termites. *Ecology,* 37:248–258.

Evans, H. E., and M. J. W. Eberhard. 1970. *The Wasps.* University of Michigan Press, Ann Arbor. 265 pp.

Frisch, K. von. 1974. *Animal Architecture.* With the collaboration of O. von Frisch. Translated by L. Gombrich. Harcourt Brace Jovanovich, Inc., New York. 306 pp.

Hancocks, D. 1973. *Master Builders of the Animal World.* Harper & Row, Publishers, New York. 144 pp., 8 pls.

Howse, P. E. 1970. *Termites: A Study in Social Behavior.* Hutchinson University Library, London, 150 pp.

MacGinitie, G. E., and N. MacGinitie. 1949. *Natural History of Marine Animals.* McGraw-Hill Book Co., New York. 473 pp.

Michener, C. D. 1974. *The Social Behavior of the Bees.* Harvard University Press, Cambridge, Mass. 404 pp.

Spradbery, J. P. 1973. *Wasps.* University of Washington Press, Seattle. 408 pp.

Sudd, J. H. 1967. *An Introduction to the Behaviour of Ants.* Arnold Ltd., London. 200 pp.

Van der Kloot, W. G., and C. M. Williams. 1953. Cocoon construction by the *Cecropia* silkworm. *Behaviour,* 5:141–157, 157–174.

Wilson, E. O. 1971. *The Insect Societies.* Harvard University Press, Cambridge, Mass. 548 pp.

Part I

HISTORICAL AND PHILOSOPHICAL BACKGROUND

Editors' Comments
on Papers 1 Through 3

1   **DARWIN**
     Instinct

2   **FABRE**
     Some Reflections upon Insect Psychology

3   **WHEELER**
     Ant-Nests

    The basic philosophy underlying modern biology, including behavioral biology, is the concept of evolution and the theory of natural selection established by Charles Darwin in the nineteenth century. The classical conception of insect behavior as stereotyped, unlearned, adaptive, and species-typical was in large part established and widely popularized by a contemporary of Darwin, J. Henri Fabre, who described with great literary talent his personal observations on the behavior of many insects, particularly solitary species near his home in southern France. Whereas Fabre was self-taught as an entomologist and was not an evolutionist, William Morton Wheeler, a specialist on social insects, is a good early example of the highly trained university zoologist, a disciple of Darwin, widely traveled, and with an essentially modern, scientific outlook. He was a world authority on the dominant group of nonhuman animals, the ants. What these three men have in common is that they were primarily naturalists and were greatly interested in animals that build remarkable nests. Nests provide a key to the more significant relations of a species with its environment, and these men relied much on the observation of nest-building species of animals in working out their respective and influential philosophies of nature.

    The story of Darwin (1809–1882) is the great and oft-told story of biology. There are many good, recent, critical accounts that evaluate the impact of his ideas on science and on human affairs (see Emerson, 1962; Mayr, 1964; Appleman, 1970). The son of a well-to-do English physician, Darwin first studied for a career in medicine, then for the ministry, but much preferred to indulge his hobbies of hunting and natural history. On graduating from Cambridge University he took an unsalaried position as the naturalist on a ship, the *Beagle,* and spent the next five years circumnavigating the globe (1831–1836). Darwin later wrote that this trip was the most important event of his life. It changed his whole outlook, and as it later happened, the outlook of much of the educated world as well. He spent the rest of his life gathering evidence for evolution, the truth of which he gradually became convinced during the voyage, and also in gathering evidence for his theory of natural selection. His industry and reflections culminated in the publishing in 1859 of the *Origin of Species,* the first edition of which sold out on the day of publication. One reason for its success, Darwin thought, was because this book was an abstract of a much larger work. To expedite publication he shortened the manuscript and unfortunately also omitted the documented references to the work of others, although he quite often mentions authorities or correspondents by name. Cambridge University Press has now published the original manuscript (Darwin, 1975), together with the references to journal articles and books, edited by R. G. Stauffer of the University of Wisconsin.

Darwin considered the existence of complex instincts to be one of the main stumbling blocks to his theory of the origin of species by natural selection, and in his chapter "Instinct" was at special pains to show how such complex instincts could have evolved. We have reproduced most of this chapter here as Paper 1. Darwin selects for his principal example the comb building of the "hive-bee" (honeybee), which he characterizes as "the most wonderful of all known instincts." This chapter well illustrates the empirical, scientific method he developed, including his inimitable ability to formulate and analyze significant problems, his powers of precise observation combined with experimentation where needed, his patient gathering and critical evaluation of all facts bearing on the solution of a problem, and his constant checking of theory against fact. Darwin seems to have relied mostly on his own observations of bees, although he acknowledges the ideas and observations of other selected authorities, especially Waterhouse in his comparisons of nests of different bees, and Tegetmeier and Huber on comb construction. In particular, the blind Huber (1814), aided by his assistant Burnens, had published a remarkably detailed description of much of the honeycomb-building process.

Darwin often uses what he calls "the principle of gradation" as a clue to evolutionary trends. For example, he describes gradation from simple rounded cells in the nest of "humble-bees" (bumblebees), through the more systematically arranged cells in the nest of a stingless bee (*Melipona*), to the highly organized and symmetrical honeycomb of the hive bee, in which a minimum of wax is used to store the maximum amount of honey because of the hexagonal shape and thin walls of the wax cells.

Darwin believed that the actual construction of the honeycomb could be shown "to follow from a few simple instincts," or behavior patterns: (1) the regular alternation of cells in different rows and on opposite faces of the two-layered honeycomb was in part explained by the bees' crowding together at certain fixed distances from each other to hollow out initial circular basins in the mass of wax secreted and deposited by the bees; (2) these initial basins served as reference points for many individual bees, each of which may work on different cells or at different times on the same cell; (3) when circles of equal size are crowded together, each circle is in contact with six other circles; the hexagonal shape of each cell then results from the tendency of each bee to use its mandibles to plane down the thicker parts of the crude cell walls, combined with (4) a tendency to stop planing and to build up the pliable walls of the cell when these become sufficiently thin. The excess wax is piled up for future use on the thick circular coping that covers the growing outer rim of each cell.

The accuracy of Darwin's observations and analysis is indicated by the fact that almost a century later a highly authoritative monograph on honeybee behavior (Ribbands, 1953) relied mainly on Darwin for a description of comb construction. The first really detailed modern studies of comb building were published exactly 100 years after Darwin's description, in two student dissertations, one in France (Darchen, 1959) and one in Germany (Lau, 1959). Each independently confirmed the essential points of Darwin's account, as well as adding new information. Lau found that, when the antennae of bees were amputated, the walls of the cells varied in thickness to an abnormal degree and that at times the opening of one cell even broke into another cell, an observation later confirmed by Martin and Lindauer (1966).

To some extent Darwin was carried away by his own enthusiasm when he declared that "the comb of the hive-bee, as far as we can see, is absolutely perfect in economising labour and wax." In the sixth edition (1872) of the *Origin of Species* he added, without further comment or details: "I hear from Professor Wyman, who has made numerous careful measurements, that the accuracy of the workmanship of the bee has been greatly exaggerated; so much so that whatever the typical form of the cell may be it is rarely, if ever realized." However, even a glance at a honeycomb verifies the great regularity, hexagonal shape, and thin walls of most of the cells; furthermore, some range of variability in cell shape does not seriously affect the general validity of Darwin's conclusions on the economic efficiency of the structure of honeycomb.

Darwin and others before and after him have by no means exhausted all the problems of comb building in the honeybee. The problem of exactly how bees measure and attain the correct and uniform size of the larger drone cells or the smaller workers cells still seems to be unknown (see Martin and Lindauer, 1966). Darwin paid little attention to the problem of how the combs in a nest of wild bees come to be parallel to one another. Some recent evidence by M. Oehmke suggests that bees may be orienting comb direction with the aid of cues received from the earth's magnetic field (see Lindauer and Martin, 1972), and if this finding is confirmed and the evidence substantiated, once again we shall have cause to marvel at the wonders of the building instinct of the honeybee.

The comb is built by the workers, which are sterile, and this fact led Darwin to the important problem of the evolution of sterile castes in the highly social insects, a problem which "at first appeared to me insuperable and actually fatal to my whole theory." He resolved the difficulty on realizing that natural selection may be applied to the family as a unit as well as to the individual and that the reproductive castes could transmit to their fertile offspring a tendency to produce sterile individuals that aid in the survival of the community as a whole. This is a form of group selection and one of the most important aspects of the theory of natural selection. We use the term "group selection" for any genetic benefit to two or more individuals. Darwin made the further important deduction that the evolution of sterile worker castes rules out any direct inheritance of the effects of use and disuse, as was held by Lamark.

Sewall Wright, building on Darwin's theory of natural selection, has developed the most widely accepted, modern genetical theory of evolution. From the statistical consequences of Mendelian heredity, he finds that the best conditions for social evolution of animals (Wright, 1945), or of evolution in general, resemble those that have proved best for continued progress in animal breeding. The establishment of many different inbred lines, or partially isolated small populations (demes), serves to multiply various characteristic genotypes. Selection among these different strains, or subpopulations, is followed by excess emigration and cross-breeding of the favored types to restore variability and to provide a basis for further progress. Selection that is too rigorous reduces the field of available variability and destroys the basis for further advance, whereas some random or sampling drift of gene frequencies in a multidimensional probability distribution permits the species population to explore a wide variety of favorable gene combinations over time. The method depends on a balance among its various factors, and Wright (1970) has termed it the "shifting balance theory of evolution." In reply to criticisms that local populatins are not generally small enough to serve as a basis for appreciable sampling drift or sufficiently isolated for enough differentiation of their genetic systems to undergo interdemic selection, Wright points out that the *effective* breeding size of natural populations and numbers of effective immigrants may often be much smaller than the actual numbers present. More information is needed on the breeding structure of populations in nature.

The nonrandom breeding distribution of animals in nature is regulated largely by their social behavior (Collias, 1951). Habitat selection and traditional locality fixation, local aggregations, restriction of reproduction to certain individuals, family life, dispersal of founder individuals to peripheral habitats, and local limitation of population densities by intraspecific conflict all favor establishment of local genetic types. Such factors tend to reduce genetic variability within demes and are balanced against sexual reproduction, the prime source of variability among living organisms.

Darwin's metaphorical use of the phrases "struggle for existence" and "survival of the fittest" occasionally lead him to make what seem to be somewhat exaggerated and inconsistent statements with respect to the theory of natural selection. Thus he concludes his chapter on instinct with the assertion that certain special instincts have originated in evolution "not as specially endowed or created instincts, but as small consequences of one general law leading to the advancement of all organic beings, namely, multiply, vary, let the strongest live and the weakest die." Surely there have been few more misleading and unfortunate statements made in the history of the human intellectual effort. This statement stands in some contrast to the concept of group selection which Darwin himself had just developed in the preceding pages of the same chapter. It also seems somewhat out of harmony with his explicit statement given earlier in the *Origin of Species* (1859, p. 62): "I use the term Struggle for Existence in a large and metaphorical sense including dependence of one being on another and including (which is more important) not only the life of the individual, but success in leaving progeny."

Interdependence of living things in the web of life or balance of nature is basic, and individual or group strength and aggressiveness in any direct sense are of course only two of the many factors involved in the "survival of the fittest." Often success in leaving progeny depends on the effective functioning of that individual as part of a social system. In turn, social groups must function effectively as units within ecological systems of a higher order. Natural selection must be operating to produce harmonious adjustments within and between all levels of organization in nature (Collias, 1944, 1951).

Allee (1931, 1945) has pointed out that no animal is really solitary throughout its whole life cycle, that interdependence is demonstrated by the fact that all living things are found in communities, and that the evolution of highly organized and complex individuals and societies is in itself evidence that the interaction between individuals has been more beneficial than harmful in the long run. He thus concludes that cooperation is the most basic principle of life. Allee has also developed evidence for his thesis that a moderate degree of crowding is beneficial or stimulating for many animal functions, and notes the widespread tendency of animals to aggregate under certain conditions. Williams (1966) believes that the biological principle of adaptation should not be invoked if one can explain the beneficial effects of aggregation by chance and by the principles of chemistry and physics alone. However, as Allee pointed out, the benefits of aggregation, irrespective of their cause, have survival value and therefore important implications for evolution of a greater degree of mutual interdependence.

Emerson (1949, 1962, 1973) has developed the concept that the progressing evolution of living systems, both as individuals and as social groups, has not been so much toward an increasing complexity as such, as toward an increase of self-regulatory control over the establishment of optimal conditions for survival and reproduction, a concept he calls "dynamic homeostasis," which includes both individual and social homeostasis. As we shall see, much of this control over habitat conditions is often exerted by external constructions or habitat modification by the animals themselves.

During the evolution of the animal kingdom there has been an ever-increasing awareness by the more advanced animals of a larger environment; the trend finds its strongest expression in man with his world of ideas, towns and cities, and exploration of outer space. Unlike all other animals, man is aware that he is a product of evolution and that he has the opportunity to modify the course of his own evolution to some degree (J. Huxley, 1943). Natural selection is now supplemented by more or less unintended human selection in evolution with a choice of goals based on human wisdom and a sense of moral values. Emerson (1962; p. 204) suggests that the important trends of cultural evolution resemble those of organic evolution in that evolving civilization similarly moves toward improved homeostasis. He further points out that interdependence of the individual and of group systems is a fact that deserves more recognition in the practical affairs of human society.

In the first paragraph of his chapter on instinct, Darwin, although he was later to write a book on the descent of man, writes: "I must premise, that I have nothing to do with the origin of the primary mental powers, any more than I have with that of life itself." In a way, just the reverse is true of our second author, J. Henri Fabre (1823–1915). Few men, if any, did more to establish the classical conception of instinct than he, and a good part of his theoretical orientation was to try to bring out the great gulf between the intellectual powers of man and those of insects.

Unlike Darwin, who voyaged around the world, Fabre spent almost all his ninety-two years in or near his home in southern France, where he was fascinated by the insects and observed them at every opportunity during his long life. Whereas Darwin was financially independent, Fabre had a lifelong battle with poverty. He was an able and talented school teacher, but he was dismissed from the lycée at the age of forty-six—it is said for some of his advanced ideas in teaching, such as admitting girls to his science classes. Henceforth, he supported himself and his family by writing popular books about insect behavior and natural science, finally earning enough money to buy a small plot of land, worthless for agriculture, but ideal for insect life. In the same year (1879) appeared the first of the ten volumes of his renowned *Souvenirs entomologiques*, which were published over a period of almost thirty years. Edwin Way Teale (1950) has written a brief but fascinating account of Fabre's life by way of introduction to a fine selection from Fabre's writings translated into English (as much of Fabre's works have been).

For presentation here (Paper 2) we have selected part of the chapter "Reflections upon Insect Psychology" from Fabre's book *The Mason-Bees,* translated by Alexander Teixeria de Mattos. We have included Fabre in this selection of articles on external construction because (1) he is representative of the majority of naturalists who, before Darwin's book was published, believed in special creation versus evolution, the great philosophical controversy of the nineteenth century (though he carried on a cordial correspondence with Darwin, who even suggested an experiment for Fabre to carry out on his mason bees); (2) by means of his precise, accurate, and numerous personal observations and ingenuous, although simple experiments on insects, especially solitary insects, together with his voluminous and popular writings, he helped mold the classical conception of instinct as stereotyped, unlearned, species-typical behavior; (3) the basic motivating force of a scientist and naturalist is curiorsity, and no one has conveyed this aspect of scientific inquiry in as articulate and beautiful a fashion, with so much enthusiasm and clarity as Fabre has done in his straight-forward style; and (4) the subject of his discussion in the selection given here is the nesting behavior of a solitary insect.

This insect is the bee *Chalicodoma muraria,* which belongs to the family Megachilidae. In this family, the female carries pollen in a brush of hairs on the underside of the abdomen instead of on the hind legs as most bees do. The female builds a nest consisting of a cluster of six to ten cells on a pebble. The cell is built of a mortar of dry clay dust mingled with a little sand and made into a paste with saliva. While the mass is still soft, angular bits of gravel are inserted one by one into the walls, hence the name "mason bee." The bee lines the inside of the cell with pure mortar and provisions it with a mixture of honey and pollen, lays a single egg on top of the mixture, and then seals the cell with a lid of pure mortar.

Fabre develops the point that the mason bee is operating mechanically with no apparent understanding of her own work. In an earlier chapter he describes how he removed an incomplete cell from a bee that had been building on it and substituted another cell, completed and provisioned but not yet capped. The bee accepted this cell but continued to build and provision until the new cell was one third higher than normal, with an excess

15

store of honey and pollen. Fabre continues this theme in the chapter presented here and again shows that the principle of fixed behavior sequences, apparently dependent on an internal reaction chain, is operating. He experiments on cells by damaging them at various times in the building cycle. Because the bee continues whatever specific behavior she has just been engaged in, whether appropriate or not, and makes the needed repairs only when she is in the relevant phase of her building cycle, he concludes that the mason bee has no insight or foresight into the results of her own behavior.

However, some later observers, working with solitary wasps, have observed that these wasps can modify their nesting behavior to some extent to meet experimentally altered circumstances. Hingston (1929) experimented with the potter wasp, *Rhynchium nitidulum,* in India. This species builds a small clay pot, covers the pot with resin, stuffs it with caterpillars as a food store for her larva, then seals the pot with a lid. When Hingston broke a hole into the side of a pot, the wasp repaired it, whether she was plastering with clay, smearing with resin, or provisioning at the time. However, if the wasp had been plastering resin before the hole was made, she first tried to repair it with resin before using any clay to patch over the hole. The wasp tends to a fixed behavior sequence, as Fabre had maintained for a mason bee, but can vary her behavior to some degree to meet emergencies. It appears that the behavior of some insects may be somewhat more variable and adaptable to emergencies than Fabre had emphasized. Hingston concluded that his potter wasps showed "rational" behavior. In a recent book on wasps, Evans and Eberhard (1970, p. 110) express the modern view: "The time is long since past when it is worthwhile to speculate whether behavior is "intelligent" or "instinctive"; what is needed are experiments to determine the causation of behavior in specific situations." The author of a modern definitive treatise on insect societies found it unnecessary even to use the term "instinct" in his book (Wilson, 1971). However, instinct is still quite often used as a convenient label for species-specific, or species-typical, behavior.

According to Michener (1974, p. 234) in his recent treatise on the social behavior of bees, whereas in typical solitary bees there is commonly an essentially unvarying sequence relating to construction and provisioning, in the social bees one finds no such limitations. The social bees seem able to omit any part of a behavioral sequence relating to construction and provisioning and can work on more than one cell in different stages of construction at nearly the same time. Michener writes: "What one would really like to know is whether in solitary bees there is an action sequence not or little controlled by sensory feedback, in contrast to a reaction sequence in which sensory stimuli control each activity to be performed, as is presumably the case in social bees." In the light of these remarks by a current leading authority on bees, Fabre's observations on the behavior of the solitary mason bee, *Chalicodoma muraria,* would seem to help fill a modern need.

William Morton Wheeler (1865–1937), one of America's best-known authorities on insect behavior during the early part of this century, was often asked his opinion of Fabre's work. His reply (1918, pp. 7–8) was that Fabre, a student of the living insect, was the greatest entomologist of his day, but because Fabre was trained as a physicist, chemist, and mathematician and did not believe in evolution, he naturally stressed and schematized the normal course of behavior in the insects. Wheeler further writes: "The variations which to Fabre were more or less negligible at once assumed great importance when biologists became evolutionists." Wheeler thought Fabre was too set in his ways of thinking when the *Origin of Species* appeared to acquire any sympathy with evolutionary theories. However, "Fabre is so preeminent in the wealth and precision of his observations, the ingenuity of his experimentation and in literary expression, that his *Souvenirs* will always endure."

Wheeler, like modern biologists generally, was a follower of Darwin. He was a specialist on the social insects, particularly ants. Ants are the dominant group of nonhuman terrestrial animals today, and unlike bees and wasps, most species of which are solitary, all species of ants are social. Wheeler's work also exemplifies the comparative method so much used by Darwin. Like many advanced American scholars of his time, he had studied at European as well as at American universities. He was well versed in foreign languages and a master of the literature in his field. A Harvard professor of entomology for some thirty years and a world traveler, he made many lasting contributions to the study of social insects. Wheeler's writings are essentially objective and modern in tone, historically minded with a wide philosophical perspective, often involve much personal observation, and are extensively documented. His book on ants, first published in 1910, was reprinted as late as 1960. His book *Social Life Among the Insects* (1923) and a later treatment, *The Social Insects* (1928), were the best available works, at least in English, on the subject, giving a comprehensive treatment on a worldwide basis, until the recent treatise by E. O. Wilson, *The Insect Societies,* appeared in 1971. Wilson was a student at Harvard of F. M. Carpenter, himself a student of Wheeler's; all three were or are ant specialists. In 1924 Wheeler was awarded the Daniel Giraud Elliot Medal of the National Academy of Sciences, the highest award of this most prestigious scientific body in the United States. Mary and Howard Evans (1970) have written a detailed, personal biography of Wheeler.

We reproduce here a selection on ant nests from Wheeler's book *Ants, Their Structure, Development and Behavior.* This selection (Paper 3) includes his classification of ant nests, one of the first such detailed, scientific classifications of the external constructions built by a major group of animals. In this chapter, Wheeler also gives a historical account of the discovery of how the arboreal weaver ant, *Oecophylla smaragdina,* constructs its leafy nests by using its silk-spinning larvae as living spools and shuttles to fasten the edges of the leaves together with silk. The building of the nest of *Oecophylla* is probably the most spectacular example known of cooperative work between individuals in construction by nonhuman animals. Wheeler's scholarly account illustrates beautifully how several independent observers are often involved in the progressive elucidation and gradual solution of a scientific problem. Finally, the last part of his chapter deals with the construction of accessory structures by ants such as roadways and aphid or coccid tents or sheds, and illustrates how building ability can be adapted to diverse and at times quite unexpected functions, in addition to the building of nests.

Each successful step in the attempt to understand complex biological phenomena leads inevitably to new problems and new solutions. Sudd (1963) has further clarified the coordination in building by *Oecophylla,* with his observation that the ants at first work independently trying to pull down or roll up leaves, but are then attracted to those individuals which first succeed, abandoning their own efforts go to help the more successful ants. From the phylogenetic viewpoint, Ledoux (1956), in his comparative survey of arboreal ant nests, has helped to clarify how the remarkable nest-building behavior of *Oecophylla* might have evolved.

REFERENCES

Allee, W. C. 1931. *Animal Aggregation: A Study in General Sociology.* University of Chicago Press, Chicago. 431 pp.

Allee, W. C. 1945. Human conflict and cooperation: the biological background. Pp. 321–367 in *Approaches to National Unity: Conference on Science, Philosophy and Religion,* L. Bryson, L. Finkelstein, and R. M. MacIver (eds). Harper & Row, Publishers, New York.

Appleman, P. 1970. *Darwin.* W. W. Norton & Co., Inc., New York. 674 pp.

Collias, N. E. 1944. Aggressive behavior among vertebrate animals. *Physiol. Zool.,* 17(1):83–123.

Collias, N. E. 1951. Problems and principles of animal sociology. Pp. 423–457 in *Comparative Psychology,* C. P. Stone (ed.). Prentice-Hall, Inc., Englewood Cliffs, N.J.

Darchen, R. 1959. Les techniques de construction chez *Apis mellifica. Ann. Sci. Nat. Zool. Biol. Anim.,* Sér. 12, 1:113–209, 8 pls.

Darwin, C. 1859. *On the Origin of Species by Means of Natural Selection, or the Preservation of Favored Races in the Struggle for Life.* John Murray, London. 502 pp.

Darwin, C. 1975. *Natural Selection.* Cambridge University Press, New York. 640 pp. (Posthumous publication, edited by R. C. Stauffer.)

Emerson, A. E. 1949. Ecology and evolution. Pp. 598–729 in W. C. Allee, A. E. Emerson, O. Park, T. Park, and K. Schmidt, *Principles of Animal Ecology.* W. B. Saunders Co., Philadelphia. 837 pp.

Emerson, A. E. 1962. The impact of Darwin on biology. *Acta Biotheoretica,* 15(4):175–216.

Emerson, A. E. 1973. Some biological antecedents of human purpose. *Zygon,* 8(3–4):294–309.

Evans, H., and M. J. Eberhard. 1970. *The Wasps.* University of Michigan Press, Ann Abor. 265 pp.

Evans, M., and H. Evans. 1970. *William Morton Wheeler, Biologist.* Harvard University Press, Cambridge, Mass. 363 pp.

Fabre, J. H. 1879–1907. *Souvenirs entomologiques.* Delegrave, Paris. 10 vols.

Hingston, R. W. G. 1929. *Instinct and Intelligence.* Macmillan Publishing Co., Inc., New York. 296 pp.

Huber, F. 1814. New observations upon bees. Translated from the French (1926) by C. P. Dadant. Published by *American Bee Journal,* Hamilton, Ill.

Huxley, J. 1943. Evolutionary Ethics. Pp. 405–422 in *Darwin,* P. Appleman (ed.). W. W. Norton & Co., New York, 1970.

Lau, D. 1959. Beobachtungen und Experiment über Entstehung der Bienenwabe (*Apis mellifica* L.). *Zool. Beitr.* 4(2):233–306.

Ledoux, A. 1956. La construction du nid chez quelques fourmis arboricoles de France et Afrique tropicale. *Proc. 10th Intern. Congr. Entomol.* (Montreal), 2:521–528.

Lindauer, M., and H. Martin. 1972. Magnetic effect on dancing bees. Pp. 559–567 in *Animal Orientation and Navigation,* S. R. Galler et al. (ed.). National Aeronautics and Space Administration Office, Washington, D.C.

Martin, H., and M. Lindauer. 1966. Sinnesphysiologische Leistungen beim Wabenbau der Honigbiene. *Z. Vergleich. Physiol.* 53(3):372–404.

Mayr, E. 1964. Introduction to a facsimile of the first edition of *On the Origin of Species* by Charles Darwin. Pp. vii–xxvii. Harvard University Press, Cambridge, Mass.

Michener, C. D. 1974. *The Social Behavior of the Bees: A Comparative Study.* Harvard University Press, Cambridge, Mass. 404 pp.

Ribbands, R. 1953. *The Behaviour and Social Life of Honeybees.* Bee Research Association, London. 352 pp.

Sudd, J. H. 1963. How insects work in groups. *Discovery,* June 1963: 15–19. London.

Teale, E. W. 1950. *The Insect World of J. Henri Fabre.* Dodd, Mead & Co., New York. 333 pp.

Wheeler, W. M. 1918. Introduction (pp. 1–8) to Phil Rau and Nellie Rau, *Wasp Studies Afield.* Princeton University Press, Princeton, N.J. 372 pp.

Wheeler, W. M. 1923. *Social Life Among the Insects.* Harcourt, Brace, New York. 375 pp.

Wheeler, W. M. 1928. *The Social Insects, Their Origin and Evolution.* Harcourt, Brace, New York. 378 pp.

Williams, G. C. 1966. *Adaptation and Natural Selection: A Critique of Some Current Evolutionary Thought.* Princeton University Press, Princeton, N.J. 307 pp.

Wilson, E. O. 1971. *The Insect Societies.* Harvard University Press, Cambridge, Mass. 548 pp.

Wright, S. 1945. Tempo and mode in evolution: a critical review. *Ecology,* 26:415–419.

Wright, S. 1970. Random drift and the shifting balance theory of evolution. Pp. 1–31 in *Mathematical Topics in Population Genetics,* K. Kojima (ed.). Springer-Verlag, New York.

1

ON

THE ORIGIN OF SPECIES

BY MEANS OF NATURAL SELECTION,

OR THE

PRESERVATION OF FAVOURED RACES IN THE STRUGGLE
FOR LIFE.

By CHARLES DARWIN, M.A.,

FELLOW OF THE ROYAL, GEOLOGICAL, LINNÆAN, ETC., SOCIETIES;
AUTHOR OF 'JOURNAL OF RESEARCHES DURING H. M. S. BEAGLE'S VOYAGE
ROUND THE WORLD.'

LONDON:
JOHN MURRAY, ALBEMARLE STREET.
1859.

Reprinted from Charles Darwin, *On the Origin of Species*, John Murray, London, 1859, pp. 207–212, 224–238, 242–244

INSTINCT

Charles Darwin

Instincts comparable with habits, but different in their origin—Instincts graduated — Aphides and ants — Instincts variable—Domestic instincts, their origin—Natural instincts of the cuckoo, ostrich, and parasitic bees — Slave-making ants — Hive-bee, its cell-making instinct—Difficulties on the theory of the Natural Selection of instincts—Neuter or sterile insects—Summary.

THE subject of instinct might have been worked into the previous chapters; but I have thought that it would be more convenient to treat the subject separately, especially as so wonderful an instinct as that of the hive-bee making its cells will probably have occurred to many readers, as a difficulty sufficient to overthrow my whole theory. I must premise, that I have nothing to do with the origin of the primary mental powers, any more than I have with that of life itself. We are concerned only with the diversities of instinct and of the other mental qualities of animals within the same class.

I will not attempt any definition of instinct. It would be easy to show that several distinct mental actions are commonly embraced by this term; but every one understands what is meant, when it is said that instinct impels the cuckoo to migrate and to lay her eggs in other birds' nests. An action, which we ourselves should require experience to enable us to perform, when performed by an animal, more especially by a very young one, without any experience, and when performed by many individuals in the same way, without their knowing for what purpose it is performed, is usually said to be instinctive.

22

But I could show that none of these characters of instinct are universal. A little dose, as Pierre Huber expresses it, of judgment or reason, often comes into play, even in animals very low in the scale of nature.

Frederick Cuvier and several of the older metaphysicians have compared instinct with habit. This comparison gives, I think, a remarkably accurate notion of the frame of mind under which an instinctive action is performed, but not of its origin. How unconsciously many habitual actions are performed, indeed not rarely in direct opposition to our conscious will! yet they may be modified by the will or reason. Habits easily become associated with other habits, and with certain periods of time and states of the body. When once acquired, they often remain constant throughout life. Several other points of resemblance between instincts and habits could be pointed out. As in repeating a well-known song, so in instincts, one action follows another by a sort of rhythm; if a person be interrupted in a song, or in repeating anything by rote, he is generally forced to go back to recover the habitual train of thought: so P. Huber found it was with a caterpillar, which makes a very complicated hammock; for if he took a caterpillar which had completed its hammock up to, say, the sixth stage of construction, and put it into a hammock completed up only to the third stage, the caterpillar simply re-performed the fourth, fifth, and sixth stages of construction. If, however, a caterpillar were taken out of a hammock made up, for instance, to the third stage, and were put into one finished up to the sixth stage, so that much of its work was already done for it, far from feeling the benefit of this, it was much embarrassed, and, in order to complete its hammock, seemed forced to start from the third stage, where it had left off, and thus tried to complete the already finished work.

If we suppose any habitual action to become inherited — and I think it can be shown that this does sometimes happen—then the resemblance between what originally was a habit and an instinct becomes so close as not to be distinguished.   If Mozart, instead of playing the pianoforte at three years old with wonderfully little practice, had played a tune with no practice at all, he might truly be said to have done so instinctively.   But it would be the most serious error to suppose that the greater number of instincts have been acquired by habit in one generation, and then transmitted by inheritance to succeeding generations.   It can be clearly shown that the most wonderful instincts with which we are acquainted, namely, those of the hive-bee and of many ants, could not possibly have been thus acquired.

It will be universally admitted that instincts are as important as corporeal structure for the welfare of each species, under its present conditions of life. Under changed conditions of life, it is at least possible that slight modifications of instinct might be profitable to a species; and if it can be shown that instincts do vary ever so little, then I can see no difficulty in natural selection preserving and continually accumulating variations of instinct to any extent that may be profitable. It is thus, as I believe, that all the most complex and wonderful instincts have originated.   As modifications of corporeal structure arise from, and are increased by, use or habit, and are diminished or lost by disuse, so I do not doubt it has been with instincts.   But I believe that the effects of habit are of quite subordinate importance to the effects of the natural selection of what may be called accidental variations of instincts;—that is of variations produced by the same unknown causes which produce slight deviations of bodily structure.

No complex instinct can possibly be produced through

natural selection, except by the slow and gradual accumulation of numerous, slight, yet profitable, variations. Hence, as in the case of corporeal structures, we ought to find in nature, not the actual transitional gradations by which each complex instinct has been acquired—for these could be found only in the lineal ancestors of each species—but we ought to find in the collateral lines of descent some evidence of such gradations; or we ought at least to be able to show that gradations of some kind are possible; and this we certainly can do. I have been surprised to find, making allowance for the instincts of animals having been but little observed except in Europe and North America, and for no instinct being known amongst extinct species, how very generally gradations, leading to the most complex instincts, can be discovered. The canon of "Natura non facit saltum" applies with almost equal force to instincts as to bodily organs. Changes of instinct may sometimes be facilitated by the same species having different instincts at different periods of life, or at different seasons of the year, or when placed under different circumstances, &c.; in which case either one or the other instinct might be preserved by natural selection. And such instances of diversity of instinct in the same species can be shown to occur in nature.

Again as in the case of corporeal structure, and conformably with my theory, the instinct of each species is good for itself, but has never, as far as we can judge, been produced for the exclusive good of others. One of the strongest instances of an animal apparently performing an action for the sole good of another, with which I am acquainted, is that of aphides voluntarily yielding their sweet excretion to ants: that they do so voluntarily, the following facts show. I removed all the ants from a group of about a dozen aphides on a dock-plant, and prevented their attendance during several hours. After this interval, I felt sure that the aphides would want to excrete. I watched them for some time through a lens, but not one excreted; I then tickled and stroked them with a hair in the same manner, as well as I could, as the ants do with their antennæ; but not one excreted. Afterwards I allowed an ant to visit them, and it immediately seemed, by its eager way of running about, to be well aware what a rich flock it had discovered; it then began to play with its antennæ on the abdomen first of one aphis and then of another; and each aphis, as soon as it felt the antennæ, immediately lifted up its abdomen and excreted a limpid drop of sweet juice, which was eagerly devoured by the ant. Even the quite young aphides behaved in this manner, showing that the action was instinctive, and not the result of experience. But as the excretion is extremely viscid, it is probably a convenience to the aphides to have it removed; and therefore probably the aphides do not instinctively excrete for the sole good of the ants. Although I do not believe that any animal in the world performs an action for the exclusive good of another of a distinct species, yet each species tries to take advantage of the instincts of others, as each takes advantage of the weaker bodily structure of others. So again, in some few cases, certain instincts cannot be considered as absolutely perfect; but as details on this and other such points are not indispensable, they may be here passed over.

As some degree of variation in instincts under a state of nature, and the inheritance of such variations, are indispensable for the action of natural selection, as many instances as possible ought to have been here given; but want of space prevents me. I can only assert, that instincts certainly do vary—for instance, the migratory instinct, both in extent and direction, and in its total loss. So it is with the nests of birds, which vary partly in dependence on the situations chosen, and on the nature and temperature of the country inhabited, but often from causes wholly unknown to us: Audubon has given several remarkable cases of differences in nests of the same species in the northern and southern United States.

[*Editors' Note:* Material has been omitted at this point.]

Cell-making instinct of the Hive-Bee.—I will not here enter on minute details on this subject, but will merely give an outline of the conclusions at which I have arrived. He must be a dull man who can examine the exquisite structure of a comb, so beautifully adapted to its end, without enthusiastic admiration. We hear from mathematicians that bees have practically solved a recondite problem, and have made their cells of the proper shape to hold the greatest possible amount of honey, with the least possible consumption of precious wax in their construction. It has been remarked that a skilful workman, with fitting tools and measures, would find it very difficult to make cells of wax of the true form, though this is perfectly effected by a crowd of bees working in a dark hive. Grant whatever instincts you please, and it seems at first quite inconceivable how they can make all the necessary angles and planes, or even perceive when they are correctly made. But the difficulty is not nearly so great as it at first appears: all this beautiful work can be shown, I think, to follow from a few very simple instincts.

I was led to investigate this subject by Mr. Water-house, who has shown that the form of the cell stands in close relation to the presence of adjoining cells; and the following view may, perhaps, be considered only as a modification of his theory. Let us look to the great principle of gradation, and see whether Nature does not reveal to us her method of work. At one end of a short series we have humble-bees, which use their old cocoons to hold honey, sometimes adding to them short tubes of wax, and likewise making separate and very irregular rounded cells of wax. At the other end of the series we have the cells of the hive-bee, placed in a double layer: each cell, as is well known, is an hexagonal prism, with the basal edges of its six sides bevelled so as to join on to a pyramid, formed of three rhombs. These rhombs have certain angles, and the three which form the pyra-midal base of a single cell on one side of the comb, enter into the composition of the bases of three adjoining cells on the opposite side. In the series between the extreme perfection of the cells of the hive-bee and the simplicity of those of the humble-bee, we have the cells of the Mexican Melipona domestica, carefully described and figured by Pierre Huber. The Melipona itself is inter-mediate in structure between the hive and humble bee, but more nearly related to the latter: it forms a nearly regular waxen comb of cylindrical cells, in which the young are hatched, and, in addition, some large cells of wax for holding honey. These latter cells are nearly spherical and of nearly equal sizes, and are aggregated into an irregular mass. But the important point to notice, is that these cells are always made at that degree of nearness to each other, that they would have intersected or broken into each other, if the spheres had been completed; but this is never permitted, the bees building perfectly flat walls of wax between the spheres which thus tend to intersect. Hence each cell consists of an outer spherical portion and of two, three, or more perfectly flat surfaces, according as the cell adjoins two, three, or more other cells. When one cell comes into contact with three other cells, which, from the spheres being nearly of the same size, is very frequently and necessarily the case, the three flat surfaces are united into a pyramid; and this pyramid, as Huber has remarked, is manifestly a gross imitation of the three-sided pyramidal basis of the cell of the hive-bee. As in the cells of the hive-bee, so here, the three plane surfaces in any one cell necessarily enter into the construction of three adjoining cells. It is obvious that the Melipona saves wax by this manner of building; for the flat walls between the adjoining cells are not double, but are of the same thickness as the outer spherical portions, and yet each flat portion forms a part of two cells.

Reflecting on this case, it occurred to me that if the Melipona had made its spheres at some given distance from each other, and had made them of equal sizes and had arranged them symmetrically in a double layer, the resulting structure would probably have been as perfect as the comb of the hive-bee. Accordingly I wrote to Professor Miller, of Cambridge, and this geometer has kindly read over the following statement, drawn up from his information, and tells me that it is strictly correct:—

If a number of equal spheres be described with their centres placed in two parallel layers; with the centre of each sphere at the distance of radius $\times \sqrt{2}$, or radius $\times 1\cdot41421$ (or at some lesser distance), from the centres of the six surrounding spheres in the same layer; and at the same distance from the centres of the adjoining spheres in the other and parallel layer; then, if planes of intersection between the several spheres in both layers be formed, there will result a double layer of hexagonal prisms united together by pyramidal bases formed of three rhombs; and the rhombs and the sides of the hexagonal prisms will have every angle identically the same with the best measurements which have been made of the cells of the hive-be(

Hence we may safely conclude that if we could slightly modify the instincts already possessed by the Melipona, and in themselves not very wonderful, this bee would make a structure as wonderfully perfect as that of the hive-bee. We must suppose the Melipona to make her cells truly spherical, and of equal sizes; and this would not be very surprising, seeing that she already does so to a certain extent, and seeing what perfectly cylindrical burrows in wood many insects can make, apparently by turning round on a fixed point. We must suppose the Melipona to arrange her cells in level layers, as she already does her cylindrical cells; and we must further suppose, and this is the greatest difficulty, that she can somehow judge accurately at what distance to stand from her fellow-labourers when several are making their spheres; but she is already so far enabled to judge of distance, that she always describes her spheres so as to intersect largely; and then she unites the points of intersection by perfectly flat surfaces. We have further to suppose, but this is no difficulty, that after hexagonal prisms have been formed by the intersection of adjoining spheres in the same layer, she can prolong the hexagon to any length requisite to hold the stock of honey; in the same way as the rude humble-bee adds cylinders of wax to the circular mouths of her old cocoons. By such modifications of instincts in themselves not very wonderful,—hardly more wonderful than those which guide a bird to make its nest,—I believe that the hive-bee has acquired, through natural selection, her inimitable architectural powers.

But this theory can be tested by experiment. Following the example of Mr. Tegetmeier, I separated two combs, and put between them a long, thick, square strip of wax: the bees instantly began to excavate minute circular pits in it; and as they deepened these little pits, they made them wider and wider until they were converted into shallow basins, appearing to the eye perfectly true or parts of a sphere, and of about the diameter of a cell. It was most interesting to me to observe that wherever several bees had begun to excavate these basins near together, they had begun their work at such a distance from each other, that by the time the basins had acquired the above stated width (*i. e.* about the width of an ordinary cell), and were in depth about one sixth of the diameter of the sphere of which they formed a part, the rims of the basins intersected or broke into each other. As soon as this occurred, the bees ceased to excavate, and began to build up flat walls of wax on the lines of intersection between the basins, so that each hexagonal prism was built upon the festooned edge of a smooth basin, instead of on the straight edges of a three-sided pyramid as in the case of ordinary cells.

I then put into the hive, instead of a thick, square piece of wax, a thin and narrow, knife-edged ridge, coloured with vermilion. The bees instantly began on both sides to excavate little basins near to each other, in the same way as before; but the ridge of wax was so thin, that the bottoms of the basins, if they had been excavated to the same depth as in the former experiment, would have broken into each other from the opposite sides. The bees, however, did not suffer this to happen, and they stopped their excavations in due time; so that the basins, as soon as they had been a little deepened, came to have flat bottoms; and these flat bottoms, formed by thin little plates of the vermilion wax having been left ungnawed, were situated, as far as the eye could judge, exactly along the planes of imaginary intersection between the basins on the opposite sides of the ridge of wax. In parts, only little bits, in other parts, large portions of a rhombic plate had been left between the opposed basins, but the work, from the unnatural state of things, had not been neatly performed. The bees must have worked at very nearly the same rate on the opposite sides of the ridge of vermilion wax, as they circularly gnawed away and deepened the basins on both sides, in order to have succeeded in thus leaving flat plates between the basins, by stopping work along the intermediate planes or planes of intersection.

Considering how flexible thin wax is, I do not see that there is any difficulty in the bees, whilst at work on the two sides of a strip of wax, perceiving when they have gnawed the wax away to the proper thinness, and then stopping their work. In ordinary combs it has appeared to me that the bees do not always succeed in working at exactly the same rate from the opposite sides; for I have noticed half-completed rhombs at the base of a just-commenced cell, which were slightly concave on one side, where I suppose that the bees had excavated too quickly, and convex on the opposed side, where the bees had worked less quickly. In one well-marked instance, I put the comb back into the hive, and allowed the bees to go on working for a short time, and again examined the cell, and I found that the rhombic plate had been completed, and had become *perfectly flat*: it was absolutely impossible, from the extreme thinness of the little rhombic plate, that they could have effected this by gnawing away the convex side; and I suspect that the bees in such cases stand in the opposed cells and push and bend the ductile and warm wax (which as I have tried is easily done) into its proper intermediate plane, and thus flatten it.

From the experiment of the ridge of vermilion wax, we can clearly see that if the bees were to build for themselves a thin wall of wax, they could make their cells of the proper shape, by standing at the proper distance from each other, by excavating at the same rate, and by endeavouring to make equal spherical hollows, but never allowing the spheres to break into each other. Now bees, as may be clearly seen by examining the edge of a growing comb, do make a rough, circumferential wall or rim all round the comb; and they gnaw into this from the opposite sides, always working circularly as they deepen each cell. They do not make the whole three-sided pyramidal base of any one cell at the same time, but only the one rhombic plate which stands on the extreme growing margin, or the two plates, as the case may be; and they never complete the upper edges of the rhombic plates, until the hexagonal walls are commenced. Some of these statements differ from those made by the justly celebrated elder Huber, but I am convinced of their accuracy; and if I had space, I could show that they are conformable with my theory.

Huber's statement that the very first cell is excavated out of a little parallel-sided wall of wax, is not, as far as I have seen, strictly correct; the first commencement having always been a little hood of wax; but I will not here enter on these details. We see how important a part excavation plays in the construction of the cells; but it would be a great error to suppose that the bees cannot build up a rough wall of wax in the proper position—that is, along the plane of intersection between two adjoining spheres. I have several specimens showing clearly that they can do this. Even in the rude circumferential rim or wall of wax round a growing comb, flexures may sometimes be observed, corresponding in position to the planes of the rhombic basal plates of future cells. But the rough wall of wax has in every case to be finished off, by being largely gnawed away on both sides. The manner in which the bees build is curious; they always make the first rough wall from ten to twenty times thicker than the excessively thin finished wall of the cell, which will ultimately be left. We shall understand how they work, by supposing masons first to pile up a broad ridge of cement, and then to begin cutting it away equally on both sides near the ground, till a smooth, very thin wall is left in the middle; the masons always piling up the cut-away cement, and adding fresh cement, on the summit of the ridge. We shall thus have a thin wall steadily growing upward; but always crowned by a gigantic coping. From all the cells, both those just commenced and those completed, being thus crowned by a strong coping of wax, the bees can cluster and crawl over the comb without injuring the delicate hexagonal walls, which are only about one four-hundredth of an inch in thickness; the plates of the pyramidal basis being about twice as thick. By this singular manner of building, strength is continually given to the comb, with the utmost ultimate economy of wax.

It seems at first to add to the difficulty of understanding how the cells are made, that a multitude of bees all work together; one bee after working a short time at one cell going to another, so that, as Huber has stated, a score of individuals work even at the commencement of the first cell. I was able practically to show this fact, by covering the edges of the hexagonal walls

of a single cell, or the extreme margin of the circumferential rim of a growing comb, with an extremely thin
layer of melted vermilion wax; and I invariably found
that the colour was most delicately diffused by the bees
—as delicately as a painter could have done with his
brush—by atoms of the coloured wax having been taken
from the spot on which it had been placed, and worked
into the growing edges of the cells all round.   The work
of construction seems to be a sort of balance struck
between many bees, all instinctively standing at the
same relative distance from each other, all trying to
sweep equal spheres, and then building up, or leaving
ungnawed, the planes of intersection between these
spheres.   It was really curious to note in cases of difficulty, as when two pieces of comb met at an angle, how
often the bees would entirely pull down and rebuild in
different ways the same cell, sometimes recurring to a
shape which they had at first rejected.

When bees have a place on which they can stand in
their proper positions for working,—for instance, on a
slip of wood, placed directly under the middle of a comb
growing downwards so that the comb has to be built over
one face of the slip—in this case the bees can lay the
foundations of one wall of a new hexagon, in its strictly
proper place, projecting beyond the other completed
cells.   It suffices that the bees should be enabled to
stand at their proper relative distances from each other
and from the walls of the last completed cells, and then,
by striking imaginary spheres, they can build up a wall
intermediate between two adjoining spheres; but, as far
as I have seen, they never gnaw away and finish off the
angles of a cell till a large part both of that cell and of
the adjoining cells has been built.   This capacity in
bees of laying down under certain circumstances a
rough wall in its proper place between two just-com-

menced cells, is important, as it bears on a fact, which seems at first quite subversive of the foregoing theory; namely, that the cells on the extreme margin of wasp-combs are sometimes strictly hexagonal; but I have not space here to enter on this subject. Nor does there seem to me any great difficulty in a single insect (as in the case of a queen-wasp) making hexagonal cells, if she work alternately on the inside and outside of two or three cells commenced at the same time, always standing at the proper relative distance from the parts of the cells just begun, sweeping spheres or cylinders, and building up intermediate planes. It is even conceivable that an insect might, by fixing on a point at which to commence a cell, and then moving outside, first to one point, and then to five other points, at the proper relative distances from the central point and from each other, strike the planes of intersection, and so make an isolated hexagon: but I am not aware that any such case has been observed; nor would any good be derived from a single hexagon being built, as in its construction more materials would be required than for a cylinder.

As natural selection acts only by the accumulation of slight modifications of structure or instinct, each profitable to the individual under its conditions of life, it may reasonably be asked, how a long and graduated succession of modified architectural instincts, all tending towards the present perfect plan of construction, could have profited the progenitors of the hive-bee? I think the answer is not difficult: it is known that bees are often hard pressed to get sufficient nectar; and I am informed by Mr. Tegetmeier that it has been experimentally found that no less than from twelve to fifteen pounds of dry sugar are consumed by a hive of bees for the secretion of each pound of wax; so that a prodigious quantity of fluid nectar must be collected and consumed by the bees in a hive for the secretion of the wax necessary for the construction of their combs. Moreover, many bees have to remain idle for many days during the process of secretion. A large store of honey is indispensable to support a large stock of bees during the winter; and the security of the hive is known mainly to depend on a large number of bees being supported. Hence the saving of wax by largely saving honey must be a most important element of success in any family of bees. Of course the success of any species of bee may be dependent on the number of its parasites or other enemies, or on quite distinct causes, and so be altogether independent of the quantity of honey which the bees could collect. But let us suppose that this latter circumstance determined, as it probably often does determine, the numbers of a humble-bee which could exist in a country; and let us further suppose that the community lived throughout the winter, and consequently required a store of honey: there can in this case be no doubt that it would be an advantage to our humble-bee, if a slight modification of her instinct led her to make her waxen cells near together, so as to intersect a little; for a wall in common even to two adjoining cells, would save some little wax. Hence it would continually be more and more advantageous to our humble-bee, if she were to make her cells more and more regular, nearer together, and aggregated into a mass, like the cells of the Melipona; for in this case a large part of the bounding surface of each cell would serve to bound other cells, and much wax would be saved. Again, from the same cause, it would be advantageous to the Melipona, if she were to make her cells closer together, and more regular in every way than at present; for then, as we have seen, the spherical surfaces would wholly disappear, and would all be replaced by plane surfaces; and the Melipona would make a comb as perfect as that of the hive-bee. Beyond this stage of perfection in architecture, natural selection could not lead; for the comb of the hive-bee, as far as we can see, is absolutely perfect in economising wax.

Thus, as I believe, the most wonderful of all known instincts, that of the hive-bee, can be explained by natural selection having taken advantage of numerous, successive, slight modifications of simpler instincts; natural selection having by slow degrees, more and more perfectly, led the bees to sweep equal spheres at a given distance from each other in a double layer, and to build up and excavate the wax along the planes of intersection. The bees, of course, no more knowing that they swept their spheres at one particular distance from each other, than they know what are the several angles of the hexagonal prisms and of the basal rhombic plates. The motive power of the process of natural selection having been economy of wax; that individual swarm which wasted least honey in the secretion of wax, having succeeded best, and having transmitted by inheritance its newly acquired economical instinct to new swarms, which in their turn will have had the best chance of succeeding in the struggle for existence.

No doubt many instincts of very difficult explanation could be opposed to the theory of natural selection, —cases, in which we cannot see how an instinct could possibly have originated; cases, in which no intermediate gradations are known to exist; cases of instinct of apparently such trifling importance, that they could hardly have been acted on by natural selection; cases of instincts almost identically the same in animals so remote in the scale of nature, that we cannot account for their similarity by inheritance from a common parent, and must therefore believe that they have been acquired by independent acts of natural selection. I will not here enter on these several cases, but will confine myself to one special difficulty, which at first appeared to me insuperable, and actually fatal to my whole theory. I allude to the neuters or sterile females in insect-communities: for these neuters often differ widely in instinct and in structure from both the males and fertile females, and yet, from being sterile, they cannot propagate their kind.

The subject well deserves to be discussed at great length, but I will here take only a single case, that of working or sterile ants. How the workers have been rendered sterile is a difficulty; but not much greater than that of any other striking modification of structure; for it can be shown that some insects and other articulate animals in a state of nature occasionally become sterile; and if such insects had been social, and it had been profitable to the community that a number should have been annually born capable of work, but incapable of procreation, I can see no very great difficulty in this being effected by natural selection. But I must pass over this preliminary difficulty. The great difficulty lies in the working ants differing widely from both the males and the fertile females in structure, as in the shape of the thorax and in being destitute of wings and sometimes of eyes, and in instinct. As far as instinct alone is concerned, the prodigious difference in this respect between the workers and the perfect females, would have been far better exemplified by the hive-bee. If a working ant or other neuter insect had been an animal in the ordinary state, I should have unhesitatingly assumed that all its characters had been slowly acquired through natural selection; namely, by an individual having been born with some slight profitable modification of structure, this being inherited by its offspring, which again varied and were again selected, and so onwards.   But with the working ant we have an insect differing greatly from its parents, yet absolutely sterile; so that it could never have transmitted successively acquired modifications of structure or instinct to its progeny.   It may well be asked how is it possible to reconcile this case with the theory of natural selection?

First, let it be remembered that we have innumerable instances, both in our domestic productions and in those in a state of nature, of all sorts of differences of structure which have become correlated to certain ages, and to either sex.   We have differences correlated not only to one sex, but to that short period alone when the reproductive system is active, as in the nuptial plumage of many birds, and in the hooked jaws of the male salmon. We have even slight differences in the horns of different breeds of cattle in relation to an artificially imperfect state of the male sex; for oxen of certain breeds have longer horns than in other breeds, in comparison with the horns of the bulls or cows of these same breeds. Hence I can see no real difficulty in any character having become correlated with the sterile condition of certain members of insect-communities: the difficulty lies in understanding how such correlated modifications of structure could have been slowly accumulated by natural selection.

This difficulty, though appearing insuperable, is lessened, or, as I believe, disappears, when it is remembered that selection may be applied to the family, as well as to the individual, and may thus gain the desired end.   Thus, a well-flavoured vegetable is cooked, and the individual is destroyed; but the horticulturist sows seeds of the same stock, and confidently expects to get nearly the same variety ; breeders of cattle wish the flesh and fat to be well marbled together ; the animal has been slaughtered, but the breeder goes with confidence to the same family. I have such faith in the powers of selection, that I do not doubt that a breed of cattle, always yielding oxen with extraordinarily long horns, could be slowly formed by carefully watching which individual bulls and cows, when matched, produced oxen with the longest horns ; and yet no one ox could ever have propagated its kind. Thus I believe it has been with social insects : a slight modification of structure, or instinct, correlated with the sterile condition of certain members of the community, has been advantageous to the community : consequently the fertile males and females of the same community flourished, and transmitted to their fertile offspring a tendency to produce sterile members having the same modification. And I believe that this process has been repeated, until that prodigious amount of difference between the fertile and sterile females of the same species has been produced, which we see in many social insects.

[*Editors' Note:* Material has been omitted at this point.]

Summary.—I have endeavoured briefly in this chapter to show that the mental qualities of our domestic animals vary, and that the variations are inherited. Still more briefly I have attempted to show that instincts vary slightly in a state of nature. No one will dispute that instincts are of the highest importance to each animal. Therefore I can see no difficulty, under changing conditions of life, in natural selection accumulating slight modifications of instinct to any extent, in any useful direction.   In some cases habit or use and disuse have probably come into play.   I do not pretend that the facts given in this chapter strengthen in any great degree my theory ; but none of the cases of difficulty, to the best of my judgment, annihilate it. On the other hand, the fact that instincts are not always absolutely perfect and are liable to mistakes ;—that no instinct has been produced for the exclusive good of other animals, but that each animal takes advantage of the instincts of others ;—that the canon in natural history, of " natura non facit saltum" is applicable to instincts as well as to corporeal structure, and is plainly explicable on the foregoing views, but is otherwise inexplicable,—all tend to corroborate the theory of natural selection.

This theory is, also, strengthened by some few other facts in regard to instincts ; as by that common case of closely allied, but certainly distinct, species, when inhabiting distant parts of the world and living under considerably different conditions of life, yet often retaining nearly the same instincts.   For instance, we can understand on the principle of inheritance, how it is that the thrush of South America lines its nest with mud, in the same peculiar manner as does our British thrush : how it is that the male wrens (Troglodytes) of North America, build " cock-nests," to roost in, like the males of our distinct Kitty-wrens,—a habit wholly unlike that of any other known bird.   Finally, it may not be a logical deduction, but to my imagination it is far more satisfactory to look at such instincts as the young cuckoo ejecting its foster-brothers,—ants making slaves, —the larvæ of ichneumonidæ feeding within the live bodies of caterpillars,—not as specially endowed or created instincts, but as small consequences of one general law, leading to the advancement of all organic beings, namely, multiply, vary, let the strongest live and the weakest die.

2

Reprinted from J. H. Fabre, *The Mason-Bees*, A. T. de Matteos, trans., Dodd, Mead & Co., New York, 1914, pp. 158–179

SOME REFLECTIONS UPON INSECT PSYCHOLOGY

J. H. Fabre

THE *laudator temporis acti* is out of favour just now: the world is on the move. Yes, but sometimes it moves backwards. When I was a boy, our twopenny textbooks told us that man was a reasoning animal; nowadays, there are learned volumes to prove to us that human reason is but a higher rung in the ladder whose foot reaches down to the bottommost depths of animal life. There is the greater and the lesser; there are all the intermediary rounds; but nowhere does it break off and start afresh. It begins with zero in the glair of a cell and ascends until we come to the mighty brain of a Newton. The noble faculty of which we were so proud is a zoological attribute. All have a larger or smaller share of it, from the live atom to the anthropoid ape, that hideous caricature of man.

It always struck me that those who held

this levelling-theory made facts say more than they really meant; it struck me that, in order to obtain their plain, they were lowering the mountain-peak, man, and elevating the valley, the animal. Now this levelling of theirs needed proofs, to my mind; and, as I found none in their books, or at any rate only doubtful and highly debatable ones, I did my own observing, in order to arrive at a definite conviction; I sought; I experimented.

To speak with any certainty, it behoves us not to go beyond what we really know. I am beginning to have a passable acquaintance with insects, after spending some forty years in their company. Let us question the insect, then: not the first that comes along, but the most gifted, the Hymenopteron. I am giving my opponents every advantage. Where will they find a creature more richly endowed with talent? It would seem as though, in creating it, nature had delighted in bestowing the greatest amount of industry upon the smallest body of matter. Can the bird, wonderful architect that it is, compare its work with that masterpiece of higher geometry, the edifice of the Bee? The Hymenopteron rivals man himself. We build towns, the Bee erects cities;

we have servants, the Ant has hers; we rear
domestic animals, she rears her sugar-yielding
insects; we herd cattle, she herds her milch-
cows, the Aphides; we have abolished slavery,
whereas she continues her nigger-traffic.

Well, does this superior, this privileged
being reason? Reader, do not smile: this is
a most serious matter, well worthy of our consideration. To devote our attention to animals is to plunge at once into the vexed question of who we are and whence we come.
What, then, passes in that little Hymenopteron brain? Has it faculties akin to ours,
has it the power of thought? What a problem, if we could only solve it; what a chapter
of psychology, if we could only write it! But,
at our very first questionings, the mysterious
will rise up, impenetrable: we may be convinced of that. We are incapable of knowing
ourselves; what will it be if we try to fathom
the intellect of others? Let us be content if
we succeed in gleaning a few grains of truth.

What is reason? Philosophy would give
us learned definitions. Let us be modest and
keep to the simplest: we are only treating of
animals. Reason is the faculty that connects
the effect with its cause and directs the act by

conforming it to the needs of the accidental. Within these limits, are animals capable of reasoning? Are they able to connect a "because" with a "why" and afterwards to regulate their behaviour accordingly? Are they able to change their line of conduct when faced with an emergency?

History has but few data likely to be of use to us here; and those which we find scattered in various authors are seldom able to withstand a severe examination. One of the most remarkable of which I know is supplied by Erasmus Darwin, in his book entitled *Zoonomia*. It tells of a Wasp that has just caught and killed a big Fly. The wind is blowing, and the huntress, hampered in her flight by the great area presented by her prize, alights on the ground to amputate the abdomen, the head and the wings; she flies away, carrying with her only the thorax, which gives less hold to the wind. If we keep to the bald facts, this does, I admit, give a semblance of reason. The Wasp appears to grasp the relation between cause and effect. The effect is the resistance experienced in the flight; the cause is the dimensions of the prey contending with the air. Hence the logical conclusion: those dimensions must be decreased; the abdomen, the head and, above all, the wings must be chopped off; and the resistance will be lessened.[1]

But does this concatenation of ideas, rudimentary though it be, really take place within the insect's brain? I am convinced of the con-

---

[1] I would gladly, if I were able, cancel some rather hasty lines which I allowed myself to pen in the first volume of these *Souvenirs;* but *scripta manent* and all that I can do is to make amends now, in this note, for the error into which I fell. Relying on Lacordaire, who quotes this instance from Erasmus Darwin in his own *Introduction à l'entomologie,* I believed that a Sphex was given as the heroine of the story. How could I do otherwise, not having the original text in front of me? How could I suspect that an entomologist of Lacordaire's standing should be capable of such a blunder as to substitute a Sphex for a Common Wasp? Great was my perplexity, in the face of this evidence! A Sphex capturing a Fly was an impossibility; and I blamed the British scientist accordingly. But what insect was it that Erasmus Darwin saw? Calling logic to my aid, I declared that it was a Wasp; and I could not have hit the mark more truly. Charles Darwin, in fact, informed me afterwards that his grandfather wrote, "a Wasp," in his *Zoonomia.* Though the correction did credit to my intelligence, I none the less deeply regretted my mistake, for I had uttered suspicions of the observer's powers of discernment, unjust suspicions which the translator's inaccuracy led me into entertaining. May this note serve to mitigate the harshness of the strictures provoked by my overtaxed credulity. I do not scruple to attack ideas which I consider false; but Heaven forfend that I should ever attack those who uphold them!—*Author's Note.*

trary; and my proofs are unanswerable. In the first volume of these *Souvenirs,*[1] I demonstrated by experiment that Erasmus Darwin's Wasp was but obeying her instinct, which is to cut up the captured game and to keep only the most nourishing part, the thorax. Whether the day be perfectly calm or whether the wind blow, whether she be in the shelter of a dense thicket or in the open, I see the Wasp proceed to separate the succulent from the tough; I see her reject the legs, the wings, the head and the abdomen, retaining only the breast as pap for her larvæ. Then what value has this dissection as an argument in favour of the insect's reasoning-powers when the wind blows? It has no value at all, for it would take place just the same in absolutely calm weather. Erasmus Darwin jumped too quickly to his conclusion, which was the outcome of his mental bias and not of the logic of things. If he had first enquired into the Wasp's habits, he would not have brought forward as a serious argument an incident which had no connection with the important question of animal reason.

[1]Cf. *Insect Life:* chap. ix.—*Translator's Note.*

I have reverted to this case to show the difficulties that beset the man who confines himself to casual observations, however carefully carried out. One should never rely upon a lucky chance, which may not occur again. We must multiply our observations, check them one with the other; we must create incidents, looking into preceding ones, finding out succeeding ones and working out the relation between them all: then and not till then, with extreme caution, are we entitled to express a few views worthy of credence. Nowhere do I find data collected under such conditions; for which reason, however much I might wish it, it is impossible for me to bring the evidence of others in support of the few conclusions which I myself have formed.

My Mason-bees, with their nests hanging on the walls of the arch which I have mentioned, lent themselves to continuous experiment better than any other Hymenopteron. I had them there, at my house, under my eyes, at all hours of the day, as long as I wished. I was free to follow their actions in full detail and to carry out successfully any experiment, however long. Moreover, their numbers allowed me to repeat my attempts until I was perfectly convinced. The Mason-bees, therefore, shall supply me with the materials for this chapter also.

A few words, before I begin, about the works. The Mason-bee of the Sheds utilizes, first of all, the old galleries of the clay nest, a part of which she good-naturedly abandons to two Osmiæ, her free tenants: the Three-horned Osmia and Latreille's Osmia. These old corridors, which save labour, are in great demand; but there are not many vacant, as the more precocious Osmiæ have already taken possession of most of them; and therefore the building of new cells soon begins. These cells are cemented to the surface of the nest, which thus increases in thickness every year. The edifice of cells is not built all at once: mortar and honey alternate repeatedly. The masonry starts with a sort of little swallow's nest, a half-cup or thimble, whose circumference is completed by the wall against which it rests. Picture the cup of an acorn cut in two and stuck to the surface of the nest: there you have the receptacle in a stage sufficiently advanced to take a first instalment of honey.

The Bee thereupon leaves the mortar and busies herself with harvesting. After a few foraging-trips, the work of building is resumed; and some new rows of bricks raise the edge of the basin, which becomes capable of receiving a larger stock of provisions. Then comes another change of business: the mason once more becomes a harvester. A little later, the harvester is again a mason; and these alternations continue until the cell is of the regulation height and holds the amount of honey required for the larva. Thus come, turn and turn about, more or less numerous according to the occupation in hand, journeys to the dry and barren path, where the cement is gathered and mixed, and journeys to the flowers, where the Bee's crop is crammed with honey and her belly powdered with pollen.

At last comes the time for laying. We see the Bee arrive with a pellet of mortar. She gives a glance at the cell to enquire if everything is in order; she inserts her abdomen; and the egg is laid. Then and there the mother seals up the home: with her pellet of cement she closes the orifice and manages so well with the material that the lid receives its permanent form at this first sitting; it has only to be thickened and strengthened with fresh layers, a work which is less urgent and will be done by and by. What does appear to be an urgent necessity is the closing of the cell, immediately after the egg has been religiously deposited therein, so that there may be no danger from evilly-disposed visitors during the mother's absence.

The Bee must have serious reasons for thus hurrying on the closing of the cell. What would happen if, after laying her egg, she left the house open and went to the cement-pit to fetch the wherewithal to block the door? Some thief might drop in and substitute her own egg for the Mason-bee's. We shall see that our suspicions are not uncalled-for. One thing is certain, that the Mason never lays without having in her mandibles the pellet of mortar required for the immediate construction of the lid of the nest. The precious egg must not for a single instant remain exposed to the cupidity of marauders.

To these particulars I will add a few general observations which will make what follows easier to understand. So long as its circumstances are normal, the insect's actions are calculated most rationally in view of the object to be attained. What could be more logical, for instance, than the devices employed by the Hunting-wasp when paralyzing her prey[1] so that it may keep fresh for her larva, while in no wise imperilling that larva's safety? It is preeminently rational; we ourselves could think of nothing better; and yet the insect's action is not prompted by reason. If it thought out its surgery, it would be our superior. It will never occur to anybody that the creature is able, in the smallest degree, to account for its skilful vivisections. Therefore, so long as it does not depart from the path mapped out for it, the insect can perform the most sagacious actions without entitling us in the least to attribute these to the dictates of reason.

What would happen in an emergency? Here we must distinguish carefully between two classes of emergency, or we shall be liable to grievous error. First, in accidents occurring in the course of the insect's occupation at the moment. In these circumstances, the creature is capable of remedying the accident; it continues, under a similar form, its actual task; it remains, in short, in the same psychic condition. In the second case, the accident is con-

[1] Cf. *Insect Life:* chaps. iii. to xii. and xv. to xvii.—*Translator's Note.*

nected with a more remote occupation; it relates to a completed task with which, under normal conditions, the insect is no longer concerned. To meet this emergency, the creature would have to retrace its psychic course; it would have to do all over again what it has just finished, before turning its attention to anything else. Is the insect capable of this? Will it be able to leave the present and return to the past? Will it decide to hark back to a task that is much more pressing than the one on which it is engaged? If it did all this, then we should really have evidence of a modicum of reason. The question shall be settled by experiment.

We will begin by taking a few incidents that come under the first heading. A Mason-bee has finished the first layer of the covering of the cell. She has gone in search of a second pellet of mortar wherewith to strengthen her work. In her absence, I prick the lid with a needle and widen the hole thus made, until it is half the size of the opening. The insect returns and repairs the damage. It was originally engaged on the lid and is merely continuing its work in mending that lid.

A second is still at her first row of bricks.

The cell as yet is no more than a shallow cup, containing no provisions. I make a big hole in the bottom of the cup and the Bee hastens to stop the breach. She was busy building and turned aside a moment to do more building. Her repairs are the continuation of the work on which she was engaged.

A third has laid her egg and closed the cell. While she is gone in search of a fresh supply of cement to strengthen the door, I make a large aperture immediately below the lid, too high up to allow the honey to escape. The insect, on arriving with its mortar intended for a different task, sees its broken jar and soon puts the damage right. I have rarely witnessed such a sensible performance. Nevertheless, all things considered, let us not be too lavish of our praises. The insect was busy closing up. On its return, it sees a crack, representing in its eyes a bad joint which it had overlooked; it completes its actual task by improving the joint.

The conclusion to be drawn from these three instances, which I select from a large number of others, more or less similar, is that the insect is able to cope with emergencies, provided that the new action be not outside the course of its actual work at the moment. Shall we say then that reason directs it? Why should we? The insect persists in the same psychic course, it continues its action, it does what it was doing before, it corrects what to it appears but a careless flaw in the work of the moment.

Here, moreover, is something which would change our estimate entirely, if it ever occurred to us to look upon these repaired breaches as a work dictated by reason. Let us turn to the second class of emergency referred to above: let us imagine, first, cells similar to those in the second experiment, that is to say, only half-finished, in the form of a shallow cup, but already containing honey. I make a hole in the bottom, through which the provisions ooze and run to waste. Their owners are harvesting. Let us imagine, on the other hand, cells very nearly finished and almost completely provisioned. I perforate the bottom in the same way and let out the honey, which drips through gradually. The owners of these are building.

Judging by what has gone before, the reader will perhaps expect to see immediate repairs, urgent repairs, for the safety of the

future larva is at stake. Let him dismiss any
such illusion: more and more journeys are undertaken, now in quest of food, now in quest
of mortar; but not one of the Mason-bees
troubles about the disastrous breach. The
harvester goes on harvesting; the busy brick-
layer proceeds with her next row of bricks, as
though nothing out of the way had happened.
Lastly, if the injured cells are high enough
and contain enough provisions, the Bee lays
her eggs, puts a door to the house and passes
on to another house, without doing aught to
remedy the leakage of the honey. Two or
three days later, those cells have lost all their
contents, which now form a long trail on the
surface of the nest.

Is it through lack of intelligence that the
Bee allows her honey to go to waste? May
it not rather be through helplessness? It
might happen that the sort of mortar which
the mason has at her disposal will not set on
the edges of a hole that is sticky with honey.
The honey may prevent the cement from adjusting itself to the orifice, in which case the
insect's inertness would merely be resignation
to an irreparable evil. Let us look into the
matter before drawing inferences. With my

forceps, I deprive the Bee of her pellet of mortar and apply it to the hole whence the honey is escaping. My attempt at repairing meets with the fullest success, though I do not pretend to compete with the Mason in dexterity. For a piece of work done by a man's hand it is quite creditable. My dab of mortar fits nicely into the mutilated wall; it hardens as usual; and the escape of honey ceases. This is quite satisfactory. What would it be had the work been done by the insect, equipped with its tools of exquisite precision? When the Mason-bee refrains, therefore, this is not due to helplessness on her part, nor to any defect in the material employed.

Another objection presents itself. We are going too far perhaps in admitting this concatenation of ideas in the insect's mind, in expecting it to argue that the honey is running away because the cell has a hole in it and that to save it from being wasted the hole must be stopped. So much logic perhaps exceeds the powers of its poor little brain. Then, again, the hole is not seen; it is hidden by the honey trickling through. The cause of that stream of honey is an unknown cause; and to trace the loss of the liquid home to that cause, to the hole in the receptacle, is too lofty a piece of reasoning for the insect.

A cell in the rudimentary cup-stage and containing no provisions has a hole, three or four millimetres[1] wide, made in it at the bottom. A few moments later, this orifice is stopped by the Mason. We have already witnessed a similar patching. The insect, having finished, starts foraging. I reopen the hole at the same place. The pollen runs through the aperture and falls to the ground as the Bee is rubbing off her first load in the cell. The damage is undoubtedly observed. When plunging her head into the cup to take stock of what she has stored, the Bee puts her antennæ into the artificial hole: she sounds it, she explores it, she cannot fail to perceive it.

I see the two feelers quivering outside the hole. The insect notices the breach in the wall: that is certain. It flies off. Will it bring back mortar from its present journey to repair the injured jar even as it did but a few minutes ago?

Not at all. It returns with provisions, it disgorges its honey, it rubs off its pollen, it

[1] .11 to .15 inch.—*Translator's Note.*

mixes the material. The sticky and almost solid mass fills up the opening and oozes through with difficulty. I roll a spill of paper and free the hole, which remains open and shows daylight clearly in both directions. I sweep the place clear over and over again, whenever it becomes necessary because new provisions are brought; I clean the opening sometimes in the Bee's absence, sometimes in her presence, while she is busy mixing her paste. The unusual happenings in the warehouse plundered from below cannot escape her any more than the ever-open breach at the bottom of the cell. Nevertheless, for three consecutive hours, I witness this strange sight: the Bee, full of active zeal for the task in hand, omits to plug this vessel of the Danaides. She persists in trying to fill her cracked receptacle, whence the provisions disappear as soon as stored away. She constantly alternates between mason's and harvester's work; she raises the edges of the cell with fresh rows of bricks; she brings provisions which I continue to abstract, so as to leave the breach always visible. She makes thirty-two journeys before my eyes, now for mortar, now for honey, and not once does she bethink herself of stopping the leakage at the bottom of her jar.

At five o'clock in the evening, the works cease. They are resumed on the morrow. This time, I neglect to clean out my artificial orifice and leave the victuals gradually to ooze out by themselves. At length, the egg is laid and the door sealed up, without anything being done by the Bee in the matter of the disastrous breach. And yet to plug the hole were an easy matter for her: a pellet of her mortar would suffice. Besides, while the cup was still empty, did she not instantly close the hole which I had made? Why are not those early repairs of hers repeated? It clearly shows the creature's inability to retrace the course of its actions, however slightly. At the time of the first breach, the cup was empty and the insect was laying the first rows of bricks. The accident produced through my agency concerned that part of the work which occupied the Bee at the actual moment; it was a flaw in the building, such as can occur naturally in new courses of masonry, which have not had time to harden. In correcting that flaw, the mason did not go outside her usual work.

But, once the provisioning begins, the cup is finished for good and all; and, come what may, the insect will not touch it again. The harvester will go on harvesting, though the pollen trickle to the ground through the drain. To plug the hole would imply a change of occupation of which the insect is incapable for the moment. It is the honey's turn and not the mortar's. The rule upon this point is invariable. A moment comes, presently, when the harvesting is interrupted and the building resumed. The edifice must be raised a storey higher. Will the Bee, once more a mason, mixing fresh cement, now attend to the leakage at the bottom? No more than before. What occupies her at present is the new floor, whose brickwork would be repaired at once if it sustained a damage; but the bottom storey is too old a part of the business, it is ancient history; and the worker will not put a further touch to it, even though it be in serious danger.

For the rest, the present and the following storeys will all have the same fate. Carefully watched by the insect as long as they are in process of building, they are forgotten and allowed to go to ruin once they are actually

The Mason-bees

built. Here is a striking instance: in a cell which has attained its full height, I make a window, almost as large as the natural opening, and place it about half-way up, above the honey. The Bee brings provisions for some time longer and then lays her egg. Through my big window, I see the egg deposited on the victuals. The insect next works at the cover, to which it gives the finishing touches with a series of little taps, administered with infinite care, while the breach remains yawning. On the lid, it scrupulously stops up every pore that could admit so much as an atom; but it leaves the great opening that places the house at the mercy of the first-comer. It goes to that breach repeatedly, puts in its head, examines it, explores it with its antennæ, nibbles the edges of it. And that is all. The mutilated cell shall stay as it is, with never a dab of mortar. The threatened part dates too far back for the Bee to think of troubling about it.

I have said enough, I think, to show the insect's mental incapacity in the presence of the accidental. This incapacity is confirmed by renewing the test, an essential condition of all good experiments; therefore my notes are full of examples similar to the one which I have just described. To relate them would be mere repetition; I pass them over for the sake of brevity.

[*Editors' Note:* Material has been omitted at this point.]

3

Reprinted from *Ants: Their Structure, Development, and Behavior*, W. M. Wheeler, Columbia University Press, New York, 1910, pp. 192–199, 216–224

ANT-NESTS

W. M. Wheeler

"Le premier objet qui frappe nos sens en commençant à étudier les mœurs des fourmis, c'est l'art avec lequel elles construisent leur habitation, dont la grandeur paroit souvent contraster avec leur petitesse; c'est la variété de ces bâtimens, tantôt fabriqués avec de la terre, tantôt sculptés dans le tronc des arbres les plus durs; ou composés simplement de feuilles et de brins d'herbe ramassés de toutes parts; c'est enfin la manière dont ils répondent aux besoins des espèces qui les construisent."—P. Huber, "Les Mœurs des Fourmis Indigènes," 1810.

Nothing is better calculated to illustrate the marvellous plasticity of ants than the study of their nesting habits. Not only may every species be said to have its own plan of nest construction, but this plan may be modified in manifold ways in order to adapt it to the particular environment in which the species takes up its abode. Even the same colony may adopt very different methods of building at different periods in its growth and development. Hence the study of formicine architecture becomes one of bewildering complexity and defies all attempts at rigid classification. Owing to this complexity it is impossible to form a correct conception of the general plan of architecture in a particular species without studying its nesting habits throughout its whole geographical range. In such a subject recourse to laboratory methods is of little avail, whereas careful and extensive observation in the field is all-important.

One remarkable peculiarity of ant-nests impresses us at the very outset when we compare them with the nests of the social wasps and bees, namely, their extreme irregularity. The ants have abandoned, if indeed they ever acquired, the habit of constructing regular and permanent cells for their brood. The advantages of such cells to the ants evidently do not outweigh the disadvantages of being unable to move their larvæ and pupæ from place to place when danger threatens or in response to the diurnal variations of warmth and moisture. In its essential features the typical nest is merely a system of intercommunicating cavities with one or more openings to the outside world (Fig. 106). Even these openings, or entrances, as they are called, are absent in the nests of hypogæic species, except at the time of the nuptial flight. The intercommunicating cavities may be excavated in the soil or in plants, and even preëxisting cavities often answer every purpose and save labor. The irregular form of the cavities is a characteristic so universal in ant-nests that it would seem to be preferred to a monotonous regularity. It may be important, in fact, in enabling the ants to orient themselves readily. The nest entrance is sometimes peculiarly modified to suit the needs of the various species. It may be left permanently

FIG. 106. Superficial galleries of *Acanthomyops latipes* as they appear on removing the stone that covers them. About ¼ natural size. (Original.)

open and guarded by workers or soldiers, or it may be closed at night; it may be enlarged or constricted for the purpose of regulating the ventilation of the cavities and preventing the inroads of enemies, it may be adroitly concealed or exposed to view and surrounded by conspicuous earth-works.

Even in this prevailing and opportunistic irregularity, however, there are singular differences of degree. The more primitive ants, like the Ponerinæ, build with a certain irregularity devoid of character. The Dorylinæ may hardly be said to build nests at all, but merely to bivouac in some convenient cavity under a stone or log, or they may temporarily occupy the nests of other ants or dig irregular runways beneath the surface of the soil. The higher ants, however, which form stationary and populous formicaries, devote a great deal of attention to architecture and work according to a more or less definite plan, which they skilfully modify to suit the conditions of a specific environment.

The nests of nearly all ants are the result of two different activities, excavation and construction.   Both of these may be simultaneously pursued by the workers, or either may predominate to the complete exclusion of the other, so that some nests are entirely excavated in soil or wood, whereas others are entirely constructed of soil, paper or silk. As the nests of the latter type resemble those of the social wasps, one might be led to suppose that they represent the original ancestral form and that the excavated are degenerate types, but the prevalence of earthen nests among ants of the most diverse genera in all parts of the world, as well as the occurrence of similar nests among the solitary bees, wasps and Mutillids, would seem to indicate that even the most

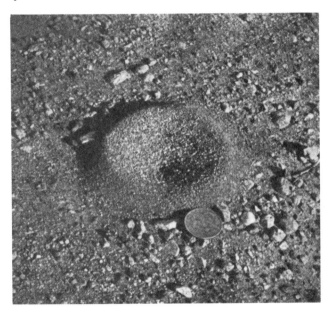

FIG. 107.   Crater of *Myrmecocystus semirufus* of the Mojave Desert ; ¼ natural size.
(Original.)

ancient ants practiced both methods of nesting.   In other words, the variable architecture of ants may be an inheritance from presocial ancestors and may have been well-established before these insects came to live in communities.

The methods employed by worker ants in making their nests are easily observed, and have been described in detail by Huber (1810) and Forel (1874). According to Forel, "They use their mandibles in two ways. When closed these organs form a kind of trowel, convex in front and above, concave beneath and behind, and pointed at the tip. This trowel is used for raking up the soft earth and also for moulding and compressing their constructions and thus rendering them more solid and continuous. This is accomplished by pushing the anterior portion of the closed mandibles forward or upward. In the second place, the mandibles, when open, constitute a veritable pair of tongues with toothed edges, at least in all of the workers of our native ants that do any excavating. They thus serve not only for transporting but also for moulding or comminuting the earth." The forelegs are used for scratching up the soil, in moulding pellets and patting them down after they have been placed in position by the mandibles, and are of so much assistance in this work that when they are cut off the insects are unable to excavate or build without great difficulty and soon abandon their work altogether.

Ants dislike to excavate in soil that is too dry and friable. When compelled to do this in artificial nests they will sometimes moisten it with water brought from a distance, as Miss Fielde (1901) has observed. She says that the workers of *Aphænogaster picea*, "like the Termites, are able to carry water for domestic uses. They probably lap the water into the pouch above the lower lip [the hypopharyngeal pocket] and eject it at its destination. A hundred or two of ants that I brought in and left in a heap of dry earth upon a Lubbock nest, during the ensuing night took water from the surrounding moat, moistened a full pint of earth, built therein a proper nest, and were busy depositing their larvæ in its recesses when I saw them on the following morning."

As even the most extensively excavated nests represent little labor compared with the nests of social wasps and bees, ants are able to leave their homes and make new ones without serious inconvenience. Such changes are often necessitated by the habit of nesting in situations exposed to great and sudden changes in temperature and moisture or to the inroads of more aggressive ants and larger terrestrial animals. Barring the intervention of such unusual conditions, however, most ants cling to their nests tenaciously and with every evidence of a keen sense of proprietorship, although there are a few species, besides the nomadic Dorylinæ, that seem to delight in an occasional change of residence. Wasmann has shown that *Formica sanguinea* often has summer and winter residences analogous to the city and country homes of wealthy people. The ants migrate from one to the other during March and April and again during late summer or early autumn (September). The summer nests are built in open, sunny places where food is abundant and the conditions most favorable to rearing the brood, whereas the winter nests are built under stumps and

FIG. 108. Nest of *Pogonomyrmex occidentalis* at Las Vegas, New Mexico: showing the basal entrance on the southeastern side. (Original.)

rocks usually in protected spots in the woods, and are used as hibernacula, or, very rarely, for protection from excessive heat during the summer.

The migration of ants from one nest to another is determined upon and initiated by a few workers which are either more sensitive to adverse conditions or of a more alert and venturesome disposition than the majority of their fellows. These workers, after selecting a site, begin to deport their brood, queens, males, fellow workers and even their myrmecophiles. The deported workers are at first too strongly attached to their old quarters to remain in the new ones and therefore keep returning and carrying back the brood. The enterprising workers, however, obstinately persist in their endeavors to move the colony till their intentions are grasped and become contagious. The indecision or indifference of many of the workers may last for days or even for weeks, during all which time files of ants move back and forth between the two nests carrying their larvæ and pupæ in both directions. But more and more workers keep joining the ranks of the radicals till the conservative individuals constitute such a helpless minority that they

FIG. 109. Large nest of *Formica exsectoides*, at Scotch Plains, N. J. Height 1 meter, basal diameter 3.25 m., circumference 10.21 m. (Original.)

are compelled to abandon the old nest and join the majority. I once observed a colony of agricultural ants (*Pononomyrmex molefaciens*) which for at least two years had occupied a nest directly in front of my house in Austin, Texas. In the autumn of the third year when certain workers decided to establish a new nest in a vacant lot about seventy feet away, I observed that it required nearly three weeks to overcome the attachment of all the workers to their old home.

Forel and Escherich (1906) distinguish two types of ant-nests, the temporary and the permanent, but this does not involve corresponding differences in architecture. The same is true of Forel's convenient distinction of monodomous and polydomous colonies. The nest of a monodomous colony is a circumscribed unit, whereas a polydomous colony, as the name implies, spreads over several nests, the inhabitants of which remain in communication with one another and may visit back and forth. This may lead to the development of accessory structures, like covered runways, but in other respects the architecture is merely a repetition of that of the simple nest. For convenience we may adopt the following classification:

A. Nests in the Soil.
  a. Small crater nests.
  b. Large crater nests.
  c. Mound or hill nests.
  d. Masonry domes.
  e. Nests under stones, logs, etc.
B. Nests in the Cavities of Plants.
  1. Nests in preformed cavities of living plants.
    a. In hollow stems.
    b. In hollow thorns.
    c. In tillandsias.
    d. In hollow bulbs.
  2. Nests in woody plant-tissues, often in cavities wholly or in part excavated by other insects.
    a. In or under bark.
    b. In twigs.
    c. In tree-trunks.
    d. In galls, pine-cones, seed-pods, etc.
C. Suspended Nests.
  a. Suspended earthen nests.
  b. Carton nests.
  c. Silken nests.
D. Nests in Unusual Sites (in houses, etc.).

E. Accessory Structures.
  a. Succursal nests.
  b. Covered runways.
  c. Tents, or pavilions.

Accurate delimitation of the foregoing categories is, of course, impossible, since two or more of them may be combined in the same nest. Thus some ants construct carton nests in dead logs or under stones, others extend their galleries from dead logs into the underlying soil. Then there are also transitional forms between the various categories, as, for example, between the small and large crater nests, and between the latter and mound nests. And lastly, a single formicary may gradually pass through a series of these categories during its growth and development.

[*Editors' Note:* Material has been omitted at this point.]

The most extraordinary ant-nests are undoubtedly the silken structures inhabited by certain species of *Œcophylla, Camponotus* and *Polyrhachis,* all genera belonging to the same subfamily. The following description of several silken nests of *Polyrhachis* and the nests of *Œcophylla* is taken from Forel (1894c):

" The nest of the *Polyrhachis jerdonii* Forel which I received from Ceylon through Major Yerbury is very interesting. This species builds upon leaves small nests, the wall of which greatly resembles in appearance the shell of many Phryganeidæ larvæ. Pebbles, and especially small fragments of plants, are cemented together by a fine web or woven together, and form a rather soft, tough web-like nest wall of a greyish-brown color. . . . We see here unmistakable small fragments of plants bound together in a web by peculiar silk threads. These silk threads are found, upon close examination, to be of very irregular thickness, often branching, and in many cases issuing from a thicker crosspiece. . . . *Polyrhachis dives*, however, no longer needs any foreign materials. It makes its nest wall out of pure silk web, exactly like coarse spun yarn or the web of a caterpillar. The web is of a brownish-yellow color and is fixed between leaves, which are lined with it and are bound together. Mr. Wroughton, of Poonah, India, sent me such a nest, simply between two leaves. A still finer, softer silk web, finer and thicker than the finest silk paper, very soft and as pliable as the finest gauze, though much thicker, of a brown color, is produced by *Polyrhachis spinigera* Mayr. . . . Here we find no more crosspieces but only silk webs. They are, however, still irregular, of varying thickness, spun

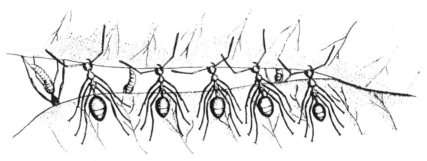

Fig. 122. Brigade of *Œcophylla smaragdina* workers drawing edges of leaves together while other workers bind them together with the silk spun by the larvæ. (Doflein.)

across each other into a web. This web is fixed in a wonderful manner in the ground, where it forms the lining of a funnel-shaped cave which is widened out into a chamber at the bottom. . . . The large nest constructed in the foliage of trees, between the leaves by *Œcophylla smaragdina* Fabr., one of the most common ants of tropical Africa, forms, however, the prototype of spun ant-nests. A great number of leaves are fastened together by a fine white web, like the finest silk stuff. This web, apart from the color, has exactly the same appearance, both to the naked eye and under the microscope, as that of *Polyrhachis spinigera*. The leaves are usually fastened together by the edges. The nest is larger, and the large, long, very vicious, reddish or greenish worker ants live in it, with their grass-green females, their black males, and the whole brood. They form very populous colonies in the branches of the trees." A similar nest is constructed by *Camponotus senex* in the forests of Brazil.

As no adult insects are known to produce silk, the question naturally arises as to how the ants manage to make these nests. Misled by some inadequate observations on *Œcophylla* published by Aitken in 1890, Forel concluded that the silk was spun by the workers from the maxillary glands. In other words, he believed it to be equivalent to the glandular component of the carton manufactured by other ants. Subsequent observations, however, have shown that it has a very different and more remarkable origin. Ridley (1890) discovered in Singapore that *Œcophylla* uses its larvæ for spinning the silk of its nest. His observations were confirmed by Holland and Green (1896, 1899) and Dodd (1902). Chun in 1903 observed that the spinning glands or sericteries of *Œcophylla* larvæ are enormously developed and Saville Kent is said to have figured the spinning larvæ of *Œcophylla* from specimens found in Australia (1897). More recently (1905) Doflein has published some interesting observations on the nest-building habits of this ant. I here reproduce his account together with three of the accompanying figures. On opening a nest for the purpose of studying its reconstruction Doflein observed " that while the majority of the workers betook themselves to defending their home, a small troop went to work to repair the rent I had made in its wall. They lined up in a very peculiar manner in a straight row as shown in the figure. They seized the edge of the leaf on one side of the rent while they fastened themselves by means of the claws on their six feet to the surface of another leaf (Fig. 122). Then they began to pull, slowly and cautiously, carefully placing one foot behind the other, while the edges of the rent were seen gradually to approach each other. It was a bizarre sight to see the animals thus working side by side with their bodies in a regular parallel series.

" Now others came and began to cut away very carefully the remnants of the old web along the edges of the rent. They bit through the web with their mandibles and tugged at it till it came away in shreds. These shreds they carried off in their mandibles to an exposed part of the nest and let them fly away in the wind, opening their mandibles whenever there was a gust. I also saw a whole row of ants carry a big piece of web to the tip of a leaf, open their mandibles as if at command and permit the piece to flutter away.

" These operations lasted nearly an hour when suddenly a strong gust of wind tore the edges of the rent out of the ants' mandibles and frustrated all their efforts. But the ants were not discouraged. Again a long row of them lined up along the slit and in half an hour they had again brought the edges pretty close together.

"I was about to despair of seeing the important part of the performance, when several workers emerged from the interior of the nest, each with a larva in its mandibles. And they did not run away with the larvæ in order to deposit them in a place of safety, but came right up to the exposed opening of the slit. There they were to be seen climbing about behind the row of workers making very odd motions with their heads. With their mandibles they held their larvæ so tightly that the bodies of the latter appeared to be compressed in the middle. Perhaps this pressure is necessary in order to excite the function of the spinning glands. It was a strange sight to see them passing between the ranks of the workers that were holding the leaves. While the latter remained on the outside, the former carried on their

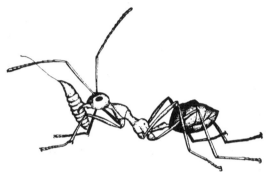

FIG. 123. *Œcophylla smaragdina* worker using a larva as a shuttle in weaving the silken tissue of the nest. (Doflein.)

work within the nest. This made it more difficult to observe what was going on. But after some time I could see very clearly that the larvæ were carried with their anterior ends directed forward and upward (Fig. 123) and were kept moving from one side to the other of the rent. At the same time each worker waited an instant on one side of the rent, as if it were gluing fast the thread spun by the larva by pressing its head against the leaf, before the head of the larva was carried across the rent and the same process repeated on the other side. Gradually, while they tirelessly pursued their task, the rent was seen to be filled out with a fine, silken web.

"There could be no doubt that the ants were actually using their larvæ both as spools and shuttles. As several workers toiled close together, they were able to cross and re-cross the threads and thus produce a rather tenacious tissue. This could be cut with the scissors and the small pieces presented a singular appearance under the microscope.

A lot of threads were seen crossing one another and in some places a number of the threads had a common direction. This agrees very well with my observations on the origin of the web. The ants are in the habit of moving to and fro with the larvæ many times in the same place before changing their position and running the thread crosswise. In this way strands are soon formed on the outside of the web and evidently save labor on the part of the ants that are holding the leaves together. Under the microscope the threads appear to be glued together at many points. This condition is very easily explained when we consider that each thread is moist and sticky for a few moments after leaving the sericteries of the larva.

" I was unable to see the thread itself while it was being spun as it is too delicate and transparent to be visible to the unaided eye. I tried to see it with a strong lens, but in a twinkling dozens of ants had covered my eyelids and after brushing them away I was only too glad to be able to see at all." Doflein (1906) has recently described and figured the voluminous spinning glands of *Œcophylla* (Fig. 124, *D*).

Dodd's account of the nest-building of *Œcophylla virescens* in Queensland, published in 1902, is also worth quoting, as it contains some interesting details not observed by Doflein :

" It is decidedly interesting to observe the insects engaged upon the construction of their domiciles. If the foliage is large or stiff, scores or even hundreds of the ants may be required to haul a leaf down and detain it in place until secured, both operations taking considerable time. It is quite a tug-of-war matter to bring the leaf into position and keep it there. The insects holding it have a chain of two or three of their comrades fastened on to them, one behind the other, each holding its neighbor by its slender waist, and all at full stretch and pulling most earnestly. What a strain it must be for poor number one ! When the leaf is far apart the ants form themselves into chains to bridge the distance and bring it down ; many of these chains are frequently required for a single leaf. I have seen a large colony at work upon a new nest, and several of these chains were from three to four inches long ; altogether there were many of them in evidence, some perpendicular, others horizontal. Up or along these living bridges other ants were passing.[1]

" Now for the web material used to build the nests. It is furnished in fine and delicate threads by the larvæ; moreover, I have only seen what appears to be half-grown examples used for the work—I have

---

[1] There can be no doubt of the accuracy of this extraordinary observation. During the summer of 1909 Prof. Ed. Bugnion showed me some fine drawings of these chains of *Œcophylla* workers, which he had recently observed in Ceylon. Prof. Bugnion did not know of Dodd's observations.

never seen a large larva being made use of. The soft and tiny grubs are held by the larger ants, who slowly move up against those pulling. Each grub is held by the middle, with head pointing forward, its snout is gently made to touch the edges of the leaves where they are joined, it is slowly moved backwards and forwards and is undoubtedly issuing a thread during the operation, which adheres to the leaf edges, and eventually grows into the web. When this web is completed it must be composed of several layers to be strong enough for the purpose of securing the leaves. Whether the larva is an unwilling instrument or not in its captor's mandibles is a point which cannot be ascertained. Maybe it is, for it cannot be comfortable in such a position. However, it supplies the web; perhaps if it were not robbed of the web for the benefit of the community it would be able to spin a cocoon for itself, in which to undergo the delicate change into the pupa state, for I have never seen a cocoon, all pupæ being quite naked.

" When contemplating the work done in these nests one cannot but marvel at the wonderful ingenuity displayed, or in endeavoring to form some idea of the vast number of larvæ which must be utilized to supply the connecting web even for a moderately sized nest, for with trees with narrow leaves, like *Eucalpytus tesselaris* for instance, many scores of leaves are required to form a nest, and each must be sewn."

Without knowing of these various observations on the Old World *Œcophylla,* Goeldi discovered the same method of nest-weaving in the Brazilian *Camponotus senex.* His observations together with a figure of a nest of this ant, have been published by Forel (1905*d*). Similar observations made by Jacobson (1905) on *Polyrhachis dives* (edited by Wasmann, 1905) and by Karawaiew (1906) on *P. muelleri* and *alexandri* prove that the silken nests of the ants of this genus are also spun by the larvæ. The latter author has described the complicated and voluminous sericteries in the larvæ of *P. muelleri* (Fig. 124*A*). Thus we have indisputable proof that the ants of three different genera, inhabiting the tropics of both hemispheres, have acquired the extraordinary habit of employing their young as instruments, or utensils, in the construction of their nests. Here, as in so many other instances, organs and functions originally developed for a very different purpose —in this particular case for spinning the pupal cocoon—have been adapted and transferred to a very different purpose.

Nests in Unusual Situations.—In this category we may include ant-nests in human dwellings, ships, etc., tenanted by such species as *Monomorium pharaonis* and *destructor, Pheidole megacephala, Solenopsis molesta* and *Prenolepis longicornis.* These, of course, originally lived in nests in the soil but on becoming house-ants they took to the crevices of the walls and woodwork. According to Forel (1874), certain European ants (*Lasius emarginatus* and others) often nest in stone masonry. In our Atlantic States *Leptothorax longispinosus*, which originally nested under small stones lying on boulders or in old nuts lying on the ground, frequently nests between the stones of the rough stone walls that enclose woods or pastures.

Accessory Structures.—Ant colonies do not always confine their constructive activities to the nest in which they are rearing their brood, but may extend their influence over the wider area on which they are accustomed to seek their subsistence. Evidences of this influence are seen in the great, bare clearings, sometimes 3–10 meters in diameter, with which *Pogonomyrmex occidentalis*, *P. barbatus* and its several varieties surround their gravel cones or discs. In addition to these clearings, *P. barbatus* also makes paths that radiate out into the surrounding vegetation, sometimes to a distance of 20 to 30 meters. These

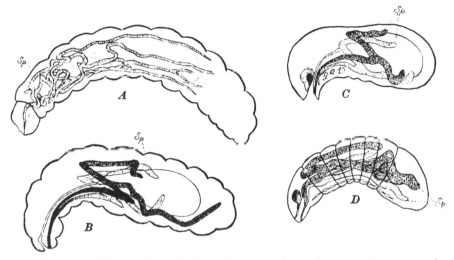

FIG. 124. The spinning glands (sericteries) of ant larvæ. (Karawaiew and Doflein.) *A*, Larva of *Polyrhachis muelleri*; *B*, of *Lasius flavus*; *C*, of *Tetramorium cespitum*; *D*, of *Œcophylla smaragdina*; the spinning glands (*sp*) are most highly developed in the two forms (*A* and *D*) which are used as shuttles in weaving the nest.

are most beautifully developed on the high plateau not far from the City of Mexico, where they are sometimes 10–20 cm. broad and resemble footpaths. More boreal species, like *Formica pratensis* of Europe and *F. integra* of the United States, often make tenuous paths which are roofed over along much of their extent with vegetable detritus and connect the different nests of a colony with one another. These and other species of *Formica* are also fond of constructing along their pathways, what Forel (1874) has called succursal nests, small excavations often resembling true nests, but used by the workers merely as places in which they can rest while foraging or escape from the heat of the sun or the pelting rain.

The tendency to construct such succursals or to establish several nests connected by run-ways is also apparent in many arboreal ants like the species of *Œcophylla, Polyrhachis, Cremastogaster* and *Liometopum*. To this habit must also be traced the construction of aphid or coccid tents, sheds or pavilions, as they are variously called. Though often at some distance from the true nests in which the brood is reared, these structures, which are usually made of carton or agglutinated earth may be regarded, nevertheless, as vestigial nests adapted to a specific purpose. Huber (1810) and Forel (1874) have described the aphid and coccid tents of European ants. Similar structures may also be seen in the tropics. On the island of Culebra I found carton tents that had been built by a variety of *Cremastogaster victima* over coccids on the lower surfaces of the leaves of *Cordia macrophylla,* and in the mountain forests of Porto Rico a yellow *Iridomyrmex* (*I. melleus*) was seen to make similar structures along the prominent ribs on the under sides of the gigantic reniform leaves of the ortegon (*Coccolobis rugosa*). Titus (1905) has shown that in Louisiana *Iridomyrmex humilis* occasionally makes coccid sheds on the surfaces of fruits like the persimmon. In our Northern States the versatile little *Cremastogaster lineolata* builds earthen or carton sheds which have been described by Osten Sacken (1862), Couper (1863), Trelease (1882) and myself (1906*b*). These structures are of small size, rarely more than 4 cm. long, more or less cylindrical or fusiform, enclosing some twig covered with plant-lice or mealy bugs (Figs. 205–209). A small round opening is left in the wall of the tent for the ingress and egress of the workers. Even the ants which spin silken nests among leaves often construct pavilions for their aphids, coccids, membracids and Lycænid caterpillars. Jacobson has observed this in *Polyrhachis dives* and Dodd gives the following description of these tents in *Œcophylla virescens:* " Not only do these strangely used larvæ provide the web to build up the nests, but they are carried considerable distances to various branches, generally near the ends, and they are there induced to furnish material for forming shelters and retreats for various scale insects, 'hoppers' and caterpillars with which the ants fraternize. Upon a tree may be seen several of these enclosures, or a dozen, occasionally many more ; as a rule a few leaves joined together. Upon large-leaved trees, like *Careya australis* or *Eucalyptus platyphylla* a single leaf doubled over and fastened down will form a sufficient cover. Upon pulling any of these apart a small flat scale in great numbers will be found adhering to the leaf. Upon another species of tree, *Acacia* say, perhaps ' hoppers ' only of a particular kind, with horned head, and their larvæ and pupæ may be found."

The striking character of the tents described in the preceding paragraph leads naturally to the question of their function and of the instincts of which they are an expression. There are several possible answers to such a question. We may suppose that the tents are built, first, for the purpose of preventing the escape of the aphids and coccids to other plants or to other parts of the same plant; second, for the purpose of protecting these insects and the ants themselves from exposure to cold, air-currents, moisture, or light; third, for the purpose of protecting the aphids and coccids from their natural enemies or from other ants. For some or all of these purposes the tents would seem to be admirable contrivances. It is probable that the aphids and coccids make the same appeal to the ants' sense of ownership as their own larvæ and pupæ. This is certainly true of some ants, like our species of *Lasius* which are fond of cultivating snow-white root aphids and coccids in their subterranean galleries. Whenever the stones covering the nests are overturned, the workers seize their charges in their mandibles and hurry away with them to a place of safety. It is natural, therefore, that ants should try to prevent the escape of their charges from a sense of proprietorship such as all ants display towards their own brood. Protection of the ants themselves from the air, and especially from the sunlight, is of great importance while they are waiting among the plant-lice for the accumulation and excretion of the honey-dew. Indeed, few of the species known to construct pavilions are at all fond of the open sunlight. This is certainly true of many of those mentioned in the preceding paragraphs. We may infer, therefore, that the ants probably build tents primarily for their own comfort and protection.

In concluding this chapter attention must be called to the fact that ants which have become parasitic on other species tend to lose completely the ability to excavate or construct nests. It is believed that even *Formica sanguinea,* which is only slightly dependent on its slaves, shows an inclination to neglect the labors of excavating. More completely parasitic genera, like *Polyergus,* though still possessing worker forms, are able to dig in the earth with their fore feet when opening up the galleries of the ants whose brood they are robbing, but they leave the construction of the nest entirely to the slaves. In highly parasitic genera, such as *Anergates, Wheeleriella,* etc., which have no worker forms, nest-building is, of course, a long-lost art.

REFERENCES

Aitken, E. H. 1890. Red Ants' Nests. *Journ. Bombay Nat. Hist. Soc.* 5(4):422.

Couper, W. 1863. Remarks on Tent-building Ants. *Proc. Ent. Soc. Phila.:* 373–374.

Chun, C. 1903. *Aus. den Tiefen des Weltmeeres.* 2(Jena):129.

Dodd, F. P. 1902. Notes on the Queensland Green Tree Ants (*Oecophylla smaragdina Fab.*). *Victorian Nat.* 18:136–142.

Doflein, F. 1905. Beobachtungen an den Weberameisen (*Oecophylla smaragdina*). *Biol. Centralb.* 25:497–507, 5 figs.

Doflein, F. 1906. *Ostasienfahrt.* Leipzig U. Berlin, Teubner.

Escherich, K. 1906. *Die Ameise Schilderung ihrer Lebenweise.* Braunschweig, Fr. Vieweg und Sohn. 223 pp., 68 figs.

Fielde, A. M. 1901. A Study of an Ant. *Proc. Acad. Nat. Sc. Phila.* 53:425–449, 2 figs.

Fielde, A. M. 1901. Further Study of an Ant. *Ibid.,* 53:521–544, 2 figs.

Forel, A. 1874. Les Fourmis de la Suisse. *Nouv. Mem. Soc. Helv. Sc. Nat.,* Zurich. 26:447, 2 pls.

Forel, A. 1894. Algunas formigas de Carnareas recogidas del Sr. Cabera y Díaz. *Ann. Soc. Espan. Hist. Nat.* series 2, Vol. 2:22.

Forel, A. 1905. Einige biologische Beobachtungen des Herrn Prof. Dr. E. Göldi an brasilianischen Ameisen. *Biol. Centralb.* 25:170–181, 7 figs.

Green, E. E. 1896. On the Habits of the Indian Ant. (*Oecophylla smaragdina F.*). *Trans. Ent. Soc. London Proc.,* pp. IX–X.

Green, E. E. 1899. Note on the Web-spinning habits of the "Red Ant" *Oecophylla smaragdina. Journ. Bombay Nat. Hist. Soc.* 13:181.

Huber, P. 1810. *Recherches sur les moeurs des fourmis indigènes.* Paris et Genève. 1 Vol.

Jacobson, Edward and E. Wasmann. 1905. Beobachtung über *Polyrhachis* dives auf Java die ihre Larven zum Spinnen der Nester benutzt. *Notes Leyden Mus.* 25:133–140.

Karawaiew, W. 1906. Systematisch-Biologisches über drei Ameisen aus Buitenzorg. *Zeitschr. Wiss. Insekt.-Biol.* 2(12):369–376, 16 figs.

Osten-Sacken, R. Von. 1882. Ants and Aphides. *Psyche* 3:343. Transl. from *Stett. Ent. Zeitg.,* 1862, pp. 127–128.

Ridley, C. V. 1890. (*On Oecophylla.*) *Journ. Straits. Branch Roy. Asiat. Soc. Singapore,* p. 345.

Saville-Kent, W. 1897. *The Naturalist in Australia.* Chapman and Hall, Ltd., London. 302 pp., 59 pls., 100 text-figs.

Trelease, W. 1882. Unusual Care of Ants for Aphides. *Psyche* 3(94):310–311.

Titus, E. S. G. 1905. Report on the "New Orleans" Ant (*Iridomyrmex humilis* Mayr). *U.S. Dept. Agri., Bur. Ent. Bull.* 52:79–84, fig. 7.

Wheeler, William Morton. 1906. The Habits of the Tent-building Ant (*Cremastogaster lineolata* Say). *Bull. Amer. Mus. Nat. Hist.* 22:1–18, 4 pls., 3 text figs.

Part II

EVOLUTION OF EXTERNAL CONSTRUCTION BY ANIMALS

Editors' Comments
on Papers 4 Through 10

4   **EMERSON**
      Termite Nests—A Study of the Phylogeny of Behavior

5   **COLLIAS**
      The Evolution of Nests and Nest-Building in Birds

6   **KASTON**
      The Evolution of Spider Webs

7   **ROSS**
      Evolution of Caddisworm Cases and Nets

8   **SCHMIDT**
      Apicotermes *Nests*

9   **MICHENER**
      Evolution of the Nests of Bees

10   **VAN LAWICK-GOODALL**
      Excerpts from *The Behaviour of Free-Living Chimpanzees in
      the Gombe Stream Reserve*

Alfred E. Emerson developed and emphasized the importance of nests of animals as diagrammatic representations for the study of the evolution of behavior. As he points out with regard to termite nests in Paper 4 (p. 97), "The nest structures are morphological expressions of behavior patterns. This quality makes aspects of behavior evolution as visible as morphological evolution and similar principles and terms may be directly applied." Termite nests express inherited behavior patterns of a population as a unit and not just of individuals. Emerson's striking examples of adaptive radiation and convergence of termite nests in response to different habitat conditions give evidence of the forces of selection acting under these differing conditions. Termites are the most ancient of the highly social insects, and all termites are highly social. Their colonies are abundant and of widespread occurrence in the tropical and warm–temperate regions of the world. They occur in diverse habitats, from rain forest to desert. In some species of termite the colonies are the largest known for any social insect, excepting army ants, and there may be up to 3 million individuals in a single colony.

Accurate classification of a group of organisms gives a firm basis for other studies, including behavior. In his treatise on insect societies, Wilson (1971) states that the current classification of termites is evidently better than that for any other group of social insects, thanks to a large and able group of taxonomists. Among these he takes special notice of Emerson. Born in 1896 in Ithaca, New York, Alfred Emerson graduated from Cornell University in 1918 and received a Ph.D. in entomology in 1925. He has devoted a lifetime to the biology of termites, has studied them in many parts of the world, and now in retirement from university work he continues their study. Largely through his efforts the American Museum of Natural History, to which he donated his personal collection, has specimens of over 90 percent of the termite species of the world, including over 80 percent of primary type specimens of the species.

Emerson is a man of broad biological and philosophical interests. He has been particularly interested in the relationships between ecology and evolution and has provided one of the first, really modern, comprehensive syntheses of the two fields (Emerson, 1949). He is a member of the National Academy of Sciences and has been president of several national societies concerned with zoology, ecology, evolution, and systematics.

In reply to a letter from N. E. Collias as to the start of his interest in termites he wrote (April 24, 1974): "When the war was over, I got a chance to go with William Beebe to British Guiana in 1919. He was an ornithologist at the time and Curator of Birds at the Bronx Zoo. He was also an enthusiast about nearly every aspect of life in the virgin rain forest. It was he who suggested I take up termites. At first I started out to work in the ecology of the forest floor, but soon found myself more and more fascinated by termite social life, their termitophiles, and their taxonomy. At first I thought I could get someone else to do the taxonomy, but could not, so I took it up myself, and have found it led into ecology, behavior, evolution, zoogeography, etc., so I have gradually branched out into several subsciences of biology using the detailed study of termites as a center for radiation. I have found myself in strange places as the result. I have lectured to psychiatrists, evolutionists, sociologists, religionists, medical societies, etc. . . . Right now I am back at termites, and am just finishing a paper dealing with a rare species discovered in 1854 that does not seem to fit into any classification well.''

In 1920 William Morton Wheeler visited Beebe's station at Kartabo, British Guiana, to further his studies of ants. Emerson, already at the station, helped him with his field work, and many years later (1964) wrote his impressions to Wheeler's biographers (Evans and Evans, 1970, pp. 275–277): "Personally, I found Wheeler a most stimulating person and he did much to steer my professional life. I had already become interested in termites before he came to Kartabo, but his enthusiasm at some of my early discoveries was most encouraging to a young man starting to feel his way into scientific research. In several instances he advised me to make certain special studies in British Guiana that I followed and with fruitful results. I also helped him in his ant studies, largely by carrying equipment, or doing the muscle work involved in finding some army ant queens. All of this intimate contact with him was a rich experience for me.''

Wheeler must have been favorably impressed with Emerson, for the latter also wrote that Wheeler's recommendation was an important factor in securing an offer to come to the University of Chicago in 1929, where Emerson remained on the faculty of the zoology department until his retirement over thirty years later.

The article reproduced here as Paper 4 appeared in 1938 and was the first extensive article by Emerson on this specific topic; we selected it because it illustrates so well the basic steps followed by later authors in dealing with the phylogeny of behavior as related to animal architecture: (1) work out the taxonomy and phylogeny of the group of animals concerned, using all available information, (2) relate variations in external construction and building behavior to the basic phylogeny of the group, and (3) relate variations in external construction to differences in ecology within the group (convergent and parallel evolution). At each stage additional information helps throw new light on other problems.

In 1964 a symposium, "The Evolution of External Construction in Animals,'' was published in *American Zoologist*. It dealt with external construction resulting mainly from overt behavior rather than just from physiological phenomena, as in the secretion of mollusk shells and coral reefs. This symposium, which seems to have been the first really comprehensive symposium on such external construction by animals that had been held by a national society, was organized by the Division of Animal Behavior of the

American Society of Zoologists and was cosponsored by three other national societies, including the Ecological Society of America, the Society for the Study of Evolution, and the Society for Systematic Zoology. The object was to call renewed attention to the promise of external construction by animals for gaining new insights into the ecology and evolution of behavior. It seems appropriate to reproduce this short symposium here, for it contains much relevant information that will serve as a basis for further study of the subject. The participants were all professors from various universities.

Various specialized groups of animals were selected to illustrate the different functions of external construction and the evolution of such construction under different types of habitat. Of all vertebrates, birds have the greatest variety of nests that aid development of the eggs and young; of all animals, spiders build the most amazing snares for unwary prey; caddisworms have evolved a varied assortment of portable cases that furnish protecting to the individual browsing in exposed sites; and finally the communal nests of termites and social bees provide excellent examples, from two taxonomically unrelated categories, of elaborate constructions by large social groups. The five reports given are reproduced here in the same order as they were originally presented. For the present volume we have added some illustrations at the end of the articles on birds, termites, and spiders.

The first report, by Nicholas Collias, is on the evolution of nests and nest building in birds (Paper 5). Darwin, in treating the subject of instinct in his *Origin of Species,* used more examples from birds and social insects than from any other groups of animals. In a less condensed version of his chapter "Instinct," published posthumously with the consent of his family (in Romanes, 1883, p. 365), Darwin states, "In the nests of birds, also, we have an unusually perfect series, from those which build none, but lay on the bare ground, to others which make a most imperfect and simple nest, to others more perfect, and so on, till we arrive at marvelous structures rivalling the weavers' art." We have been studying nests and nest building of weaverbirds (Ploceidae) and other birds over a period of almost twenty years; however, our initial stimulus for this work came not from reading the above statement by Darwin, which we were pleased to find but only recently discovered, but rather from Emerson at the University of Chicago, where N. E. Collias was a student in the zoology department for the seven years just prior to World War II.

This report on the evolution of the nests of birds attempts to trace possible sequences from simple to complex nests, but, in comparison with the reports on other groups of animals, the phylogenetic relationships of the orders and families of birds are as yet poorly understood. Hence the sequences given are often to be looked on as representing evolutionary grades or levels of progressive change, rather than as reflecting specific genealogies. There has been a great deal of parallel evolution and convergence among birds, making the job of the taxonomist in ornithology unusually difficult. However, the very extent of such parallel or convergent evolution helps bring out the similarity of the selection pressures that have repeatedly produced similar types of nests under similar conditions. One can readily find correlations between nest type and such factors as habitat type, size of bird, and the like. Then, too, the taxonomy of the weaverbirds is better known than that of many other passerine families, since one of the best avian taxonomists, James P. Chapin (1954), paid special attention to this family and gave us a relatively firm foundation from which to study their nests. In general, we found that variations in weaverbird nests confirmed and paralleled Chapin's 1954 taxonomy (Collias and Collias, 1964).

The second article in the symposium (Paper 6) is on spiders. Spiders have a greater variety of uses for silk than do any other animals; they use their spinning ability not only to capture prey but also to spin retreats, to make webs that aid courtship and mating activities, to construct a silken case for the eggs and young, and to spin out long threads that enable young spiders to ride the wind and disperse far and wide. B. J. Kaston, who wrote this report, is also the author of a well-known and recently revised field guide to American spiders (1972; originally published in 1952) that after more than 20 years is still popular. For many years he was a professor of biology at Central Connecticut State College, after which he joined the department of zoology at San Diego State University in California.

The third article in the symposium (Paper 7) is on caddisworms. Many of the aquatic larvae of the caddisflies (Trichoptera) fashion a portable case, often weighted down with small stones, sand grains, bits of plant debris, or other flotsam, which furnishes additional protection to the larva. Herbert H. Ross, author of the article, has written extensively on caddisflies, their evolution and classification. Twenty years before the article reproduced here was published he realized that the phylogeny of the Trichoptera was badly in need of study. The adult morphology gave little information on the early phylogeny of the group, so he turned to a more detailed study of larval morphology. This proved highly rewarding, and at the same time it became apparent that the important morphological features of the larvae were correlated with their type of external construction, suggesting that changes in behavior may have preceded changes in morphology. Professor Ross is also the author of a well-known textbook of entomology and more recently of a book on evolution. He has been president of the Entomological Society of America and of the Society for the Study of Evolution, and has been acting chief of the Illinois State Natural History Survey.

The fourth article (Paper 8), by Robert S. Schmidt, deals with the *Apicotermes,* a group of termites which make some of the most spectacular subterranean nests built by any animal. The building behavior of this genus of termites is scarcely known, but their nests provide some of the best examples of evolutionary sequences in behavior. Schmidt received his Ph.D. in zoology from the University of Chicago under the guidance of Alfred Emerson for a study of the phylogeny of the nests of *Apicotermes,* rapidly became an authority on the subject, and wrote several papers subsequently on the nests of this group. For eleven years he held a career development award from the National Institute of Neurological Diseases and Blindness. He is now on the faculty of the Stritch School of Medicine of Loyola University in Chicago, where he has demonstrated his versatility by doing pioneering research on the neuroethology of vocalization in frogs. Recent reviews of *Apicotermes* nests have been written by C. Noirot (1970) and A. Bouillon (1970) in *The Biology of Termites,* edited by Krishna and Weesner.

The fifth and last article (Paper 9) is by Charles D. Michener on the evolution of the nests of bees. In reply to a letter from N. E. Collias, he wrote (May 13, 1974): "I became interested in the nest structure of bees for two reasons: First, the structure varies among the groups of bees and is often relatively conservative. It therefore frequently provides supplementary characters of systematic interest that help to clarify or reinforce our views of the major groups of bees. Of broader general interest, I believe, is the point that the nest structures have evolved in such a way that they provide for the well-being and survival of the bees that inhabit them. The ways in which they do this are of considerable interest; the interaction of nest structure with the environment to benefit the insect is usually much easier to understand than the relationship between body structure and the environment."

Michener is professor of entomology at the University of

Kansas in Lawrence. Among his many honors, he has been president of the American section of the International Union for the Study of Social Insects, of the Society for the Study of Evolution, and of the Society for Systematic Zoology.

Unlike ants and termites the great majority of bees are solitary, and one can build up sequences that illustrate early stages in the evolution of social life and the associated nests. Michener is the author of a recent treatise (1974) on social behavior in the various families of bees. He has been particularly interested in the nesting behavior and social life in some of the more primitive bees, particularly the sweat bees (Halictidae), as well as in the relatively advanced stingless honeybees (Meliponini, Apidae). His article brings out the complexities in the phylogenetic interrelations of bee nests, and also helps throw some light on the evolution of the classic honeycomb of the common honeybee, *Apis mellifera.*

The most general conclusion drawn from a summary of the symposium was that the most elaborate constructions made by animals in the various major terrestrial groups of the animal kingdom can all be traced back eventually through various intermediate stages to a simple burrow or hole in the ground (Collias, 1964). Many water animals of diverse types make burrows, and with the invasion of the land during early evolution, burrows, natural crevices, and holes were probably used for shelter and to conserve moisture. Digging of a burrow in which the eggs are placed is common today among the more primitive terrestrial animals. For example, the most primitive living spiders, the Liphistiidae of eastern Asia, which still have segmented abdomens, build silk-lined burrows from which signal threads radiate. The birds, which are so noted for their diversity of nests, originated from reptiles, many of which bury their eggs in the soil, as do the megapodes among birds.

Various principles of evolution are as well illustrated by animal architecture, and therefore by behavior, as by animal morphology, including the principles of gradation, adaptive radiation, parallel and convergent evolution, the exploitation of new ecological opportunities, increase in homeostasis and the efficient use of energy, principles relating ontogeny to evolution, and principles of group selection and regressive evolution.

In considering the evolution of highly specialized and complex external constructions, it is often possible to arrange a graded series of simpler structures that, in a general way, probably represents the evolutionary steps concerned. Certain weaverbirds make a finely woven, roofed nest with the entrance at the bottom of a long entrance tube that projects downward from the nest. Such a nest can be traced back through more crudely woven nests without any entrance tube and with a side entrance, to nonwoven, semidomed, or unroofed and bowl-like nests in trees, or on the ground, and finally to mere hollows or scrapes in the ground having little if any special nest materials.

Among spiders, Kaston shows how the beautifully symmetrical vertical web of the typical orb weavers, with its sticky spiral, can be traced back through horizontal orbs to crude platforms or sheets without viscid threads, to irregular simple meshworks in vegetation, and thence to radiating signal threads from a silk-lined burrow in the ground. Similarly, as Schmidt shows for *Apicotermes* and Michener for bees, the most elaborate nests built by members of these groups have evolved from nests consisting of a few cells in the ground.

Most bees are solitary, and many primitive bees build nests that are branching burrows, each branch ending as a cell lined with impervious material and furnished with a store of food for the larva, which makes no cocoon. This lining, which is painted on with the mouth parts of the adult bee and may or may not be partly soluble in wax solvents, helps maintain the proper humidity in the closed cell and probably helps protect the food store from fungi. In sphecoid wasps, from which bees evolved, the most common type of cell is an unlined cavity in the ground.

Bees had two main adaptive radiations: one, by the more primitive, mostly solitary, short-tongued bees, such as the Halictidae, occurred at a time when most of the angiosperms had shallow flowers; the other, by the long-tongued bees, coincided with the evolution of flowers with deeper corollas.

Many stages in the evolution of nest construction are obvious in the various halictid bees. Solitary species have one to a few cells in a burrow. In some social halictids the nests are excavated by several bees, which cooperate in lining and provisioning the cells. Cells increase in number, especially with an increase in group size, and labor is saved by shortening lateral branches to concentrate the cells. Ventilation and drainage of cell clusters may be improved by excavating an air space in the soil around the clusters. This may be followed in evolution by a shift from excavated to constructed cells of the same shape.

The most social of the long-tongued bees are the Apidae, which include the orchid bees, the bumblebees, the stingless honeybees, and the true honeybees. In the Apidae the relatively thick walls of the cells are made of mud, excrement, resin, wax, or a mixture of resin and wax. Cell shape varies from round in most Apidae to hexagonal in *Apis*. Stingless bees have mass provisioning and cap the cells after the egg has been laid in each cell, unlike the *Apis* and some bumblebees, in which the origin of progressive provisioning has helped solve the problem of susceptibility of the pollen mass to mold.

Parallel or convergent evolution in the origin of various adaptive features of external construction is very common, just as it is in the origin of morphological and physiological features. But in the case of external constructions we are dealing more directly with a record of behavioral evolution. The roofed nest so common among tropical passerine birds has evolved repeatedly and independently in different families. The orb web seems to have evolved independently in the evolution of two main groups of spiders, the cribellate and ecribellate (noncribellate) families. This phenomenon, Kaston believes, is a case of convergence or parallelism, such as occurs with respect to the morphology of these two groups, and is analogous to the adaptive similarities between certain marsupial and placental mammals. The cribellum is a sievelike transverse plate on the abdomen just in front of the spinnerets. From it, a special row of curved setae (the calamistrum) on the fourth legs pulls silken bands, each covered with a pubescence of fine filaments that entangle the prey. As Emerson pointed out, among the termites, rain-shedding devices in arboreal species have evolved convergently in three different subfamilies, while Schmidt shows that at least the circular galleries in the underground nests within the genus *Apicotermes* evolved independently. In bees the vertical comb with its two opposed layers of cells has evolved independently in a stingless bee (*Dactylurina*) and in the true honeybee (*Apis*).

The existence of convergent evolution calls for a closer analysis of the significant principles and ecological forces involved. The principle of economy, as Michener suggests, has resulted in the clustering of cells in a regular arrangement, and simple combs are found in many kinds of bees. We see again the principle of economy of energy in the construction of the scarcely visible snare of the advanced orb weavers which covers a maximum of area with a minimum of material.

Energy may require gaseous exchange, often a problem for animals that live deep in the soil. Evolution toward optimal spacing of ventilation pores in the subterranean nests of different

90

species of *Apicotermes* permits maximum gaseous exchange with minimum interference by diffusion from adjacent pores with each other. Schmidt calculates that the presence of optimally spaced pores, combined with the location of the nest in an air space in the soil, or in a loose envelope of organic and inorganic material, may increase by some fifty times the total surface available to the nest for gaseous exchange.

Exploitation of a new, major, ecological opportunity is evident in the evolution of spiders from predominantly ground-dwelling species to types that live and hunt among vegetation, which coincides in general with the origin and evolution of flying insects, and culminating in the orb web and such highly specialized aerial snares as the springing web, the retiarus, and the bolas. Similarly, the origin of the portable case made it possible for the more advanced caddisworms to browse on diatoms, their chief food, in open and exposed situations. There followed a secondary evolution of a great variety of such cases.

Some apparent instances of ontogeny recapitulating some earlier stages in phylogeny of an external construct may help provide clues to this phylogeny. The *Uloborus* spiderling has neither cribellum nor calamistrum until after its first molt, and during this early period it weaves its web without the characteristic spiral. In the nests of certain species of *Apicotermes*, different parts of the same nest may illustrate different stages in both ontogeny and phylogeny of the nest.

In the course of evolution specific ontogenetic events can be displaced or projected forward from a later to an earlier stage of development or life history. This phylogenetic principle of precocious shift of specific events in the life history is illustrated in caddisworms by Ross. Originally, caddisworms had no portable case but constructed a cocoon for pupation in a fixed spot. In certain lines of evolution the young larva anticipates this stage by a premature construction of the outer cocoon, which is still made in more primitive species just before pupation. This cocoon may be weighted with small stones or other debris and is used as a portable case by the larva. Ross applies the principle that genetic factors often exert their effects by influencing the relative timing of developmental events to a theory of how the portable case of caddisworms might have evolved from a precocious spinning of the outer cocoon by the young larva. To support and elaborate this theory, he draws on the work of Van der Kloot and Williams on certain lepidopterous larvae, close relatives of caddisworms, and he suggests that such premature spinning might have resulted from greater inhibition of the juvenile hormone secreted by the corpora allata.

In vertebrates, an example of the principle of precocious shift or of the anticipatory projection of events in the life history is seen in the evolution of courtship behavior of many birds. The practice of manipulating materials used in nest construction or of going through nest-building movements has in many species been projected forward in time to become part of the courtship display of the male to the female.

External constructions are relatively constant species-specific features. The role of individual experience in construction is generally assumed to be minimal, but the nature of the interaction between genetic and experiential factors has been little investigated. Nor is much known yet about the genetic basis of external construction in animals. However, since the elaborate nests of the highly social insects are made by workers normally sterile, it has been emphasized that natural selection in these cases has been working on the community as a whole, with the survival of changes in the genetic capacity of the species.

Regressive evolution is illustrated by loss of nest building in various parasitic birds and in parasitic bees. Ventilation pores in nests of *Apicotermes* have phylogenetically regressed in drier soils. Kaston indicates that, among spiders, loss of ability to build a web has evolved independently in three different ecribellate genera (*Euryopis, Pachygnatha,* and *Celaena*).

The symposium of which we have been summarizing some highlights made no mention of the evolution of external construction by early man. This human talent is foreshadowed by certain capabilities among the great apes—the building of nests in which to sleep or rest each day (three species) and the construction and use of simple tools, especially by the chimpanzee. Jane Van Lawick-Goodall has made some outstanding contributions to our knowledge of both these phenomena in her own classic studies of the chimpanzee in its native habitat in Africa. George Schaller (1963) has studied the mountain gorilla in central Africa and John MacKinnon (1974) the orangutan in Borneo and Sumatra, and their observations on nest building help round out the picture for the great apes. Whereas the adult male gorilla often sleeps on the ground in a crudely made nest, the females and young males make better-constructed platforms in trees, as the chimpanzee does regularly. The orangutan, most arboreal of the group, inhabits heavy rain forest country, and often builds an overhead shelter for its nest, piling loose branches over itself. It also uses such overhead shelters for shade, for camouflage, and in play. In contrast, roofed nests seem to be rare among chimpanzees, although one was observed by Jane Goodall.

The great apes build their nests by hand, and this means that they are preadapted to the evolution of tool use. The ability to make and use a variety of simple tools is well established now for the chimpanzee, and Goodall's discoveries and observations of tool use in this manlike creature are the most noted of her many remarkable observations of this species in its wild state. Excellent films illustrating this behavior were made by her husband and are available from the University of California film library, among many other university libraries.

In 1960, the anthropologist, L. S. B. Leakey, well known for his key discoveries of fossil man and fossil apes, encouraged Goodall, then working for him in Nairobi, to undertake a long-term study of chimpanzees in the wild in the hope of shedding some light on the behavior of human ancestors, and it was he who obtained funds to finance her early fieldwork. Accompanied only by her mother and her African assistants, with some help from the local authorities, Goodall set up camp in the remote Gombe Stream Game Reserve in Tanzania on the shores of Lake Tanganyika in a malaria-ridden district. At first, the chimpanzees were very shy and elusive and kept at a distance. Only after many months of remarkably persistent, patient, and tactful daily effort by Goodall did the chimpanzees gradually lose their early fear to the extent that they sometimes even visited the campsite, and she was able to make many close observations of the details of chimpanzee life. Early in the project Leakey sent a young photographer from the National Geographic Society, Baron Hugo Van Lawick, to make documentary films of the work. Hugo became much interested in the research, helped her in all aspects of it, and eventually the couple married.

Not merely devoted but well-organized effort from a scientific viewpoint characterized Goodall's studies, which were accepted for a Ph.D. thesis at Cambridge University in England and published in 1968 as part of Volume I of *Animal Behaviour Monographs*. The sections on nest-making behavior for sleeping and resting and the section on aimed throwing and tool using are reproduced here as Paper 10. This work was initially financed by the Wilkie Foundation and later by the National Geographic Society.

The program having proved itself admirably is now aided by

several other funds and foundations and, with the help of a series of assistants and independent investigators, it still continuing at the Gombe Stream Research Center, where Goodall is the scientific director. She has summarized much of the more recent studies and sequelae to her thesis work in a widely read book, *In the Shadow of Man* (1971), and other recent work, with special reference to tool using by chimpanzees, in two articles (1970, 1973).

REFERENCES

Bouillon, A. 1970. Termites of the Ethiopian region. Pp. 154–280 in Vol. 2, *Biology of Termites,* K. Krishna and F. M. Weesner (eds.). Academic Press, Inc., New York.

Chapin, J. P. 1954. *The Birds of the Belgian Congo.* Part IV, Bull. Am. Mus. Nat. Hist., 75:1–846, 27 pls.

Collias, N. E. 1964. Summary of the symposium on the evolution of external construction by animals. *Am. Zool.,* 4(2):241–243.

Collias, N. E., and E. C. Collias. 1964. *Evolution of Nest-building in the Weaverbirds (Ploceidae).* Univ. Calif. Publ. Zool., 73:1–162, 38 pls.

Emerson, A. E. 1949. Ecology and Evolution. Pp. 419–435 and 598–729 in W. C. Allee, A. E. Emerson, O. Park, T. Park, and K. Schmidt, *Principles of Animal Ecology.* W. B. Saunders Co., Philadelphia. 837 pp.

Evans, M., and H. Evans. 1970. *William Morton Wheeler, Biologist.* Harvard University Press, Cambridge, Mass. 363 pp.

Kaston, B. J. 1972. *How To Know the Spiders.* 2nd ed. William C. Brown Co., Dubuque, Iowa. 290 pp.

MacKinnon, J. R. 1974. The behavior and ecology of wild orangutans (*Pongo pygmaeus*). *Anim. Behav.,* 22(1):3–74.

Michener, C. D. 1974. *The Social Behavior of the Bees: A Comparative Study.* Harvard University Press, Cambridge, Mass. 404 pp.

Noirot, C. 1970. The nests of termites. Pp. 73–125 in Vol. 2, *Biology of Termites,* K. Krishna and F. M. Weesner (eds.). Academic Press, Inc., New York.

Romanes, G. J. 1883. *Mental Evolution in Animals.* With a posthumous essay on instinct by Charles Darwin. Kegan Paul, Trench and Co., London. 411 pp. Appendix by Darwin, pp. 353–384.

Schaller, G. B. 1963. *The Mountain Gorilla: Ecology and Behavior.* University of Chicago Press, Chicago. 431 pp.

Van Lawick-Goodall, Jane. 1970. Tool-using in primates and other vertebrates. *Advan. Study Behav.,* 3:195–250.

Van Lawick-Goodall, Jane. 1971. *In the Shadow of Man.* Houghton Mifflin Co., Boston. 297 pp.

Van Lawick-Goodall, Jane. 1973. Cultural elements in a chimpanzee community. In *Precultural Primate Behaviour,* E. W. Menzel (ed.). Karger Publs., Basel.

Wilson, E. O. 1971. *The Insect Societies.* Harvard University Press, Cambridge, Mass. 548 pp.

4

Reprinted from *Ecol. Monogr.*, **8**(2), 248–284 (1938)

TERMITE NESTS—A STUDY OF THE PHYLOGENY OF BEHAVIOR

Alfred E. Emerson

Hull Zoological Laboratory, University of Chicago

CONTENTS

	PAGE
INTRODUCTION	249
TERMITE NESTS AS ILLUSTRATIONS OF BEHAVIOR EVOLUTION	250
PRE-ISOPTERAN NESTING BEHAVIOR	251
NESTING BEHAVIOR OF THE KALOTERMITIDAE	252
Excavating Behavior of the Kalotermitidae	253
Construction Behavior of the Kalotermitidae	254
NESTING BEHAVIOR OF THE MASTOTERMITIDAE AND HODOTERMITIDAE	254
NESTING BEHAVIOR OF THE RHINOTERMITIDAE	256
NESTING BEHAVIOR OF THE TERMITIDAE	258
Nest Materials	259
Structures	260
CASTES INVOLVED IN NEST CONSTRUCTION	265
ECOLOGICAL FUNCTIONS OF THE NEST	266
SELECTION OF THE NESTING SITE	271
NEST DIVERGENCE WITHIN A GENUS	272
CONVERGENT EVOLUTION WITHIN A SIMILAR HABITAT	276
DISCUSSION	278
SUMMARY	281
LITERATURE CITED	282

INTRODUCTION

Almost every naturalist who travels in tropical countries observes the striking nests of termites and many figures of these nests have been published (Hegh 1922). The significance of termite nests to biological theory, however, has been only partially stressed, although Darwin (1859) places considerable emphasis upon the evolution of "instinct" as illustrated by "neuter and sterile insects." He states:

But I must confess, that, with all my faith in natural selection, I should never have anticipated that this principle could have been efficient in so high a degree, had not the case of these neuter insects led me to this conclusion. I have, therefore, discussed this case, at some little but wholly insufficient length, in order to show the power of natural selection, and like-wise because this is by far the most serious special difficulty which my theory has encountered. The case, also, is very interesting, as it proves that with animals, as with plants, any amount of modification may be effected by the accumulation of numerous, slight, spontaneous variations, which are in any way profitable, without exercise or habit having been brought into play. For peculiar habits confined to the workers or sterile females, however long they might be followed, could not possibly affect the males and fertile females, which alone leave descendants. I am surprised that no one has hitherto advanced this demonstrative case of neuter insects, against the well-known doctrine of inherited habit, as advanced by Lamarck.

Subsequent writers have referred to Darwin's statement and have used the sterile castes as an argument against the Lamarckian concept (Ball, 1890, 1894; Weismann 1893; Holmes 1911), but the rich recent information made available through greater exploration and systematic study of tropical faunas has not yet been adequately interpreted. I propose, in the following pages, to present the best cases known to me illustrating the principles of evolution as applied to termite nesting behavior.

I am indebted to Mr. G. F. Hill, Senior Entomologist in the Division of Economic Entomology in Australia, an eminent contributor to our knowledge of termites, for the photographs of Australian termite nests. The African nests were photographed by Mr. Herbert Lang while on the Lang-Chapin expedition of the American Museum of Natural History during which he made a splendid collection of termites and took many valuable field notes. The other photographs were taken by the author or under the author's direction.

Studies of north African and European termites were made by the author during the tenure of a John Simon Guggenheim fellowship, during which time Dr. F. Santschi and Dr. F. Silvestri were particularly helpful. Studies in British Guiana were made at the Tropical Research Station of the New York Zoological Society at Kartabo, and I am greatly indebted to the director, Dr. William Beebe, for the opportunities he afforded. Studies in Panama were largely made at the Barro Colorado Island Laboratory of the Institute for Research in Tropical America. The author is indebted to Mr. James Zetek for many courtesies extended during his sojourn on the island. My thanks are due Mr. George Lee for the opportunity to study savannah termites in Panama. For the opportunity to study Californian termites, I am indebted to the Termite Investigations Committee and particularly to Dr. C. A. Kofoid and Dr. S. F. Light.

I am also grateful to many men interested in this problem with whom I have had conversations or correspondence, although they should not be held responsible for any erroneous views expressed in the following pages. Among them all, I should particularly mention Dr. W. C. Allee, Mr. G. F. Hill, Dr. N. A. Kemner, Dr. H. Klüver, Dr. T. E. Snyder, Dr. W. M. Wheeler and Dr. Sewall Wright.

TERMITE NESTS AS ILLUSTRATIONS OF BEHAVIOR EVOLUTION

The nests of termites have certain attributes which make them objects of great biological interest. These may be listed as follows:

(1) The nest structures are morphological expressions of behavior patterns. This quality makes aspects of behavior evolution as visible as morphological evolution and similar principles and terms may be directly applied.

(2) The nest results from the activity of a large number of individuals cooperating in building, organizing and enlarging the structure. Individual variation is thus practically cancelled and the nest stands as an expression of the behavior of a population.

(3) The nest-forming behavior is predominantly an inherited species-pattern. The nests of a given species show striking similarity in material, general shape, internal organization and ecological position. In most instances, the workers which construct the nest have had no contact with workers of other colonies. The reproductive castes, which do not exhibit complicated nest-building behavior, fly from the parental nest, pair, and lay eggs which develop into workers which again establish a nest with the specific characteristics of the parental nest.

(4) The caste which builds the nest in all the higher termites is the sterile worker helped, in some cases at least, by the larger soldier nymphs. Thus we have a complete control over any possibility of the inheritance of an acquired character. No nest-building habit or structural modification of a sterile caste acquired during ontogeny could be transmitted to the succeeding generation. This control over any Lamarckian influence is even better than that found in other social insects. Workers of wasps, bees and ants, although without

socially functional young in most cases, not infrequently lay eggs which, being unfertilized and haploid, develop into fertile males. This is not known among termites. Workers of termites may, therefore, be considered as somatic individuals, physiologically separated from the germ plasm of the reproductive individuals, making a directive influence upon the genetic determiners a practical impossibility.

(5) The nests are definitive enough and at the same time specific enough to give us excellent evolutionary sequences. These may be correlated with the known morphological evolutionary relationships of the species. We have nest sequences of species in the same genus available, information that should give us a clue to the influence of species divergence upon these behavior patterns. In numerous instances we also find nest characteristics common to many species within a single genus or a group of related genera, thus indicating a long stability of the inherited patterns.

(6) Astonishing examples of adaptive modification of the nests together with convergent evolution of nest structure in certain environments present evidence of the force of selection acting upon the inherited patterns.

(7) The evolution of the nesting behavior is one aspect of the evolution of polymorphism. The colony of polymorphic insects has many attributes of the individual multicellular organism and exhibits many interesting parallels to organismal coordination mechanisms and interrelations of parts. Natural selection probably acts upon the colony as a whole more than upon the individual termite.

PRE-ISOPTERAN NESTING BEHAVIOR

The available indications point toward excavation of wood as the primitive nesting behavior. Certain wood-eating roaches, such as *Cryptocercus punctulatus* Scudder, probably represent the closest approach to the activities of the blattoid ancestors of the termites. Cleveland (1934, p. 190) describes the excavations of this roach as follows:

The wood is honeycombed with galleries which, for the most part, run parallel with the grain. In some of the sounder logs, particularly chestnut, which is often very hard, the roaches are seldom found near the outside. * * There is little evidence that they ever leave the log and enter the ground. * * They pass well-formed pellets of dry, woody material which is not utilized in any way for building purposes or for the construction of passage ways, differing in this respect from many species of more highly specialized termites.

With reference to excavations for the ootheca, Cleveland (p. 207) states:

a groove had been made in the wood where none existed before and the ootheca had been carried approximately six inches, placed in the groove, and sealed off so completely that only a portion of one end was visible.

These observations and others indicate the probability that habitations in excavated wood and the care of the eggs, as well as wood-feeding, symbiotic relationship with gut-inhabiting protozoans, and development of a family organization, antedated the origin of the Isoptera.

NESTING BEHAVIOR OF THE KALOTERMITIDAE

Among the living termites, *Mastotermes darwiniensis* Froggatt (Mastotermitidae) is universally admitted to be the most primitive morphologically. However, the descriptions of its nesting activities (Hill 1921, 1925) would seem to indicate an advance over the behavior of kalotermitids usually considered more advanced from a morphological standpoint. As Imms (1919) and Emerson (1926, p. 92) have remarked, in certain morphological details *Archotermopsis* (Kalotermitidae) is more primitive than *Mastotermes*. One may either conclude that *Mastotermes* has undergone evolution toward more intricate behavior after its divergence from the ancestral isopteran stock, or else that degenerative evolution of the behavior patterns has occurred in the Termopsinae and other kalotermitids. With only meager evidence, I am inclined toward the former hypothesis.

From morphological considerations, *Archotermopsis wroughtoni* (Desneux) is the most primitive living member of the Kalotermitidae. The following description of the nest is taken from Imms (1919, p. 126), who also figures the galleries:

Any structure which might be designated a termitarium or nest is absent, and the bulk of the members of a colony are to be found in irregular chambers situated in the decayed portions of a tree trunk. The wood is perforated in various directions by large galleries or tunnels. Certain of these passages pass outwards in a radial direction terminating just beneath the bark, if the latter be present. The majority, however, run in a longitudinal direction following the grain of the wood. The insect does not construct tunnels of cemented material on the surface of the tree trunk or the ground, as is the custom among a very large number of Termites. There is usually, therefore, no outward manifestation of its presence, which probably accounts for the insect having so long remained a rarity. When necessary, however, it closes up crevices by means of a cement of masticated ligneous material, or of excrementous matter held together by salivary secretion. The centre of a colony is occupied by the ova and very young larvae, and in their immediate vicinity the queens and kings are to be found.

The above description also applies to the nests of species in the related genus, *Zootermopsis*, found in the western United States. These species have been classed by Light (1934, v. 136) as "damp-wood termites" because they seem to be more dependent upon moisture than the majority of the Kalotermitidae which are often classed as "dry-wood termites." Both the damp-wood and dry-wood termites may live wholly within excavated wood and usually do not invade the soil, although in certain instances, notably *Kalotermes (Paraneotermes) simplicicornis* Banks, soil excavations may be found (Light 1934, p. 140; 1937).

EXCAVATING BEHAVIOR OF THE KALOTERMITIDAE

Although the excavation of wood is probably the most primitive behavior for the construction of a protected nesting site, the responses to environmental factors are quite intricate. A cork in a test-tube containing a captive colony of *Kalotermes flavicollis* (Fabricius) was invaded by the termites through small openings about the diameter of the larger nymphs. The end of the cork was cut off with a knife and revealed the excavations of the

FIG. 1. Cork excavated by *Kalotermes flavicollis.*

termites (Fig. 1). Through examination of the figure, it will be noted, in the first place, that the galleries are elongated and are strikingly correlated with the spring growth rings, only occasional passageways being found in the summer growth rings. This is probably a reaction to mechanical stimuli, although these rings doubtless differ in chemical composition as well. Secondly, it will be noted that the termites excavate close to the surface, but do not penetrate the surface. The layer between the excavation and the outside is so thin that light shines through when the cork is held up to the light. One might classify this phenomenon as a reaction to light, to changes of humidity or to a weakness in the thinner layers of cork. If pieces of wood within a test tube are subject to the attack of termites, no such thin layer is left, thus indicating that the weakened material is not the only factor, although the termites would seem to be sensitive to such differences. Likewise, termites do not avoid light shining through glass test tubes or containers, providing the light is not too hot and the humidity is not lowered materially. Toleration experiments and experiments with a humidity gradient conducted by Williams (1934) indicate sensitivity to humidity in most cases studied, although some of the Kalotermitidae do not seem to be highly sensitive to low humidities. Pending further experimentation, it would seem to me that a change in humidity was the factor dominantly responsible for the separation of the excavations from the exterior. Pillars of corky material are to be seen in the photograph. These have been left around the hollow lenticel pores in the cork running at right angles to the growth rings. Such pillars are probably best interpreted as reactions to changes in humidity.

To what extent the excavating termites respond to stresses and strains in the wood has not yet been satisfactorily determined, although the problem is open to experimental attack. Observations, particularly by those investigating damage to building construction wrought by termites, seem to indicate that termites will not weaken timbers sustaining weights to the point of collapse. If such timbers collapse, extra strains have been placed upon them through storms, earthquakes, or additional weights to which the termites have had no opportunity to react while making their excavations. Spatial factors may also play a rôle but have not been much investigated.

There is little to indicate that the presence of the king, queen, young or eggs influence the excavation activities of the Kalotermitidae. These primitive termites do not have grossly enlarged queens and the reproductive castes are usually found in galleries approximating the size and general appearance of the other galleries. Eggs may be found in the cells occupied by or near the reproductive castes. They may be picked up and gathered into small groups by the nymphs as has been observed by the author in captive colonies of *Zootermoposis angusticollis* (Hagen) and *Kalotermes flavicollis*.

CONSTRUCTION BEHAVIOR OF THE KALOTERMITIDAE

Although excavation as the result of feeding activities supplies termites with a protected nesting site, positive constructions supplement the burrows in producing habitations even in the most primitive kalotermitids.

After the colonizing flight and subsequent pairing, the reproductive couple of *Kalotermes minor* Hagen excavates a small hole or cell in available wood and plugs the entrance. Harvey (1934, p. 225) states: "This plug is a mixture of partially chewed wood and a secretion of the termites which acts as a cement."

The worker-like nymphs of a mature colony of *Zootermopsis angusticollis* kept in a quart mason jar built extensive partitions separating cells in the excavated wood and also connecting the wood with the glass. These partitions were made largely from the pellets of excrement which were cemented by saliva, and by liquid excrement which was extruded upon the pellets worked into place by the mouth-parts. Small experimental holes made in the metal cover of the jar were plugged in a similar manner. The nymphs would seem to react directly to changes in humidity, although other factors may also contribute to the cooperative action involved in building a partition.

Such building activities as those described for *Kalotermes* and *Zootermopsis* are probably to be found with little variation throughout the Kalotermitidae. In a few instances, permanent runways may be excavated through soil which enable the termite colony to exploit food resources outside the nest. Such runways are not common in the Kalotermitidae, but have been reported by Light (1934a, p. 311; 1937).

NESTING BEHAVIOR OF THE MASTOTERMITIDAE AND HODOTERMITIDAE

Hill's (1921, 1925) accounts of the nests of *Mastotermes darwiniensis* leave little doubt that the nesting behavior must be more specialized than that found in the Kalotermitidae. The nests are subterranean and are not always closely associated with the feeding excavations. Extensive construction and covered galleries are reported; the colonies contain over a million individuals, and foraging termites may destroy wooden materials over a hundred yards from the nest.

Even though this account gives the impression of rather specialized nesting behavior, it would seem to differ quantitatively rather than qualitatively from that of the Kalotermitidae.

The hodotermitids seem to be an offshoot of primitive kalotermitids or possibly pre-kalotermitids. Their social organization indicates considerable specialization beyond that observed in the Kalotermitidae. They have become harvesters and feed largely upon grass; they have an active, pigmented caste with compound eyes which forages on the surface in the day time and is usually considered an adult worker, a caste which seems to be lacking in the Mastotermitidae and Kalotermitidae.

Accompanying the evolution of structure and social integration, we find elaborate nesting behavior transcending anything observed among other primitive termites. Fuller (1915) has given excellent descriptions and figures of the nests of the South African species of this family (also Hegh 1922, pp. 232, 233). The nests of *Hodotermes (H.) mossambicus transvaalensis* Fuller are built in excavated chambers under the surface of the earth. The soil is brought to the surface and dropped in small piles which are distributed by the rain. Fuller (p. 350) states:

> The hive-cavities, with one exception, were all sub-spherical, having a horizontal diameter of 24 in. and a perpendicular height of 18 in. The cavities are partitioned by very numerous horizontal and close-set shelves. These are constructed of a thin and very papery substance which does not dissolve in water or in alcohol. The shelves lie one above another with striking regularity, and are attached to a series of clay brackets projecting from the walls. The shelves are not equi-distant apart throughout the cavity, but range from 6 to 15 mm. Innumerable little cylindrical columns of wooden texture, spread over the field of each shelf, hold the whole fabric together. These little columns are not stairways; the insects pass up and down from storey to storey of the hive by short inclines.

Of particular interest is the construction of the supporting columns. It would seem necessary to assume an intricate reaction to strains in the nest material to account for such elaborate architecture.

Nests of *Hodotermes (Anacanthotermes) ochraceus* (Burmeister) which I examined on the outskirts of Kairouan, Tunisia, consisted of soil excavations without surface indications except that the soil particles had been cemented together over the cells near the surface forming a brittle cover which gave off a hollow sound when lightly tapped. Some of these cells were filled with strips of plant epidermis from the stems of surrounding scrub vegetation. A tiny termitid, *Eremotermes indicatus* Silvestri, was found living in these storage cells feeding upon the food gathered by *H. ochraceus*. Extensive chambers contained all stages of nymphs and all castes of *H. ochraceus*.

A mason jar was filled with soldiers and various sizes of "workers." The smallest was about ½ cm. long and the largest about 1 cm. long. Soil together with strips of plant epidermis collected by the termites was also included. The termites immediately started to construct passageways and galleries and to accumulate the food. The construction was performed by working pieces of dirt moistened by saliva into place with the mouth-parts. No abdominal excretion was observed during the construction activities. All sizes of "workers" including the next to the smallest engaged in the labor.

These data indicate a much more developed nesting behavior in the Hodotermitidae than has been found among the Kalotermitidae. The emancipation from a wood diet, the construction of subterranean galleries, the storage of food, and the subtle manipulation of materials in the construction of complex layers of cells, passageways and supports; all indicate greater elaboration of the inherited behavior patterns than is found elsewhere below the Termitidae.

NESTING BEHAVIOR OF THE RHINOTERMITIDAE

The family Rhinotermitidae was derived, according to the available morphological evidence (Hare, 1937), from a kalotermitid stock fairly closely related to *Stolotermes*. The social development is much more marked than in the Kalotermitidae and a true sterile adult worker caste has become differentiated. Specialization of the worker is accompanied by larger numbers of individuals in the colonies, enlargement in size and reproductive capacity of the queen, incorporation of certain specialized termitophiles into the social community—all indicative of more complex social organization.

Excavations and passageways in the soil enable these insects to reach soil moisture which seems to be necessary for their existence in nature and also enables them to exploit food resources remote from their nests. Because of their adjustment to soil conditions, they are typical "subterranean termites" and have been so classed by Snyder (1920, p. 89).

The development of more elaborate nesting behavior might be expected in this family, but in reality the nests are still largely to be found in excavated galleries in wood with somewhat more complex cells, partitions and covered tunnels than are characteristic of the Kalotermitidae.

The more elaborate nest structures among the Rhinotermitidae are made by certain species of *Coptotermes*. Oshima (1919) has published some excellent figures of the nests of *C. formosanus* Shiraki. He states (p. 333):

The nest consists of a mixture of abdominal excreta and clay or sand, pasted together with a special secretion of the salivary glands. Sometimes it is rigid and compact and seems like a piece of rock. However, it is inflammable and burns rapidly, leaving a small amount of ash.

Ehrhorn (1934, p. 327, 329) shows photographs of the nests of the same species and Light (1934, p. 142) figures the nests of *C. vastator* Light. Hill (1915, p. 92) gives an excellent description of the nests of *C. acinaciformis* (Froggatt) accompanied by photographs. This species builds conspicuous mounds usually found at the base of a tree or enveloping a stump. The mounds may reach a height of six to eight feet. The walls of these nests are constructed of fine particles of earth and sand firmly cemented together. The walls vary in thickness from two inches near the top to twelve inches near the ground or on the sides. The interior is composed of triturated wood molded into curious forms. Near the ground are found thin-walled horizontal cells serving as the "nursery." Hill also states:

The queen is generally found about three inches from the ground, and about the middle of the nest, in a low domed cell with more or less level floor, from which she cannot excape.

It is apparent that the nests of *Coptotermes* illustrate the use of various materials, the organization around social functions, and reaction to the presence of the queen resulting in the construction of a "royal cell."

During the colonizing period it is necessary that exits from the nest be made in order to allow the imagoes to emerge. Holes are excavated by the workers in all cases observed and the soldiers guard the exits during the emergence of the imagoes. The exits are plugged by the workers at the end of the flight. Two instances of building activity associated with the colonizing flight have been observed. Wood of a beam in a small house three miles from Kartabo, British Guiana, was infested with *Coptotermes testaceus* (L.). On the morning of July 2, 1924, Dr. S. C. Crawford and the writer noticed workers constructing a small shelf about an inch long, a half inch wide and a half inch high, at the mouth of a small excavated hole in the wood on the vertical face of the beam. Soldiers guarded the operations. About two hours later a light rain fell and as it ceased, flying imagoes of the species were noticed. They were soon discovered emerging from the hole in the beam and using the shelf as a platform from which they took to the air. The next morning only a plugged hole was visible, the workers having dismantled the shelf during the night. Mr. James Zetek observed a somewhat similar construction during a flight of *C. niger* Snyder from the base of a tree in Ancon, Canal Zone, on August 26, 1923.

It is extremely difficult to analyze such behavior in terms of responses to simple stimuli. Although our ignorance is great, one must assume responses to intricate social stimuli. Anthropomorphic explanations, however, do not assist our understanding.

NESTING BEHAVIOR OF THE TERMITIDAE

The Termitidae are morphologically and socially the most specialized of the termites and have doubtless evolved from a rhinotermitid stock. The abundance of nest types are too numerous to describe in this report and only a few of the more outstanding and significant examples will be mentioned.

Many nests of the Termitidae are not more elaborate than the simplest nests of the Rhinotermitidae. In fact, some nests seem even less complex. The author dug into a subterranean nest of *Syntermes snyderi* Emerson in the rain-forest of British Guiana (Fig. 2) and found galleries extending

FIG. 2. Subterranean excavated nest of *Syntermes snyderi*.

below the surface to a depth of about four feet and covering a horizontal circular area with a diameter of about twelve feet. The galleries were large and seemed to be simple excavations in the sandy-clay soil, although there were small lumps lining the galleries in places that might have been excretions of the termites. The excavated dirt was deposited on the surface in loose piles above the nest. No covered tunnels were constructed on the surface, the termites moving in exposed trails. They cut out pieces of dead leaves which served as their food. Leaf fragments about one centimeter in diameter were stored in some compartments of the nest.

The humidity and consistency of the material composing the walls of the excavated galleries must be fairly constant in such a nest. The excavating termites would seem to be reacting to spatial factors as well as to strains in the supporting walls, rather than to humidity or differentiation of material as in the case of *Kalotermes flavicollis* described in the preceding pages.

NEST MATERIALS

Materials composing the nests of termitids (see Hegh, 1922) are similar to those already described, but one often finds a given group specializing in the use of certain substances. Most of the species use particles of dirt or sand. The nest of *Anoplotermes (A.) silvestrii* Emerson in British Guiana was of such extreme hardness that sparks flew from the hatchet when the nest was opened. In laboratory colonies, no abdominal excretion was observed during the nest building activities. Each particle of dirt was moistened only with a salivary secretion as the worker placed it in position. Laboratory colonies of *Microcerotermes arboreus* Emerson, however, revealed the workers placing pieces of dirt moistened with saliva in place and then turning and excreting a drop of thick dark fluid upon the newly inserted particle. *Crepititermes verruculosus* Emerson exhibited still different actions. Practically all the building observed in the laboratory was constructed through the use of abdominal excretions of thick dark fluid. The intestines of the workers seemed to be filled with this muddy material. They were sometimes observed to bring pieces of material for building, but no salivary secretion was seen nor did they work the material into place in the manner so characteristic of most termites.

On the other hand, wood derivatives are used almost exclusively by the majority of the species of *Nasutitermes, s. str.* (Fig. 3), a tropicopolitan group of termites composed of a large number of species. These termites can build typical nests in such places as the branches of standing dead trees which have been killed by the rise of dammed waters in Gatun Lake during the construction of the Panama Canal. Such nesting sites are not available for species dependent upon dirt in the construction of their nests.

Beaumont (Dudley, 1889, p. 91) described the construction activity of a species of Panamanian termite (either *N. corniger* (Motschulsky) or *N. ephratae* (Holmgren). The behavior is essentially similar to that described for *Microcerotermes arboreus* except that particles of wood or carton are used instead of dirt. Bugnion (1927, p. 18) also gives a fine detailed account of similar behavior of *Nasutitermes (= Eutermes) ceylonicus* (Holmgren) which uses both wood, sand and abdominal excretions in its construction. An

interesting variation was observed by the author while watching captive
colonies of *Nasutitermes guayanae* (Holmgren) in British Guiana. Instead
of placing the piece of material in its position with saliva and then excreting

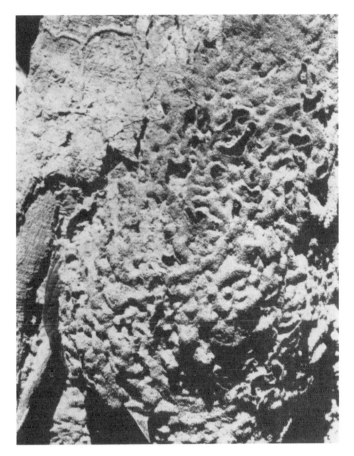

FIG. 3. New cells in process of construction during the
enlargement of the nest of *Nasutitermes guayanae*.

abdominal fluid, the termite invariably excreted the abdominal substance
first, and then turned and worked the piece of building material into place
with its mandibles.

STRUCTURES

The structures vary considerably in form and function (see Hegh, 1922),
thus indicating a wide variety of stimuli affecting the behavior pattern. Covered tunnels may lead out from the nest. These are built along odor trails
followed by the termites at times of high humidity. One of these covered
tunnels built by *N. guayanae* led from a nest thirty feet up in a palm tree,
down some vine stems to the ground, and around by a sandy clearing in the

forest to a small dead tree. In all, the distance was 162 feet, although the dead tree was only 35 feet in a straight line from the palm tree.

The covered tunnels may lead down into the ground where the excavations are lined with carton (materials cemented by saliva or excrement). Covered tunnels on the walls of Chilibrillo Cave, Panama, made by *N. corniger*, were estimated to be about twenty feet below the surface of the ground.

William Beebe observed *Macrotermes carbonarius* (Hagan) at Penang, Malaya, walking in open trails on carton roadways which the termites constructed over rough ground (See Emerson, 1937a, p. 247). I assume these roads to be a response to mechanical factors.

Termites will use the same materials used in the construction of their nest or tunnels for burying foreign insects introduced into their colony. Dudley (1889) has figured this action in *Nasutitermes* and the author has witnessed it many times following the attack upon an introduced foreign termite or termitophile.

The size of the nest is quite often fairly characteristic of the species. For example the nests of *N. pilifrons* (Holmgren) were quite often much larger than those of *N. corniger* or *N. ephratae* in Panama. Likewise in British Guiana, the nests of *N. surinamensis* (Holmgren) were typically larger than those of *N. guayanae*, *N. costalis*, or *N. ephratae* and, in turn, these were larger than the nests of *N. gaigei* Emerson. The huge dirt nests of *N. pyriformis* (Froggatt) reaching a height of 18 feet would seem to be characteristic of the species (Froggatt, 1905). The size of the nest may be proportional to colony size and reproductive capacity. The number of individuals in one nest (Emerson, 1937a, p. 247) of *N. surinamensis* was estimated quite carefully to be three million in round numbers. The nest measured six feet in height and three and one-half in greatest diameter. The queen from this nest was 24 mm. long and 8 mm. wide. She laid 2938 eggs in 18 hours. A small nest of *Microcerotermes arboreus* in British Guiana seemed small enough to count the entire colony (excepting foragers). It measured 6 in. long, 2 in. wide and 2 in. thick. The volume was approximately 110 cu. cm. There were 5876 termites and 2109 eggs in the nest. 4006 of the termites were mature and 1870 were nymphs. The mature individuals were composed of 1 queen, 1 king, 114 soldiers, and 3890 workers divided into 2624 large light-headed types, 939 small dark-headed types and 327 intermediates. Two larger nests of this same species contained about 250,000 eggs and 300,000 eggs respectively. A queen measuring 21 mm. in length laid 1680 eggs in 24 hours. The highest rate of oviposition which I have measured was 357 eggs in one hour in the case of a queen of *Anoplotermes silvestrii* which measured 50 mm. in length. These tropical queens continue to lay eggs steadily without much variation during diurnal or seasonal cycles and are doubtless often quite old. Through consideration of nests in Africa known to be at least

40 or 50 years old and each occupied by a first form queen, I think such an age for some of the reproductive castes is not impossible.

Enlargement of the nest may indicate some of the stimuli to which the workers respond. In the case of *Nasutitermes guayanae* (Fig. 3) a hole in

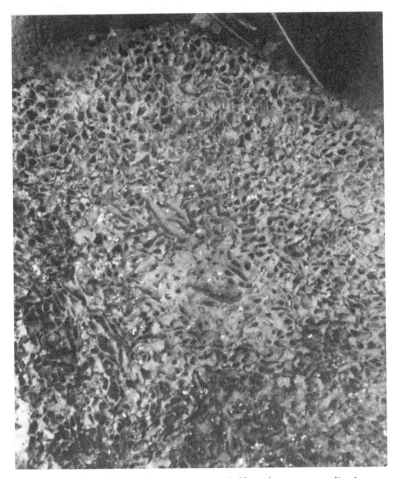

Fig. 4. Vertical section of a nest of *Nasutitermes costalis* showing the thick walls in the vicinity of the royal cell.

the exterior of the nest is made and a new cell is built over the opening. Workers build the walls on all sides until they meet at the top. *Constricto-termes cavifrons* (Holmgren) enlarges its nest by roofing the area between parallel surface ridges (Fig. 14). The point of junction between the walls is perfectly even and seems to be explained only in terms of subtle sensitivity to spatial relations. The subsequent thickening of the walls of the cells takes place below the surface. The walls in the vicinity of the royal cell are greatly thickened in the nests of many species of termites (Fig. 4) evidently stimulated by the presence of the queen. Grassé (1937) gives an interesting and detailed account of the organizat'on of the nest of *Macrotermes (Bellicositermes) natalensis* including consideration of the royal cell.

Holdaway (1933) gives interesting data upon the chemical differences in materials composing the parts of the mound nests of *Nasutitermes (= Eutermes) exitiosus* (Hill). An analysis of the inorganic material shows that the outer wall averages 64 per cent, the inner wall 15 per cent, and the "nursery" 11 per cent. There seems little doubt that social and environmental factors influence the use of materials in the same nest.

FIG. 5. Mound nest of *Amitermes medius* Banks. Aguadulce, Panama.

By far the most astounding structural detail in the construction of nest walls is to be found in the subterranean nests of the various species of the African genus *Apicotermes*. Careful descriptions and figures of these nests have been published (Desneux 1918; Hegh 1922, pp. 385-394, 712; Sjöstedt 1923, 1926, Taf. 13; Reichensperger 1923). The determination of the known species is to be found in Sjöstedt (1926, p. 153). The simplest of these nests has external walls perforated by tiny pores or channels too small for the termite to walk through and probably functioning as ventilation pores. The next most complex nest has the external openings of the pores emerging into small pits or holes considerably wider than the pores, and the most complex nests have circular galleries within the walls connected to the outside by small pores and to the inside by small pores of approximately equal diameter. In the latter case, it would seem difficult not to assume that the pits or holes of the second type of nest had evolved into the circular galleries of the third type. Such behavior evolution would seem to present a sequence which reminds one of the ascon, sycon and leucon types of canal systems in the sponges. The nests of *Apicotermes* also illustrate most clearly the inherent nature of the hereditary influences upon the behavior which are generic in character rather than distinctive of species alone. Of course we might expect to find certain types of behavior characteristic of larger groups than genera, and such is the case. The finest example of distinctive construction behavior characteristic of an entire subfamily (Macrotermitinae) is discussed later.

Kinsey (1936, pp. 8, 56, 57) reviews current concepts of higher taxonomic categories. Among others, the concept that genera, subfamilies and families are arbitrary groupings merely for the convenience of the taxonomist and without objective reality has often been stated. The correlation of many morphological characteristics together with behavioristic, physiological, ecological and geographical characteristics as seen in the genera, subfamilies and families of termites has convinced the author that the groups of species are indicative of a relationship which may be scientifically symbolized in the names of the higher categories which thus rest upon sound objective evidence with statistical significance of correlation.

As has already been pointed out, food is sometimes stored in the cells of the nests of *Hodotermes* and *Syntermes*. Food storage may involve construction activity also. In the nest of *Constrictotermes cavifrons* (Fig. 14) the basal hanging portion is composed of a dark pasty material which I interpret to be stored food. It is possible, however, that this is segregated excrement. In any case, the material has sufficient nutrient value to be used by a large number of other insects including tineid caterpillars probably belonging to the genus *Amydria* or *Exoncotis* (det. W. T. Forbes), larvae of an acanthocerid of the genus *Acanthocerus*, elaterid larvae, larvae of the sciarid genus *Sciara*, and finally another species of termite, *Termes (= Microtermes) inquilinus* (Emerson). *T. inquilinus* has never been found in any other place and I am confident that it is confined to the nests of *C. cavifrons*. Its galleries are separated from those of the host termite and are lined with carton of its own construction, but the nesting activities largely consist of excavations in the stored organic material deposited by the host termite. If the two species come together when the nest is opened, a violent battle ensues. Although the nests of the genus *Termes* are not as definitive as those of most species of *Nasutitermes*, it seems safe to assume that this particular species of *Termes* has undergone degenerative evolution of the nesting behavior in association with its social semi-parasitism. The closest related species, *Termes fur* (Silvestri), seems to have the same relationship to *Constrictotermes cyphergaster* (Silvestri) (Silvestri 1903, p. 128). These cases remind one of the evolutionary degeneration of the nesting behavior among the parasitic birds and the genus *Psithyrus* of the bumble bees.

Kemner (1929) interprets the existence of carton nodules in the nests of *Microcerotermes depokensis* Kemner as food storage activity. Similar nodules were observed in nests of *Nasutitermes pilifrons* during my studies of Panamanian termites, but I am not sure of their function.

Probably the most remarkable construction activities associated with nutrition are to be found in the fungus gardens of all species of the subfamily Macrotermitinae. The excrement of the termites or finely pulverized plant material (see Grassé, 1937) is built into elaborate convoluted structures so organized as to give a maximum surface for the growth of the fungus (Wheeler 1907; Sjöstedt 1907; Hegh 1922; Bugnion 1927). The gardens are quite distinct from the rest of the nest, which is constructed in quite a different manner. They present a picture of complex reactions to complex stimuli. Fungus-growing behavior is confined to this subfamily of termites. Examples of nest-building behavior patterns characteristic of higher taxonomic categories of birds are given by Chapin (1917).

CASTES INVOLVED IN NEST CONSTRUCTION

The nests of the primitive Kalotermitidae are excavated or constructed by the nymphs of the soldiers and reproductive castes except for the original cell of the colonizing pair. It is probable that the Mastotermitidae follow

FIG. 6. Mounds of *Amitermes vitiosus* (?) Hill between Camooweal, Queensland, and Newcastle Waters, Northern Territory. Photograph through the courtesy of G. F. Hill.

the same rule. The situation among the Hodotermitidae is more obscure because the exact status of the darkly pigmented, eyed "worker" has not been definitely established. Among the Rhinotermitidae and Termitidae, the adult sterile worker has become differentiated through neoteinic evolution from the soldier nymph (Emerson, 1926, 1935; Hare, 1934). However, it is quite common to find smaller worker-like forms, fully pigmented and with the abdomens containing the same materials as those found in the adult workers. These individuals seem to be nymphs of workers or soldiers and their behavior is the same as that of the adult workers as far as is known (Emerson 1926). It is thus probable that the worker-like nymphs of the soldiers have the same complex behavior patterns as the workers, and function in the construction and nutrient activities. Mature soldiers, however, never seem to assist in nest building. Snyder (1920, p. 190) states that the nasute soldier uses the fluid exuded from the beak in making shelter tubes, but I am convinced that this statement is an error. It has not been demonstrated that the nymphs of the reproductive castes of the Rhinotermitidae or Termitidae ever take an active part in gathering much food or constructing the elaborate nests. The construction of the original cell in wood or dirt by the colonizing pair, however, indicates that nesting behavior has not been wholly inhibited in the reproductive castes.

Spencer (1893) attempted to explain the differences between the behavior of the reproductive caste and the worker in the Hymenoptera as a loss of instinct on the part of the reproductive caste, the primitive species having the instincts before caste divergence took place. Holmes (1911) points out

FIG. 7. East face of mounds of *Amitermes meridionalis.*
Photograph through the courtesy of G. F. Hill.

the weakness of this contention, which is further demonstrated in the behavior evolution occurring long after the establishment of the adult sterile worker caste among the termites.

ECOLOGICAL FUNCTIONS OF THE NEST

The nest would seem to be of value to the termites as a means of controlling certain variables in the environment or for allowing sufficient elasticity of conditions to enable the insects to avoid certain environmental extremes. In order to test the variation in temperature within nests in relation to variations in the environment, two nests of *Nasutitermes corniger* on Barro Colorado Island were chosen. One was located in a clearing on a small stump, the base in contact with the ground and the top two feet from the ground. The stump was under a banana tree and was forty feet from the edge of the forest. The greatest diameter of the nest was one foot. A short thermometer (4 in.) was placed in a hole stoppered by a cork. The hole was about 1 inch wide and the bulb of the thermometer was 5½ inches from the surface. The other nest was attached to a small vine in the forest about twenty feet from the edge of the clearing. The forest roof was about fifty feet high. The bottom of the nest was 1½ feet from the ground. The nest was 2 feet high and 1½ feet in greatest diameter. A hole about 1 inch wide was dug into the nest so that the bulb of the thermometer (matched with the

TABLE 1. Temperature and light measurements through a daily cycle in the environment and in clearing and forest nests of *Nasutitermes corniger* in Panama.

Time	Clearing				Forest				Remarks
	Nest Temp. °F.	Shade Temp. °F.	F. C. Light on Nest	F. C. Light in Sun	Nest Temp. °F.	Shade Temp. °F.	F. C. Light on Nest	F. C. light in Sun Fleck	
10 A.M.	79.0	83.5	1300	10500	80.5	82.0	25	200	Sunny
11 A.M.	86.0	86.0	11500	11500	81.0	81.0	20	5000	Sunny
12 M.	87.0	85.0	8500	8500	81.5	79.5	30	5500	Sunny
1 P.M.	90.0	86.0	9500	9500	82.5	79.5	10	30	Partly cloudy
2 P.M.	92.0	79.5	800	800	83.0	77.5	½	—	Cloudy
3 P.M.	92.0	79.5	950	950	83.0	77.0	5	—	Cloudy
4 P.M.	91.0	80.0	195	195	83.0	78.0	0	—	Light rain
5 P.M.	91.0	79.0	160	160	83.0	78.0	0	—	Light rain
6 P.M.	89.0	77.5	50	50	83.0	78.0	0	—	Light rain
7 P.M.	88.5	77.5	0	—	82.5	77.0	0	—	Rain stopped
8 P.M.	87.0	74.5	0	—	82.5	74.5	0	—	Stars out
9 P.M.	84.5	74.5	0	—	82.5	74.0	0	—	Stars out
10 P.M.	83.5	74.0	0	—	81.0	74.0	0	—	Stars out
11 P.M.	82.0	74.0	0	—	80.5	74.0	0	—	Stars out
12 P.M.	82.0	74.0	0	—	80.5	74.0	0	—	Stars out
1 A.M.	82.0	73.5	0	—	81.0	74.0	0	—	Stars out
2 A.M.	81.0	74.5	0	—	79.0	74.8	0	—	Cloudy
3 A.M.	81.0	74.5	0	—	79.0	75.0	0	—	Stars out
4 A.M.	80.5	74.5	0	—	78.5	75.0	0	—	Stars out
5 A.M.	79.0	74.5	0	—	78.5	75.0	0	—	Stars out
6 A.M.	79.5	75.8	5	5	78.5	75.0	0	—	Cloudy
7 A.M.	80.0	77.0	450	450	78.5	76.0	10	—	Cloudy
8 A.M.	80.5	78.0	600	1800	78.0	77.5	50	300	Sunny
9 A.M.	81.0	79.0	1100	1200	78.5	77.0	15	15	Cloudy
10 A.M.	82.0	80.5	1600	1650	78.5	78.0	20	25	Cloudy
11 A.M.	83.0	83.0	2500	3000	79.0	79.0	25	35	Dim sun

thermometer in the clearing nest) was 6½ inches from the surface and the hole was stoppered. Measurements (Table 1) were taken approximately on the hour on September 2d and 3d, 1935, for a period of over 24 hours. Light readings (in foot candles) were taken by means of a Weston illuminometer. In general the measurement of illumination in the sun in the clearing showed little difference from that on the nest. A light rain fell intermittently from 2.30 to 7.00 P.M. and doubtless influenced the temperature and illumination records.

One concludes from these measurements (Table 1) that (1) the direct sun rays on the nest raise the internal temperature above the shade temperature of the surroundings, and (2) that the temperatures follow the daily rhythm of the external temperatures, but lag behind the rising and falling external temperatures and do not reach the extremes found outside the nest. Of course, the sun temperatures outside the nest which reached maxima above 120°F. (the limit of the thermometers available) were far in excess of any internal nest temperature, and the nest temperatures measured above the maximum shade temperatures were doubtless due to the direct rays of the sun.

It is thus possible to say that the nest structure partially protects the termites from the temperature extremes found outside of the nest. Cowles (1930) measured external and internal nest temperatures in the nests of *Trinervitermes trinerviformis* in Natal and his table shows the same tendency in these mound nests. Subterranean nests probably follow the temperature cycles of the surrounding soil very closely and this feature may be considered one advantage of such a habitat. Although it is possible that there is some migration of termite workers and soldiers influenced by tem-

FIG. 8. South end of mounds of *Amitermes meridionalis*.
Photograph through the courtesy of G. F. Hill.

perature variations within the nest (Holdaway, 1935) and between the nest and the ground, the queen is of necessity usually confined to the royal cell and the young nymphs are seldom found far from the center of the nest.

Of probably greater importance than temperature is the control of humidity within the nest as compared to the extremes to be found in the external air. No measurements of nest humidities have been made, but it is safe to assume that the occupied nests of the Rhinotermitidae and Termitidae have almost a saturated air humidity (See Cowles, 1930, p. 23). The reaction of these termites to humidity gradients (Williams 1934; Emerson, unpublished experiments) indicates that they move away from dry air toward saturated air and that they die from even a brief exposure to dry atmosphere. Species of the Kalotermitidae often show more toleration to dry atmosphere and also often do not move from dry air to saturated air when exposed to humidity gradients. It is this dependence upon saturated or nearly saturated humidity that is probably the most important reason why the rhinotermitids and termitids are typically soil termites and, when above the surface of the soil, they typically construct nests and tunnels which maintain humid conditions not very different from the subterranean habitat. Slight differences in the moisture requirements between *Reticulitermes hesperus* Banks and *R. tibialis*

FIG. 9. Rain-shedding dirt-carton nest of *Amitermes excellens*. A wood-carton nest of *Nasutitermes guayanae* is visible on the right side of the trunk.

Banks demonstrated by Williams (1934) make it reasonable to suppose that such differences may often determine the ecological distribution of the species. Even the desert termites, however, so control the humidity of the air surrounding them that they live in an atmosphere close to saturation. The case is different in the Kalotermitidae, however, and the ability of *Cryptotermes*, for example. to live in the wood of dry furniture is doubtless linked with a greater toleration to dry atmosphere than is found in the species of Rhinotermitidae and Termitidae.

Other factors that may be of some importance to termites are the oxygen necessities and the elimination of carbon dioxide. Williams (1934) reports positive experiments upon these factors. The ventilation pores in the nests of *Apicotermes* may function for the exchange of gases. It may well be that the elimination of carbon dioxide may determine the site of the nest to some extent. Soil heavily soaked in water or with the surface flooded would certainly not offer a favorable site for a termite nest and the tendency to construct mounds and arboreal nests may enable termites to live in otherwise unfavorable localities.

The termite literature abounds in references to the avoidance of light. The only definite orientation with response to light of which I am aware among termites is the photopositive reaction of the winged imagoes during the first part of the colonizing flight. I have myself reported (Emerson 1929) a "negative phototropism" of the imagoes before the flight starts and after it is finished, but more accurate experimentation indicates that the pho-

topositive reaction is only present for a short time and the insect is neutral
to light at other times. This neutrality does not apply to heat and to increased evaporation, however, and the actions usually interpreted as negative
responses are probably the result of these other factors which are so often

Fig. 10. Detail of rain-shedding projections of nest of *Amitermes excellens.*

closely associated with increase in the intensity of light. I should consequently not interpret the construction of nests or tunnels as the result of
photonegative behavior.

Another factor emphasized in the literature, particularly by Bugnion
(1927), is the construction of the nests for the protection of the inhabitants
from predatory enemies. It is certainly true that exposed termites form
ready prey for wasps, ants, spiders, lizards, and other predators. Ants in
particular have doubtless been enemies of termites for many ages and some
species prey only upon termites (Wheeler 1936). Experiments as yet unpublished indicate the ability of the soldiers to combat these enemies in comparison with the defensive ability of the workers. The evolution of the
soldier caste has doubtless been guided by the selection of efficient means of
defense against predators and it is quite conceivable that the nest-building
patterns have also been selected because of their value in protecting the colony from attack. I have never observed a direct influence of the presence of
predators upon construction which cannot better be explained as a response
to other factors, however, and I am of the opinion that any such direct re-

sponse to ants as that postulated by Bugnion (1927, pp. 14, 30) needs critical verification.   A few predators are especially adapted to overcome the protection the nest affords.   The modified forefeet of the spiny anteater, pangolin, aard vark, and New World anteaters enable these animals to invade the hard nests of termites.   The cylindrical sticky tongues of these phylogenetically diverse mammals are astonishingly efficient in penetrating the complex system of cells in the termite nest.   The prehensile tail, found in the tamandua, silky anteater, and some pangolins, enables these predators to reach the arboreal nests.   The remarkable adjustments of these specialized termitophagous mammals only serve to emphasize the value of the nests as protection from general predators and indicate how strong the selection pressure was to influence the evolution of such striking convergent adaptations.

Not only is the nest a protection against animal predators, but it is also a protection from harmful fungi.   Laboratory colonies removed from nests often succumb to molds which seem to be effectively controlled under natural conditions.

SELECTION OF THE NESTING SITE

One might suppose that the selection of the nesting site involved no more behavior response than is found in seeds scattered by the wind.   It is true that the flying termites scatter in all directions and a very large proportion alight in unfavorable locations.   I have seen thousands entrapped in the surface of a river where fish were rapidly devouring them.   I have seen thousands die of heat and evaporation on the hot sands of dunes.   I have seen paired couples entering crevices in a wharf which was soon to be inundated by a rising tide.   Certainly the mortality is high during the colonizing flight.   On the other hand, the colonizing pair in *Termes inquilinus* and *T. fur* would seem to select the nests of *Constrictotermes cavifrons* and *C. cyphergaster* respectively for their nesting site and there are other indications that the imagoes are not wholly subject to chance.   There is strong evidence that considerable powers of selection are manifested, not so much by the colonizing pair as by the workers after the colony has developed.   The migration of an entire colony of *Nasutitermes costalis* including the royal couple and many termitophiles has been reported (Emerson, 1929).   The process of moving the nesting site will rarely be seen by human eyes, but migrations must be assumed when nests are found in locations where the royal pair could not have excavated their original cell.   Such is the case for every nest of *Constrictotermes cavifrons* seen in the British Guiana forest (Fig. 14).   These nests were invariably found on living trees with smooth bark, and in nine out of ten cases the tree slanted at an angle to the ground.   The living bark was not invaded by the termites.   The evidence seems convincing that all these nests were established through the migration of the entire colony.   The uniformity of the nesting site in this species indicates behavior reactions

of which we know practically nothing. How blind, rather small insects can
select a tree with smooth bark rather than rough, the under side of a slanting
surface rather than the upper, and then, after at least preliminary construction by hundreds of individuals, stimulate the entire colony including the
grossly physogastric queen to move to the new site, is difficult to analyze. It
is not surprising that uncritical commentators fall back upon anthropomorphic explanations of such facts.

Especially in arboreal nests, height above the ground may be characteristic of the species or genus. Of the nests which I have studied in the
New World, *Nasutitermes costalis, N. ephratae, N. surinamensis, N. wheeleri,
N. acajutlae, N. corniger, N. pilifrons, N. guayanae, N. similis, N. colum-
bicus, Constrictotermes cavifrons, Anoplotermes (Speculitermes) arboreus,
Amitermes excellens* and *Microcerotermes arboreus* build at varying heights
from the ground. The last four species build dirt carton nests and the others
build of wood carton. Other species seem to be limited to sites at most only
a few feet above the ground, such as *Anoplotermes silvestrii, A. banksi, A.
brevipilus* and various species of *Termes* and *Armitermes*. Every gradation
from these sites to subterranean locations may be found, but the range for
each species seems to be fairly characteristic, although there may be considerable variation within each species.

Nests may be built quite rapidly. A nest of *Amitermes excellens* Silvestri was constructed about 50 feet from the ground on a tree near the
Kartabo laboratory between September 15, 1919, and June 1, 1920. This
nest was about 3 feet long, 1 foot wide and ½ foot thick. A colony of
stingless bees had partly occupied the interior. I assume that the colony
migrated to a new site in this case. Nests enlarged in proportion to the
growth of the colony would doubtless grow more slowly.

NEST DIVERGENCE WITHIN A GENUS

It has often been observed that adaptive modifications are more conspicuous as one ascends in the ranking of the animal group. Adaptations between families are more obvious than between genera, and adaptations between genera are more obvious than between species of the same genus. One
explanation of this difference in different categories may be that genera arise
through further speciation of adapted species, while less adapted species may
not survive through the ages of competition. Striking adaptive divergence
probably takes a long time under fairly strong selection pressure, while speciation mechanisms may operate with slight selection (Wright, 1932, p. 363).
If the mechanisms of evolution are operating to produce adaptive behavior
in the same way that they operate to produce adaptive physiological interaction and structural growth, we may expect to find this rule illustrated in
the nesting behavior of termites.

The general outline of family differences in nests has already been presented. Many genera show similar nesting tendencies between various species, notably the dirt carton nests of *Microcerotermes*, the wood carton arboreal nests of most species of *Nasutitermes, s. str.* (Figs. 3, 4, 9), the mushroom-shaped nests of *Cubitermes* (Fig. 11), and the perforated nests of *Apicotermes*.

It is also possible to show striking divergence between the nests of species of the same genus, notably in the genera *Nasutitermes, Anoplotermes,* and

FIG. 11. Mushroom-shaped dirt-carton nest of *Cubitermes loubetsiensis* Sjöstedt, Medje, Belgian Congo. The cap with peripheral projections sheds rain from the stem. The height measured 44 centimeters. The vertical section shows the connecting holes between the cells and the queen in the middle just above the narrow part of the stem. Photographed by Herbert Lang.

Apicotermes. I have selected the genus *Amitermes* as the best example known to me which illustrates such divergence as well as striking adaptive nest modifications.

Amitermes is a genus found in every zoogeographical realm, but in general it is found in· the tropics and only in the warmer portions of the temperate realms. As a genus, it seems particularly adjusted to dry regions, many more species being found in such regions as the southwestern semi-deserts of the United States than in the rain-forests of the tropics.

The majority of the species do not build definitive nests above the ground which lend themselves to photography. Examples of such inconspicuous nests are those of *A. beaumonti* Banks in Panama, *A. wheeleri* (Desneux) in Texas, and *A. santschii* Silvestri in north Africa.

A number of species, particularly in Australia, build mounds above the ground which may be conspicuous features of the landscape (Figs. 5, 6). Illustrations of such nests of *Amitermes* are to be found in Hill (1922, 1922a, 1935), Froggatt (1905), Mjöberg (1920), Hegh (1922), Snyder and Zetek (1934), Emerson (1937a). It will be noted upon examination of these various photographs that there is a fair amount of variation in the shapes of the nests within a given species, but it is also obvious that characteristic distinction between the different species exists and in several instances the species may be easily recognized by means of the nest alone.

From many viewpoints, the most remarkable nest known is that of *A. meridionalis* (Froggatt). This nest is the famous "magnetic" nest found in north Australia (Figs. 7, 8) which is figured by Froggatt (1905), Hegh (1922), Hill (1922, 1935), and Emerson (1937a). Hill (1935) questions the determinations of the nests described and figured by Saville-Kent (1897, 1897a) and Mjöberg (1920). The largest of the "meridian" nests are about 12 feet high with a north-south length of about 10 feet and an east-west width of about $3\frac{1}{2}$ feet at the base in the middle. Hill (1935) gives the most complete and accurate account of the geographical and ecological distribution, the shape and orientation of the nest, and a discussion of the theories to account for the peculiar orientation. Hill states (private correspondence):

I have thought that the probable reasons for the mounds being built with the long axis approximately north and south is to obtain the maximum total amount of solar radiation during the winter months, and the minimum during the heat of the day in summer.

It would seem to me that the explanation probably will ultimately be found in the control of temperature, although theories placing emphasis upon humidity and wind have already been proposed. Not only is the explanation of the adaptive value of such a nest important, but a knowledge of the factors to which the termites are reacting directly would add much to our concept of nest evolution.

Andrews (1927) and Dreyer and Park (1932) discuss ant nests of the genus *Formica* which are oriented with reference to light and give measurements of light and temperature. These ant nests have their broadest face toward the south and the most direct rays of the sun. The difference in the orientation of the nests of *A. meridionalis* may very likely be owing to their tropical location and possibly to the greater extremes of temperature to which they are subjected.

In sharp contrast to the mound nests of the more arid savannahs, *A. excellens* constructs nests on the sides of surrounding trees in the rain-forest of British Guiana. Galleries covered the entire trunk of a dead tree in an old clearing near the Kartabo laboratory (Figs. 9, 10). The tree was covered to a depth of about 8 inches and to a height of about 45 feet. The material was a sandy-dirt carton which easily crumbled in the hand. The exterior was covered by numerous finger-like projections extending downward and outward. Upon breaking these projections they proved to be hollow and were always occupied by a few termites. The function seems to be for the shedding of rain water during heavy tropical showers. In this particular

FIG. 12. Rain-shedding dirt-carton nest of *Cubitermes subarquatus* on a tree at Medje, Belgian Congo, composed of a series of caps with peripheral projections. Photographed by Herbert Lang.

FIG. 13. Nest and rain-shedding, chevron-shaped structures of *Procubitermes niapuensis* on a tree at Niapu, Belgian Congo. Photographed by Herbert Lang.

locality the annual rain fall was around 110 inches. Heavy rain fall (5 to 8 inches) of short duration was not uncommon.

When these structures are first started on the side of a tree, a long vertical covered gallery is built on the trunk, and side branches are built in a peculiar parallel arrangement inclined downward on the tree at an angle of about 45° to form a "herring bone pattern" (Emerson, 1937a, p. 246). The function of this construction is also probably the efficient shedding of rain which descends the tree trunk.

CONVERGENT EVOLUTION WITHIN A SIMILAR HABITAT

The adaptive significance of a modification is more convincing if it is possible to show that organisms from stocks not possessing the modification have separately evolved a similar adjustment when subjected to natural selection within a similar habitat. The function of the finger-like projections on the nest of *Amitermes excellens* becomes more clear if similar nest structures appear in unrelated termites typically found in or near rain-forests. Convergence of rain-shedding structures may be demonstrated in three different subfamilies of the Termitidae including the Amitermitinae already discussed. Without doubt these subfamilies had no common ancestor from which they could have inherited such a pattern.

The Termitinae include several African genera which can be arranged in a series possibly illustrating an evolutionary sequence. *Cubitermes* often constructs a unique mushroom-shaped nest (Fig. 11), the specific variations of which have been well illustrated by Sjöstedt (1913), Hegh (1922), and Emerson (1928). These nests show definite generic similarity but at the same time show specific differences. The majority are capped by an umbrella-like structure, convex on top and projecting outward at the periphery in such a way as to shed rain from the top of the nest. Finger-like projections, quite similar to those described on the exterior of the nest of *Amitermes excellens*, often extend around the edge of the cap. The nest of *Cubitermes subarquatus* Sjöstedt (Fig. 12), found in the Congo rain-forest, deviates from the mushroom-shaped form found in more open country by appearing to be a succession of caps, one above the other. This nest type might easily have evolved from the mushroom type. The remarkable nest (Fig. 13) of a species from the Congo rain-forest, *Procubitermes niapuensis* Emerson, belongs to a closely related genus. The termites build a series of hollow, chevron-shaped, dirt-carton structures extending for some distance above the nest proper on the side of the tree trunk. Mr. Herbert Lang, who examined and photographed the nest, is of the opinion that the ridges function for the shedding of rain (Emerson 1928).

In the rain-forest of British Guiana, still another nest exhibits rain-shedding modifications (Fig. 14). The species is *Constrictotermes cavifrons* belonging to the Nasutitermitinae. The entire structure sometimes reaches a length of about 3 feet. It is attached for most of its length to the side of a smooth-barked live tree. A discussion of other features of this nest is to be found in the preceding pages. Pertinent to our present consideration is the series of solid, sharp carton ridges which extend over the surface of the nest and continue downward at an angle along the tree trunk at the sides of the nest. These ridges are also built some distance above the nest and form a series of chevron-shaped structures which remind one of similar structures in the nest of *Procubitermes niapuensis* (Fig. 13). In the case of the nest of *C. cavifrons*, however, the ridges are more numerous, are sharp and solid, and are built over the surface of the nest. The surface of the bark is cleaned of lichens and similar outgrowths between the ridges, and an open passageway is left above the nest which allows ease of travel up the tree. I have observed the nest during a heavy shower and there is no question that the

FIG. 14. Profile and detail of edge of nests of *Constrictotermes cavifrons* on smooth-barked slanting trees at Kartabo, British Guiana. Sharp, solid, rain-shedding ridges extend above the nest on the tree trunk as well as over the surface of the nest and down the sides of the trunk. The termites have bridged over the ridges as they enlarged the nest (upper right). The hanging basal part of the nest (lower left) is composed of stored food and is occupied by the semi-parasitic termite, *Termes inquilinus*.

ridges deflect the sheet of water descending the trunk of the tree in such a manner as to keep the nest relatively dry and moistened only by the drops which fall directly upon it. Hingston (1932) describes this nest and arrives at the conclusion that the ridges "shoot the water forward and outward clear of the nest." My observations did not indicate that the water was "shot" outward, but that the stream was deflected to the side and ran down the trunk without wetting the nest. Hingston also describes an experiment in which he cut 18 ridges above the nest, allowing the water to run into the nest. He states that the water moistened and softened the earth carton of which the nest is composed. In the fourth week following the injury to the ridges, the termites started to repair and replace the ridges and had completed eight during the sixth week when the observations ceased. This experiment of Hington's is most interesting because it indicates that the termites may react to the percolation of the rain-water into their nest.

DISCUSSION

Because the insects in the experiment outlined above have successfully met an emergency, Hingston classifies this behavior as intelligence (also see Imms 1931, p. 16). A cut in the finger is an unusual event for the cells involved and adaptive clotting of blood followed by the growth of tissues reacting to complex factors result (Arey 1936), but the physiologist would hardly be inclined to say the reacting cells were intelligent. Such somatic physiological activity is known to be influenced by hereditary factors as evidenced by the genetics of haemophilia. The termite behavior recorded by Hingston seems to me to offer a significant parallel.

Unless the nesting behavior be largely hereditary, there is no more phylogenetic significance in its evolution than in the evolution of human architecture. Human architecture seems to rest wholly upon a capacity for "conditioning" and intelligent response to environmental and esthetic factors. From the standpoint of behavior analysis, it is in sharp contrast to termite architecture. Both types of behavior, of course, may be of fundamental importance to the biological success of the species.

The tendency to resort to anthropomorphic explanations is noteworthy in observers who are astonished by the remarkable complexity of termite life. Maeterlinck (1926, 1928) is especially prone to such generalizations. Bugnion (1927) and Imms (1931, p. 102) resort to the "lapsed intelligence" theory of Lewes with its Lamarckian implications to explain the origin of "instincts". All of the references to termite "intelligence" known to the author carry strong anthropomorphic connotations.

Kemner (1929) has given a classification of the termite nests which he studied in Java and also discusses the systems of classification used by Holmgren (1906). In certain respects I find myself in agreement with Kemner's phylogenetic arrangement—namely in treating the kalotermitid-type nest as the most primitive and postulating the evolution of the rhino-termitid-type from the kalotermitid-type. In the arrangement of the phylogenetic sequence of the various types of termitid nests, however, I cannot agree with Kemner because the arrangement is opposed to phylogenetic sequence based upon a study of the morphological characters of all the castes (Hare 1937). It would seem to me that all correlated evidences of phylogeny would have to be considered and certainly comparative morphology cannot be ignored, not because behavior or physiological characteristics are

less important than morphological characteristics, but the significantly correlated characters which have been studied from a phylogenetic viewpoint are
largely morphological at the present state of our knowledge. Without breaking away from the deduced phylogenetic history of the major groups of
termites, one may assume that the most primitive termitids were ground and
log dwellers not differing greatly from the rhinotermitids in their nesting
behavior. Many morphologically specialized termitids have not deviated
much from this basic behavior type. Radiation, however, may conceivably
have occurred in several directions leading to simple ground excavations,
mound nests, fungus-garden construction, arboreal dirt nests, arboreal wood-
carton nests, and other more specialized types, a few of which have been
discussed in the preceding pages. It is not necessary to assume that any of
the more basicly specialized termitid-type nests were derived from any other
specialized type and there is much evidence against this hypothesis. Among
certain closely related groups such as the species of *Apicotermes* and species
of *Cubitermes* and *Procubitermes*, one may postulate evolutionary sequences
which fit the morphological data, but a postulate that *Microcerotermes*-type
nests evolved from *Nasutitermes* (= *Eutermes*)-types and that the fungus-growing behavior was an aftermath of the type of nest-building behavior of *Microcerotermes* seems to me wholly out of line with a considerable body of facts upon which the phylogenetic arrangement of these
groups is based.

The separation of analogous from homologous morphological structures
through the maze of genetic modifications, physiological influences, growth
patterns, degenerative changes, and convergent adaptations, is a difficult task.
With increasing knowledge of connecting links and a better understanding of
the principles of embryology and of evolution, we are enabled to rectify many
of the mistaken conclusions of the past. However, in spite of the fact that
the principle of homology has been applied to the evolution of termite nests,
I think that the best available evidence is a parallel series of homologous
morphological structures. Whether homology is assumed for structure, for
behavior or for physiological action, it would seem to rest upon a similar
genetic basis manifesting itself through successions of enzymic effects and
physiological actions and interactions (Wright 1934, p. 33). That behavior
may be dependent at times upon local chromosomal influences has been demonstrated by Whiting (1932) in his studies of the reproductive reactions in
sex mosaics of the parasitic wasp, *Habrobracon*.

I have cited one instance of degenerative evolution of nesting behavior
in the case of *Termes inquilinus*. Degenerative evolution of behavior may
have occurred in numerous instances, but few cases are associated with sufficient data to justify a reasonable conclusion. Degenerative evolution is an
example of a negative and usually non-adaptive directional tendency that
has been a stumbling block for many evolutionary theorists. Modern ge-

126

netics, however, has shown that mutations are often degenerative in their manifestations, that they may occur at statistically predictable rates, that species characters are usually an outgrowth of gene and chromosome combinations, and that natural selection is probably responsible for keeping wild populations from exhibiting degenerative tendencies of functional characteristics. "Mutation pressure" resulting in degenerative effects might produce a degenerative evolution if selection pressure for a particular structure or function were removed (Wright 1929; Darlington 1936). Degenerative mutations would seem to have more effect in small interbreeding populations. Another positive influence has been postulated (Wright, 1929) in the probable fact that "each character is affected by many genes and each gene affects many characters" (Wright, 1934, p. 30). Thus selection probably does not act upon the single effect, but selects genes and gene combinations which result in a sum total of favorable effects. Consequently a gene or system of genes causing degeneration may be selected for other beneficial effects, particularly if a former beneficial effect has been lost or decreased through a change in the environment.[1] It is thus possible to explain known evolutionary trends on the basis of modern genetic theory without recourse to Lamarckism. Nonadaptive evolution is possible (Wright 1932, p. 363) and it is not necessary to assume adaptive value for every specific character. The multiple effects of genes also give a reasonable explanation of non-utilitarian vestigial structures and recapitulative tendencies. Sudden complete losses of the hereditary basis of complicated structures would probably indicate such gross genetic changes that other vital functions would be affected. It is more probable under natural conditions that the old heredity has merely become overlaid by new heredity suppressing or modifying structures or actions (Wright 1935, p. 105). Needham (1930) emphasizes the opinion that "organs are only recapitulated in so far as they are necessary for the development of the ones which are required in the ontogeny in question, and all the other old ones disappear." Although recapitulative structures and actions may often have functions, it seems difficult to believe all cases are functional and Wright's explanation allows for the possibility of recapitulation of non-functional organs or actions.

I have searched my experience and the literature for a case illustrating vestigial behavior among termites, but I am unable to offer a single instance which is worthy of consideration. In a sphere far removed from the insect world, however, I wonder whether the rather futile action of dogs in scratching dirt following defecation may not be considered an example of behavior which has undergone phylogenetic degeneration, but is still visible as a vestigial pattern (see Enders 1935 for an account of specific differences in faecal behavior of Panamanian cats).

[1] Wright, in commenting upon this statement, adds: "The type allele in each series is that which has the most favorable net effect on all characters. If one character loses in importance relative to others, there will be a shift in the alleles in many series with degeneration of the character losing importance as a consequence of the increased development of the others."

The foregoing discussion emphasizes the phylogeny of inherited species behavior patterns, but I see no reason why individual behavior involving modifiability through experience (Emerson, 1933) may not be characteristic of termites. Such behavior seems to be relatively unimportant in the phylogenesis of the species patterns discussed in this paper. Ranges of variation in the inherited pattern may also be demonstrated under identical environments, and different environments will doubtless bring about variation in the patterns with the same heredity. A better understanding of the ontogenetic development of both genetically and environmentally induced behavior is highly pertinent. These problems await future investigation.

SUMMARY

1. Termite nests may be used as examples of behavior evolution because they are morphological indications of behavior patterns, they express the behavior of a population, the patterns are hereditary, there is a natural control over any Lamarckian influence, evolutionary sequences are available, adaptive modifications may be demonstrated, and coordination mechanisms may be partially analyzed.

2. Wood-eating roaches excavate galleries in wood but make no constructions.

3. The Kalotermitidae excavate wood and construct partitions, indicating responses to humidity and mechanical or chemical factors.

4. The Mastotermitidae exhibit a quantitative advance in nest construction compared to the Kalotermitidae.

5. The Hodotermitidae show a further advance with subterranean nests, elaborate carton construction, and food storage.

6. The Rhinotermitidae have separately evolved subterranean adjustment and in some species show building activities in response to social factors as well as physical factors.

7. Excavated subterranean nests of the Termitidae exhibit the influence of mechanical and spatial factors.

8. Materials used for construction may be dirt, wood, or excrement, cemented by saliva or anal excretions.

9. Structures may include covered tunnels, roads, rain-shedding projections and ridges, nests of characteristic size and differentiation, ventilation pores in the walls, stored food, and fungus gardens.

10. Sterile workers and nymphs of sterile soldiers and workers construct the nests in the Rhinotermitidae and Termitidae.

11. The ecological functions of the nest are control of temperature, control of humidity, and protection from predators and harmful fungi, all enabling the termites to live in otherwise uninhabitable niches.

12. The nesting site may be selected partly or wholly by the colonizing pair, but often is selected by the workers and is followed by a colony migra-

tion. Height of the nest from the ground may be fairly characteristic of the species.

13. Different species within a genus show great divergence in nesting behavior. Species of the genus *Amitermes* have subterranean nests, mound nests, arboreal nests, nests oriented with reference to the sun, and rain-shedding constructions.

14. Convergent evolution of rain-shedding constructions has occurred in the Amitermitinae, Termitinae, and Nasutitermitinae.

15. A discussion of "intelligence," correlation of morphological and behavior homologies, degenerative evolution and "vestigial" behavior is included.

LITERATURE CITED

Andrews, E. A. 1927. Ant-mounds as to temperature and sunshine. *Jour. Morph. Physiol.* **44**: 1-20.

Arey L. B. 1936. Wound healing. *Physiol. Rev.* **16**(3) : 327-406.

Ball, W. P. 1890. Are the effects of use and disuse inherited? London. 1894. Neuter insects and Lamarckism. *Nat. Sci.* **4**: 91.

Bugnion, E. 1927. The origin of instinct. *Psyche Monogr.*, No. 1, London. (also in appendix in Forel, A. 1928. The social world of the ants).

Chapin, J. T. 1917. The classification of the Weaver-Birds. *Bull. Amer. Mus. Nat. Hist.* **37**: 243-280.

Cleveland, L. R. 1934. The wood-feeding roach *Cryptocercus*, its protozoa, and the symbiosis between protozoa and roach. *Mem. Amer. Acad. Arts and Sci.* **17**(2) : x + 185-342, 60 pls.

Cowles, R. B. 1930. The Life History of *Varanus niloticus*. *Jour. Ent. Zool.* **22**: 1-31.

Cunningham, J. T. 1894. Neuter insects and Darwinism. *Nat. Sci.* **4**: 91.

Darlington, P. J., Jr. 1936. Variation and atrophy of flying wings of some Carabid beetles (Coleoptera). *Ann. Ent. Soc. Amer.* **29**(1) : 136-179.

Darwin, C. 1859. The origin of species.

Desneux, J. 1918. Un nouveau type de nids de Termites. *Rev. Zool. Afric.* **5**: 298.

Dreyer, W. A., and Park, T. 1932. Local distribution of *Formica ulkei* mound-nests with reference to certain ecological factors. *Psyche* **39**: 127-133.

Dudley, P. H. 1889. Observations on the termites or white ants of the Isthmus of Panama. *Trans. N. Y. Acad. Sci.* **8**: 85-114.

Ehrhorn, E. M. 1934. The termites of Hawaii, their economic significance and control, and the distribution of termites by commerce. Chap. 27 in Kofoid, C. A. Termites and termite control. 2d edit.

Emerson, A. E. 1926. Development of a soldier of *Nasutitermes (Constrictotermes) cavifrons* (Holmgren) and its phylogenetic significance. *Zoologica* **7**: 69-100. 1928. Termites of the Belgian Congo and the Cameroon. *Bull. Amer. Mus. Nat. Hist.* **57**: 401-574. 1929. Communication among termites. *Trans. Fourth Intern. Congr. Ent., Ithaca* **2**: 722-727. 1933. Conditioned behavior among termites (Isoptera). *Psyche* **40**: 125-129. 1935. Termitophile distribution and quantitative characters as indicators of physiological speciation in British Guiana termites (Isoptera). *Ann. Ent. Soc. Amer.* **28**(3) : 369-395.

1937. Termite nests—a study of the phylogeny of behavior (Abstract). *Science* **85**(2193) : 56.

1937a. Termite Architecture. *Natural History* **39**(4) : 241-248.

Enders, R. K. 1935. Mammalian life histories from Barro Colorado Island, Panama. *Bull. Mus. Comp. Zool.* **78**: 385-502.

Forel, A. 1928. The social world of the ants. American edit. 1929. New York.

Froggatt, W. W. 1905. White ants (Termitidae). *Dept. Agric. New South Wales, Misc. Publ.* **874**. 47 pp.

Fuller, C. 1915. Observations on some South African termites. *Ann. Natal Mus.* **3**: 329-504.

Grassé, P. P. 1937. Recherches sur la Systématique et la Biologie des Termites de l'Afrique occidentale française. *Ann. Soc. Ent. France* **106**: 1-100.

Hare, L. 1934. Caste determination and differentiation with special reference to the genus *Reticulitermes* (Isoptera). *Journ. Morphol.* **56**(2) : 267-293.

1937. Termite Phylogeny as Evidenced by Soldier Mandible Development. *Ann. Ent. Soc. Amer.* **37**(3) : 459-486.

Harvey, P. A. 1934. Life history of *Kalotermes minor*. Chap. 20, part II, in Kofoid, C. A. Termites and termite control. 2d. edit.

Hegh, E. 1922. Les Termites. Bruxelles.

Hill, G. F. 1915. Northern Territory Termitidae. Part i. *Proc. Linn. Soc. New South Wales* **40**: 83-113.

1921. The white ant pest in northern Australia. *Commonwealth Inst. Sci. Ind. Bull.* **21**: 1-26.

1922. On some Australian termites of the genera *Drepanotermes, Hamitermes* and *Leucotermes. Bull. Ent. Res., London* **12**(4) : 363-400.

1922a. Descriptions and biology of some north Australian termites. *Proc. Linn. Soc. New South Wales* **47**(2) : 142-160.

1925. Notes on *Mastotermes darwiniensis* Froggatt (Isoptera). *Proc. Roy. Soc. Victoria* **37**: 119-124.

1935. Australian *Hamitermes* (Isoptera), with descriptions of new species and hitherto undescribed castes. *Commonwealth Council for Sci. Ind. Res. Pamphlet No.* **52**. *Syst. Ent. Contrib.* **1**: pp. 13-31.

Hingston, R. W. G. 1932. A naturalist in the Guiana forest. New York. xiii + 384 pp.

Holdaway, F. G. 1933. The composition of different regions of mounds of *Eutermes exitiosus* Hill. *Jour. Council Sci. Ind. Res.* **6**: 160-165.

1935. The termite population of a mound colony of *Eutermes exitiosus* Hill. *Jour. Council Sci. Ind. Res.* **8**: 42-46.

Holmes, S. J. 1911. The evolution of animal intelligence. New York.

Holmgren, N. 1906. Studien über südamerikanische Termiten. *Zool. Jahrb. Abt. Syst.* **23**: 521-676.

Imms, A. D. 1919. On the structure and biology of *Archotermopsis. Philos. Trans. Roy. Soc. London.* **209**(Ser. B) : 75-180.

1931. Social behaviour in insects. New York.

Kemner, N. A. 1929. Aus der Biologie der Termiten Javas. *Xe Congrès international de Zool., Budapest.* Deuxième partie, pp. 1097-1117.

Kinsey, A. C. 1936. The origin of higher categories in *Cynips. Indiana Univ. Publ. Sci. Ser.* **4**: 1-334.

Kofoid, C. A. *et al.* 1934. Termites and termite control. Univ. Calif. Press. 2d edit.

Light, S. F. 1934. Habitat and habit types of termites and their economic significance. Chap. 12 in Kofoid, C. A. Termites and termite control. 2d edit.

1934a. The desert damp-wood termite, *Paraneotermes simplicicornis*. Chap. 25 in Kofoid, C. A. Termites and termite control. 2d edit.

1937. Contributions to the Biology and Taxonomy of *Kalotermes* (*Paraneotermes*) *simplicicornis* Banks (Isoptera). *Univ. Calif. Publ. Entomology* **6**(16) : 423-464.

Maeterlinck, M. 1926. La Vie des Termites. Paris.

1928. The Life of the White Ant. New York.

Mjöberg, E. 1920. Isoptera. Results of Dr. E. Mjöbergs Swedish scientific expeditions to Australia 1910-1913. *Arkiv. f. Zool.* **12**(15) : 1-128.

Needham, Joseph. 1930. The biochemical aspect of the recapitulation theory. *Biol. Rev.* **5** : 142-158.

Oshima, M. 1919. Formosan termites and methods of preventing their damage. *Phil. Jour. Sci.* **15** : 319-383.

Reichensperger, A. 1923. Die Bauten des *Apicotermes occultus Silv.* *Bull. Soc. Nat. Luxembourg n.s.* **17** : 52-59.

Saville-Kent, W. 1897. The naturalist in Australia. London.

1897a. Remarkable termite mounds in Australia. *Nature* **57** : 81-82.

Silvestri, F. 1903. Contribuzione alla conoscenza dei Termitidi e Termitofili dell' America meridionale. *Redia* **1** : 1-234.

Sjöstedt, Y. 1907. Wissenschaftliche Ergebnisse der Schwedischen zoologischen Expedition nach dem Kilimandjaro, dem Meru und den umgebenden Massaisteppen, Deutsch-Ostafrika, 1905-1906. Termitidae. 37 pp.

1913. Über Termiten aus dem inneren Kongo, Rhodesia und Deutsch-Ostafrika. *Rev. Zool. Afric.* **2** : 354-391.

1923. Über das unterirdische Nest einer bisher unbekannten Termite aus Kongo. *Arkiv f. Zool.* **15** (No. 20) : 1-8.

1926. Revision der Termiten Afrikas 3. Monographie. *Kungl. Svenska Vet.-akad. Handl.* (Tredje Ser.) **3**(1) : 1-419.

Snyder, T. E. 1920. A revision of the Nearctic termites. Part 2. Biology. *U. S. Nat. Mus. Bull.* **108** : 87-211.

Snyder, T. E. and J. Zetek. 1934. The termite fauna of the Canal Zone, Panama, and its economic significance. Chap. 30 in Kofoid, C. A. Termites and termite control. 2d edit.

Spencer, H. 1893. A rejoiner to Professor Weismann. *Contemporary Rev.* **64** : 893.

Weismann, A. 1893. The all-sufficiency of natural selection. *Contemporary Rev.* **64** : 309, 596.

Wheeler, W. M. 1907. The fungus-growing ants of North America. *Bull. Amer. Mus. Nat. Hist.* **23** : 669-807.

1936. Ecological relations of Ponerine and other ants to termites. *Proc. Amer. Acad. Arts and Sci.* **71**(3) : 159-243.

Whiting, P. W. 1932. Reproductive reactions of sex mosaics of a parasitic wasp, *Habrobracon juglandis. Journ. Comp. Psych.* **14** : 345-363.

Williams, O. L. 1934. Some factors limiting the distribution of termites. Chap. 4 in Kofoid, C. A. Termites and termite control. 2d edit.

Wright, S. 1929. Fisher's theory of dominance. *Amer. Nat.* **63** : 274-279.

1932. The rôles of mutation, inbreeding, crossbreeding and selection in evolution. *Proc. 6th Intern. Congr. Genetics* **1** : 356-366.

1934. Physiological and evolutionary theories of dominance. *Amer. Nat.* **68** : 24-53.

1935. A mutation of the Guinea Pig, tending to restore the pentadactyl foot when heterozygous, producing a monstrosity when homozygous. *Genetics* **20** : 84-107.

ADDENDUM BY A. E. EMERSON

Nomenclatural Changes

Today the family Kalotermitidae is more restricted to a group of genera that include *Kalotermes flavicollis* (Fabricius); *Incisitermes minor* (Hagen), which includes *Kalotermes minor* Hagen in its synonymy; and *Paraneotermes simplicicornis* (Banks), which includes *Kalotermes (Paraneotermes) simplicicornis* Banks, in its synonymy.

In 1938 the Kalotermitidae included primitive Hodotermitidae, recognized in 1975 as the subfamily Termitopsinae, which now includes *Archotermopsis* and *Zootermopsis,* and the subfamily Stolotermitinae, which includes the single genus *Stolotermes.*

All the references to the Hodotermitidae in 1938 are now classified under the subfamily Hodotermitinae, which includes *Hodotermes mossambicus* Hagen with *Hodotermes (H.) mossambicus transvaalensis* Fuller in its synonymy, and *Anacanthotermes ochraceus* (Burmeister), which includes *Hodotermes (Anacanthotermes) ochraceus* (Burmeister) in its synonymy.

The families Mastotermitidae, Rhinotermitidae, and Termitidae as used in 1938 are still valid in 1975 with some modifications not pertinent to the 1938 discussion. In 1975, it is thought that the primitive Mastotermitidae and Kalotermitidae may have a common ancestor that branched from the more primitive unknown isopteran stem and that the primitive Hodotermitidae represented by the living *Archotermopsis* probably arose from a more primitive isopteran stem that also gave rise to the Rhinotermitidae, and thence independently to the Serritermitidae and Termitidae. This phylogeny of families is paralleled by the evolution of nesting behavior, with some specialized behavior appearing in otherwise primitive termites.

Some of the generic and specific names of the Termitidae mentioned in 1938 are invalid in 1975. *Syntermes snyderi* Emerson is a synonym of *Syntermes spinosus* (Latreille). Both *Nasutitermes pilifrons* (Holmgren) and *N. acajutlae* (Holmgren) are synonyms of *Nasutitermes nigriceps* (Haldeman). The genera and subgenera *Anoplotermes (A.)* and *Anoplotermes (Speculitermes)* in South America in the 1938 article are both unnamed new genera. *Amitermes medius* Banks is a synonym of *Amitermes foreli* Wasmann. The subgenus *Bellicositermes* is a synonym of *Macrotermes,* and the species *natalensis* referred to in the 1938 article is probably *Macrotermes bellicosus* (Smeathman). *Cubitermes loubetsiensis* Sjostedt is a synonym of *Cubitermes finitimus* Schmitz.

Changes in the Biological Basis of Termite Ethology

The following references provide a better understanding of the genetics of behavior, homology, and analogy of behavior and, in particular, of the regressive evolution of vestigial behavior than that discussed in 1938.

Bouillon, A. 1964. Structure et accroissement des nids d'*Apicotermes* Holmgren (Isoptera, Termitidae). Pp. 295–326 in Bouillon, A. (ed.), *Études sur les termites africaines*. Un Colloque international Univ. Lovanium, Léopoldville, 11–16 mai, 1964, sous les auspices de l'UNESCO. Éd. de l'Université, Léopoldville. 419 pp.

Desneux, J. 1948. Les Nidifications souterraines des *Apicotermes*. Termites de l'Afrique tropicale. *Rev. Zool. Bot. Afr.*, 41(1):1–54, 22 pls.

Desneux, J. 1952. Les Constructions hypogées des *Apicotermes*. Termites de l'Afrique tropicale. Étude descriptive et essai de phylogénie. *Ann. Mus. Roy. Congo Belge*, Sér. 8, *Sci. Zool.*, 17:1–98, 50 pls.

Desneux, J. 1956. Le Nid d'*Apicotermes rimulifex* Emerson. *Rev. Zool. Bot. Afr.*, 53:91–97.

Emerson, A. E. 1953. The African genus *Apicotermes* (Isoptera: Termitinae). *Ann. Mus. Roy. Congo Belge*, Sér. 8, *Sci. Zool.*, 17:99–121.

Emerson, A. E. 1956a. A new species of *Apicotermes* from Katanga. *Rev. Zool. Bot. Afr.*, 53:98–101.

Emerson, A. E. 1956b. Ethospecies, ethotypes, taxonomy, and evolution of *Apicotermes* and *Allognathotermes* (Isoptera: Termitidae). *Am. Mus. Novitates 1771*, pp. 1–31.

Emerson, A. E. 1958. The evolution of behavior among social insects. Chap. 15 in Anne Roe and G. G. Simpson, *Behavior and Evolution*, pp. 311–335. Yale University Press, New Haven, Conn.

Emerson, A. E. 1961. Vestigal characters of termites and process of regressive evolution. *Evolution*, 15:115–131.

Grassé, P.-P., and C. Noirot. 1948. Sur le nid et la biologie du *Sphaerotermes sphaerothorax* (Sjostedt). *Ann. Sci. Nat. Zool.*, Ser. 11, 10:149–166.

Grassé, P.-P., and C. Noirot. 1954. *Apicotermes arquieri* (Isoptere): ses constructions, sa biologie. Considérations générales sur la sous-famille des Apicotermitinae nov. *Ann. Sci. Nat. Zool.*, Sér. 11, 16:345–388.

Schmidt, R. 1955. The evolution of nest-building behavior in *Apicotermes* (Isoptera). *Evolution*, 9(2):157–181.

5

Reprinted from *Am. Zool.*, **4**(2), 175–190 (1964)

THE EVOLUTION OF NESTS AND NEST-BUILDING IN BIRDS

Nicholas E. Collias

Department of Zoology, University of California at Los Angeles

External construction is a phenomenon found throughout the animal kingdom from ameba to man. Such things as the nests of birds, the webs of spiders, the cases of caddisfly larvae, and the nests of termites and bees, are often mentioned as examples of highly specialized instinctive behavior. Their study is an old subject, but one that needs to be reviewed in a critical and comprehensive way in relation to modern biology. These structures play a central role in the lives of the animals concerned, and help bring to a focus their habitat requirements.

The object of the present survey is to attempt to formulate some general problems, and to stimulate renewed interest and continuing analysis in a promising field of scientific endeavor that has been somewhat neglected in recent years.

My own interest in the evolution of birds' nests began years ago when I took a graduate course on the principles of speciation given by Dr. Alfred E. Emerson at the University of Chicago. Emerson is well known for studies of the evolution of termites and termite nests (1938). He has also called attention to the promise of the nests of birds for investigating the evolution of behavior (Emerson, 1943).

A nest may be defined as an external construction that aids the survival and development of the eggs and young. The following table gives a summary of the main types of birds' nests. Good photographs illustrating them may be found in Stresemann's treatise on ornithology (1927-1934) and in many other more recent books about birds.

A CLASSIFICATION OF NEST-BUILDING IN BIRDS

I. Incubation by environment.
    Eggs generally buried in the ground. Megapodes.

II. Incubation by parents. Most birds.
    1. Dig or use nest cavity
    2. Nest is not enclosed in a cavity
       a. Open nests
       b. Roofed nests
       c. Compound nests

III. Eggs placed in nests of other birds. Brood parasites.

At the outset, in a discussion of nest evolution it is necessary to keep in mind certain principles in order to maintain perspective and orientation.

Competition between species has often resulted in great differences in the habitats and nest sites occupied by related species. In turn, differences in the nature of the substrate for the nest has imposed special engineering requirements on nests with regard to materials, form, structure, and placement. Building of a nest requires a considerable expenditure of energy, and it is common for many birds to make 1000 or more trips to gather and bring all the necessary materials. Natural selection may therefore be expected to generally favor anything that tends to economize on effort, so long as undue sacrifice of any crucial advantage of the species is avoided.

Nests are so closely related to habitat and habits in any given species that there has been a tremendous amount of recurrent, convergent, and parallel evolution of different nest types in birds, making it difficult to delineate particular phylogenies. In any particular line, evolution may lead to either an increasing complexity or, conversely, to an increasing simplification of nests, depending on conditions. The emphasis and object of this report is not to attempt to develop special family geneologies, but rather to attempt to analyze the ecological nature of the selection pressures that have led to the evolution of the main types of nests.

The primary, basic, and most general functions of a nest are to help insure

---

Preparation of this review was aided by Grant 22236 from the National Science Foundation.

warmth and safety to the developing eggs and young. But growth and survival of the young depend on the total biology of the species, and to fully understand the forces in evolution of the nest of any species one must often be familiar with other aspects of its life history as well. Conversely, selection of a given type of nest-site influences other aspects of species behavior, as Cullen (1957) has shown for the Kittiwake Gull, and von Haartmann (1957) for hole-nesting birds in general.

The problems of securing warmth and safety are most acute for small birds and their young, explaining why birds of small body size generally build nests that are as a rule more elaborate and better concealed than are those of larger birds. Reduction in size and energy content of eggs may have helped make possible evolution of small birds (Witschi, 1956; Dawson and Evans, 1960), but have also required increased parental care of the young which hatch in a helpless well-nigh embryonic state. There are all degrees between precocial and altricial young and a parallel development of parental care (Nice, 1962), related in a general way to nesting habits.

ORIGIN OF NEST-BUILDING IN BIRDS

Ability of birds to maintain a high constant body temperature no doubt developed gradually, probably coincident with evolution of the ability to fly. Even today there are a few birds that are known to pass into a state of torpidity under certain conditions, and their body temperature falls drastically (Bartholomew, Howell, and Cade, 1957). If birds passed into a torpid state during cool nights of the breeding season, they could scarcely incubate their eggs effectively. During the transitional period of early avian evolution, when temperature regulation mechanisms were being perfected, some birds probably continued to bury their eggs, leaving incubation to the sun or to decaying vegetation after the reptilian fashion, and as megapodes do among living birds today. The probably low nocturnal body temperature of ancestral birds is an argument in favor of the idea that

in an early stage of avian evolution, incubation of the eggs may have depended in part or wholly upon some source of heat other than that furnished by the parental body.

NESTS OF MEGAPODES (MEGAPODIIDAE)

Nests of living megapodes, recently summarized by Frith (1962), have a tremendous range of variation. Within the confines of one genus, *Megapodius* (perhaps in need of revision), the nest may vary from a simple, small pit dug in the sand, large enough for just one egg, to a gigantic mound of soil and decaying vegetation as much as 35 feet in diameter and 15 feet high, and probably the largest structure made by any one bird. Frith points out a striking parallel between bird and reptile on certain sunlit coral beaches where the turtles heave up out of the sea to dig holes on the beach to lay their eggs, and the megapodes walk out of the bush to dig pits in the sand for their eggs. Similarly, in places where dark tropical forests fringe the rivers, female crocodiles build mounds of leaves in which their eggs are laid, in close proximity to the leafy mounds of megapodes.

Change in climate, especially after the close of the Mesozoic era, from humid tropical or subtropical to drier conditions with greater extremes of temperature, was probably met in early avian evolution by two different solutions with respect to the problem of incubation of the eggs. Some birds developed the method of incubation by direct application of heat to the eggs from the body of the parent. Other birds developed into mound builders, burying the eggs deeply in the ground, safeguarding them from the harsh conditions, and at the same time evolving considerable efficiency in regulating the temperature of the mound about the eggs.

It is possible that modern megapodes once had ancestors that sat on their eggs as other birds do, and evolved their now seemingly peculiar mode of incubation by a sort of regressive evolution. But such is the advantage of direct parental incubation as in most modern birds, that the opposite theory of mound-building in megapodes as a primitive retention of a reptilian habit might well be true. Frith suggests that one thing is certain. Some megapodes have very greatly improved in the ability to regulate the temperature of the mound compared with any present day reptiles. Thus, the Mallee Fowl (*Leipoa ocellata*) lives in arid regions of Australia where the temperatures may range from below freezing to above 38°C, and even in midsummer the night temperature may be 17° lower than the day temperature. Nevertheless, this bird manages to maintain the incubation temperature about the eggs buried in its mound relatively constant between 32° and 35°C.

ORIGIN OF DIRECT PARENTAL INCUBATION OF EGGS

The open nest may well have evolved in ancient birds from a type like the simplest nest built by modern megapodes. In some parts of Australia, *Megapodius* merely lays a single egg in a small pit in the sand or in a crevice in a rock, covers the egg with leaves, and departs, leaving the task of incubation to the sun (Frith, 1962). Many precocial birds having open nests on the ground cover their eggs with either plant materials, downy feathers, or earth, when away from the nest.

Once early birds had evolved the ability to maintain a high temperature through the night independently of environmental temperature, there would be establishd coincidentally a tremendous selection pressure favoring direct parental incubation of the eggs. At the same time the spread of a species into new cooler habitats relatively free from reptilian predators would be facilitated. The danger of predation on the eggs from various nocturnal enemies, especially from contemporary mammals which were small and probably nocturnal, provided considerable value to the habit of staying with the eggs and defending them if necessary during the night. Total predation on the eggs would diminish with shortening of their developmental period.

Herrick (1911) suggested that the origin of incubation by sitting on the eggs probably arose from the tendency of birds to conceal their eggs with the body as a protection from potential predators.

CAVITY-NESTING BIRDS

Cavity nesting has evolved repeatedly in birds at virtually every stage of evolution. Initial use of natural cavities may be followed by modification of the cavity, and ultimately may lead to the evolution of special excavated nesting cavities in the ground, in banks, or in trees. About half the orders of birds recognized by Mayr and Amadon (1951) contain some species that nest in cavities. Whole orders of cavity nesters are represented by the kiwis, parrots, trogans, coraciiform, and piciform birds.

Cavity nesting provides evident shelter from the elements and conserves energy (Kendeigh, 1961). In addition, there is much direct statistical evidence from various studies (see Nice, 1957) that it is safer for altricial birds of the North Temperate Zone to nest in holes than in open nests. Nice's summary showed that only about ½ of some 22,000 eggs of various species with open nests resulted in fledglings, whereas ⅔ of 94,400 eggs were successful in hole-nesting birds. Populations of small birds like the House Wren (*Troglodytes aëdon*) (Kendeigh, 1941), have often been greatly increased by putting out a good supply of nest boxes.

The shelter and safety furnished by cavities has resulted in intense competition for these cavities. Conversely, birds building an open nest, involving as it does a greater chance of nest failure, are subject to relatively strong selection pressure to build an adequate nest or to develop markedly efficient concealing coloration.

The Prothonotary Warbler (*Protonotaria citrea*) is one of the few American warblers that build their nests within natural cavities or birdhouses. A fledgling rate for this species of only 26% (of 413 eggs) was found by Walkinshaw (1941) in Michigan. In contrast, in Tennessee the fledgling rate was 61% (of 163 eggs). This difference in success of fledgling was largely attributed to a high population of House Wrens competing for nesting places with the warblers in Michigan, whereas House Wrens were absent from the study locality in Tennessee. The wren destroys the warbler's eggs by puncturing them with its bill in the absence of the warbler.

Intense aggressive competition for tree holes has been a profound force in evolution of different size classes among such birds as woodpeckers, corresponding to different sized entrance holes typical of each species. Along with each size class of woodpecker goes a host of other species of birds that compete with the woodpecker for the corresponding size of nest cavity. The European Starling is notorious in this regard, and I have seen a starling in Ohio seize a flicker by the tail and cast it out of the flicker's freshly dug tree cavity in which a pair of starlings subsequently nested and reared a brood. Sielmann (1959) gives a number of graphic instances of direct observation in Germany of birds competing aggressively for tree holes. One of his most interesting examples shows how a small bird can compete successfully with a larger one for a nest cavity in a tree. When the European Nuthatch (*Sitta europaea*) takes over a tree cavity, it forestalls its chief rivals, the starlings, by collecting mud from nearby puddles and plastering the mud around the entrance to the tree hole, making the entrance so small and narrow that while the nuthatch can slip through, the larger starling cannot.

In the classic case of the hornbills, the male is often said to provide for the safety of his mate and young ones by imprisoning the female in her nest cavity in a tree, walling up the entrance with mud, leaving just enough room for her to put her beak out so he can feed her during the prolonged period of incubation and care of the young, releasing her at the end of that time. But, in the few species studied in detail, Moreau (1937) points out that it is the female who plasters herself in, using such materials as mud brought by the male and mixed with saliva as an adhesive, and whenever she is ready to leave it is the female who pecks her way out. The Ground Hornbill (*Bucorvus*) generally nests in depressions in trees such as those formed at the tip of a big broken branch, but it does not wall them up with mud, illustrating one possible step in the evolution of the very specialized nesting habits of other modern hornbills.

Different stages in evolution of the ability to excavate nest cavities in trees by birds probably were: (1) use of natural cavities; (2) modification of these cavities in various ways by the bird; (3) excavation in decaying or very soft wood as in some titmice (Hinde, 1952) and the tiny Olivaceous Piculet (*Picumnus olivaceus*) of South America does (Skutch, 1948); and finally, (4) chiseling of nest cavities in hard, living trees, as the large Black Woodpecker (*Dryocopus martius*) of Europe does (Sielmann, 1959).

Many birds nest in holes in the ground, and several steps in the evolution of excavated burrows in the ground may be suggested: (1) a shallow scrape such as is made by many ground nesting birds; (2) a relatively short burrow like that of the Rough-winged Swallow (*Stelgidopteryx ruficollis*), a bird which often merely nests in crevices; and (3) the eggs laid at the end of a burrow which may reach a length of over six feet in a bank, as in the Bank Swallow (*Riparia riparia*) (Bent, 1942). Furthermore, this swallow forms its tunnel in something of an upward course, insuring protection from driving rain.

The evolutionary climax of excavated nests is the construction of nesting cavities by certain birds inside the nests of social insects. The distribution of the Orange-fronted Parakeet (*Aratinga canicularis*) in Mexico and Central America closely approximates that of the colonial termite, *Eutermis (Nasutitermes) nigriceps,* in the nests of which the parakeet breeds, apparently using only nests still occupied by the termites (Hardy, 1963). Hindwood (1959) points out that 49 species of birds, including kingfishers, parrots, trogons, puff-birds, jacamars, and a cotinga, are known to breed in either terrestrial or arboreal nests of termites. In fact, some 25% of the species of kingfishers of the world nest in termites' nests. As the excavation by a bird progresses, the termites seal the exposed portions of their nests so that there is no actual contact between the birds and the insects.

Many birds that breed in termites' nests no not normally eat the insects in the colonies in which they nest (ibid.). The Rufous Woodpecker (*Micropternus brachyurus*) of the Orient seems to nest almost exclusively in occupied carton-like ant nests, of the genus *Crematogaster,* which have a painful bite. The woodpecker feeds on the ants of the colony in which it nests (Smythies, 1953). The Gartered Trogon (*Chrystrogon caligatus*) of Central America has been recorded breeding in wasps' nests, eating out the adult wasps before doing so (cf Hindwood, 1959, ibid.).

Birds that breed in nests of social insects are all from taxonomic groups characterized by nesting in cavities, and cavities in old and deserted termite nests are at times used by birds that normally breed in earth-banks or tree-holes. These facts show how the habit of consistently breeding in nests of social insects might have evolved.

Concerning all birds that nest either in natural or modified cavities or in cavities excavated by themselves or by other birds, we can make two concluding statements regarding the general significance of this habit. First, various specialized types of excavation have evolved, and secondly, nesting in a cavity goes a long way toward meeting the essential functions of a nest for warmth and safety, and thereby tends to block further evolution of truly elaborate increment nests of the type that are built up from specific materials. In fact, such nests as are placed within cavities may undergo a regressive evolution. All degrees of increasing simplification and reduction to a mere pad are seen in the case of Old World sparrows (Passerinae) that nest in tree holes (Collias and Collias, 1964b).

EVOLUTION OF OPEN NESTS ON THE GROUND

When direct parental incubation began, it was no longer necessary to dig a pit for the eggs. However, most birds that nest on the surface of the ground today, still begin their nest by making a circular scrape with the feet while crouching low and rotating the body (Dixon, 1902). This hollow may then be more or less lined with various materials protecting the eggs from the cold, damp ground.

A rim of materials around the body of the incubating parent serves to provide insulation for the eggs in many ground-nesting birds with simple nests, the materials being pushed to the periphery and built up into a circular form by much the same type of movements of the feet and body as are involved in making the initial scrape in the ground. An added feature, preventing the flattening down of the nest rim, seen in many ground-nesting birds, is the act of repeatedly reaching out with the bill, drawing in materials to the breast or passing them back along one side of the body before dropping. All of these patterns of making a ground nest are clearly seen, for example, in the Canada Goose (Collias and Jahn, 1959). But the Canada Goose will not walk or fly to the nest with materials in its bill, as cormorants or gulls and many other birds will do.

Open nests on the surface of the ground are more exposed to the elements and to predations than are enclosed nests. For example, eggs of the Horned Lark (*Otocoris alpestris*) laid early in the season in the northern United States may sometimes become frozen. Horned Larks place the nest in open country in a shallow scrape protected from the chill of prevailing winds by a clod or tussock of grass (cf Bent, 1942).

Parental behavior may supplement or substitute for a nest under particularly severe conditions of exposure. In the Arctic, persistent close incubation by the parent bird seems to be a major adaptation and occurs regardless of whether or not the nest is well insulated (Irving and Krog, 1956).

For example, it has been found that the Semipalmated Sandpiper (*Ereunetes pusillus*) with no nest keeps its eggs as warm as do various other birds nesting in the Arctic and having a substantial nest. The Emperor Penguin (*Aptenodytes forsteri*), which breeds in the Antarctic winter, rests its single egg on the feet, covers it with a fold of belly skin and incubates it against the body. Perhaps no other animal breeds under such trying conditions. The fury of recurrent severe storms is met by the birds huddling together in a close mass (Rivolier, 1956).

Eggs or nestlings exposed to strong tropical or subtropical sun in open situations are customarily shaded by the body and wings of the parent bird, as in the case of the Sooty Tern (*Sterna fuscata*) of Midway Island, in which the nest is a mere scrape in the coral sand (Howell and Bartholomew, 1962).

Nests on the surface of the ground are especially liable to be flooded, and ground-nesting birds often build their nests on any slight elevation and may build their nests up during a flood. For example, in the Adelie Penguin (*Pygoscelis adeliae*) of the Antarctic, the nest which is made of small stones functions chiefly to raise the eggs and incubating bird above ground level, thus lessening the dangers of flooding during thaws or of being buried by snow during blizzards (Sladen, 1958). During a thaw, Sladen noticed one nest had a stream of ice cold water running through it. The incubating male, his eggs half submerged, kept reaching forward, collecting and arranging stones around him. Next day the eggs were above water though the stream passed on either side of the nest. Eventually these eggs hatched. The Painted Snipe (*Rostratula benghalensis*) in Australia may lay its eggs on the bare ground when the ground is dry, but if water is lying on the ground, a solid nest of rushes and herbage is made. (Serventy and Whittell, 1962).

There is evidence that it is safer for a bird to nest on a platform over water in a marsh than to nest on the surface of dry land. Kiel (1955) found that 50% of 149 nests of dabbling ducks, which nest on the uplands, were successful. In contrast, during the same 4-year period of study in Manitoba, he found that 73% of 227 nests of diving ducks, which nest over water, were successful. Kiel attributed the greater nesting success of the diving ducks to greater protection from terrestrial mammalian predators and to greater safety from waves and floods by virtue of their nest sites.

The high degree of exposure to predation to which birds nesting on the surface of the ground are often subject is no doubt the reason why such species of birds provide some of the classical examples of concealing coloration, such as the incubating ptarmigan or the coloration of eggs in many shore birds. In these instances concealment by coloration and behavior acts as a substitute for concealment by a nest and a safe nesting site. In fact, in certain cases where the color pattern of the eggs and young and parents closely matches the surroundings, as in the European Stone Curlew (*Burhinus oedicnemus*) (Welty, 1962, photo, p. 264) and the Whip-poor-will (*Antrostomus vociferus*; Bent, 1940, Plate 23), the nest may virtually disappear, presumably because a nest itself would be too conspicuous on the surface of the ground.

EVOLUTION OF OPEN NESTS ON TREES AND CLIFFS

The dangers of ground nesting and the intense competition for tree holes have apparently provided a strong selection pressure leading to the evolution of increment nests placed in trees. The Tooth-billed Pigeon (*Didunculinae*) of Samoa is said to have abandoned its ground-nesting habits and taken to nesting in trees (Austin, 1961) after cats had been introduced by whaling ships. There is some direct evidence that nesting in trees is generally safer than nesting on the ground. In the prairie country of northwestern Oklahoma where there are few trees, Downing (1959) found in a

Mourning Dove (*Zenaidura macroura*) population 49% success of 167 tree nests but only 29% success of 130 ground nests.

Conversely, under nesting conditions safe from predators, certain species of birds that normally nest in trees elsewhere may return to ground nesting (Preston and Norris, 1947), thereby conserving the energy that would be required to constantly fly up into a tree carrying nest materials or food for nestlings. On Gardiner's Island in New York, in the absence of mammalian predators, such birds as the Osprey and the Robin have often nested on the ground, although normally nesting in trees elsewhere in the United States.

Species of birds with precocial young are generally ground nesters; species with altricial and therefore relatively helpless young generally nest either on the ground or in trees and bushes. It is probable that passerine birds, with their perching foot structure and altricial young, evolved first in relation to arboreal life in trees and bushes, and secondarily reinvaded ground habitats where they continue to construct well-rounded bowl-shaped or cup-like nests.

The nature of the materials used to help solve the problem of securely placing and attaching a nest in a tree vary with the body size of the bird and its lifting power. Large birds use twigs and branches that are not readily blown out of the tree by ordinary winds. Medium-sized birds use small twigs or grasses or both, sometimes adding mud to help attach and bind the nest materials. A great many small birds are known to use spider silk or insect silk as a binding material both for the attachment of the nest and for fastening together various other materials. Some birds of a given species may use herbaceous plant material when they nest on the ground, and twigs when they place their nest in a tree e.g. certain herons (cf Palmer, 1962).

Other problems of placing nests in trees are related to specific sites in the tree. Among the crudest twig nests are those of doves. In Oklahoma, Nice (1922) found that 39 Mourning Dove nests in crotches were almost twice as successful in producing young large enough to fly as were 59 nests on branches.

Practically all birds line their nests with finer and softer materials than are used for the foundation and outer shell of the nest.

The platform nests of some large birds such as the American Bald Eagle (*Haliaeetus leucocephalus*) and the European White Stork (*Ciconia ciconia*) are largely constructed of twigs and branches, and these nests are added to year after year, thus providing some economy of effort. Such nests may become very large and very old. Herrick (1932) gives an age of 36 years for a Bald Eagle nest, and Haverschmidt (1949) dates back to 1549 one White Stork nest which was still in use in 1930. The limiting factor to continued nest growth is often the weakening and increased liability to windfall of the nest-tree. The particular eagle nest described by Herrick, perhaps the largest eagle nest on record, was 12 feet tall, 8½ feet across the top and its weight was estimated at over 2 tons. It was situated in a tall tree and in its 36th year this nest fell with the tree in a storm.

The altricial nestlings of small birds, such as passerines (which comprise more than half of the living species of birds) hatch in a blind, relatively naked, and almost embryonic state. Compared with platform nests, the cup nest built by small birds provides more adequate protection to the young, being so constructed as to resist stresses that might cause the nest to collapse inward or outward. According to Nickell (1958), rootlets used as nest linings are the most constant feature of the nests of Catbird (*Dumatella carolinensis*) and Brown Thrasher (*Toxostoma rufum*). These rootlets are moist and flexible when placed in the nest and become like small wire springs when dry, serving as an inner bracework which preserves the shape of the nest basket.

It is probable that every type of material characteristic of the nest of a given species of bird has a definite function according to the physical properties of the material rather than its taxonomy, and that the proportions of materials of different types that are used vary not only with availability but with the requirements of particular substrate and habitat situations (Horváth, 1963). Horváth finds that robin nests contain more mud when the birds have to use short materials, more tough flexible rootlets when the nest is in an especially windy spot, and more moss when in a relatively cold microclimate.

Cup nests of the smallest species of birds, particularly when in cool or exposed environments, are likely to be heavily insulated, e.g., nests of most species of hummingbirds have a thick lining of downy material (Ruschi, 1949). The tightly constructed cup provides effective insulation of the ventral surface of the incubating female, i.e., from that part of her body from which the greatest heat loss probably occurs, as Howell and Dawson (1954) have demonstrated for the Anna Hummingbird (*Calypte anna*). There is some evidence that compared with lowland species, the species of hummingbirds that nest in high mountains may build nests having relatively thick walls (Wagner, 1955), or seek the protection of caves (Pearson, 1953).

The building movements of birds that make open nests in trees are similar in certain basic ways to those used by birds that make simple ground nests, particularly with regard to methods of shaping the nest concavity by means of scraping movements of the feet combined with rotation of the body and pushing movements of the breast. Additional movements more characteristic of tree nesters than of ground nesters are the thrusting of twigs or grasses into the nest mass with trembling movements of the bill, and in the case of many small birds the habit of wiping cobwebs from the bill in fastening materials on to the growing rim of the nest, as Van Dobben (1949 has described for the Icterine Warbler (*Hippolaius icterina*) in the Netherlands. The sticky threads of cobweb are then stretched out and passed back and forth across the rim of the nest.

Elevated nests attached to vertical faces of cliffs, buildings, or caves furnish protection against non-avian predators, but pose special problems of nest attachment. For this purpose the swifts have generally specialized on adhesive saliva (Lack, 1956; Medway, 1960), while the swallows have evolved toward a more general use of mud probably with some admixture of saliva (cf Bent, 1942). Different species of swifts or swallows, respectively, can be arranged in graded series from those swifts like *Collocalia francica* making nests of pure saliva (source of the ideal "birds' nest soup" of the Chinese), or from those swallows making nests of nearly pure mud, like some Cliff Swallows, through various admixtures with plant and other materials to more conventional types of bird nests (ibid). The nest cement secreted by the cave swiftlet, *Collocalia fuciphaga*, is sparse and soft, and the nest, which is built up from moss and other plant materials, can only be placed on an irregularity in the cave wall which will take all or a good part of the weight of the nest, unlike the nest of other cave swiftlets which can be glued to vertical walls in a cave (Medway, 1960).

EVOLUTION OF ROOFED NESTS

Small birds in particular require the protection from cold, rain and predation furnished by enclosed nests. Building of a roofed increment nest is very rare among non-passerine birds, whereas half of some 82 families and distinctive subfamilies of passerine birds recognized by Mayr and Amadon (1951), in their classification of birds of the world, build roofed nests or contain representatives that do so. At the same time roofed nests, aside from use of natural cavities, are unusual among passerine birds of the north temperate zone, but roofed nests are very common among small tropical birds with altricial young, being typical of many tropical families and genera (Collias and Collias, 1959, 1964b).

The roofed nest very probably evolved from a nest type that was open above. This conclusion is suggested by the fact that in most instances birds having non-pensile roofed nests start with the basal platform and then build up the sides and roofs, as for example, in the Wattled Starling (Liversidge, 1961), Australian Grassfinches (Immelmann, 1962), European Wren (Armstrong, 1955), Cliff Swallow (Emlen, 1954), and the tropical tanager, *Chlorophonia occipitalis* (Skutch, 1954).

Among predators of nestling birds, snakes are more numerous and varied in the tropics than in colder regions, and roofed nests perhaps help deter snakes as well as other enemies. Pitman (1958) has recently summarized many instances of snake and lizard predation on birds. However, the exact techniques used by snakes or any other predators of nesting birds do not seem to have been investigated. Probably knowledge of such techniques would help explain certain nest specializations. The true weaverbirds (Ploceinae) belong to an old world family in which all of the species build a roofed nest. In many of the species that place their nests in herbaceous vegetation near the ground, the nest has a side entrance, whereas in species that place their nests in trees the nest generally has a bottom entrance. Domed nests of weaverbirds placed in trees tend to evolve a firm pensile or pendulous attachment and long entrance tube about the ventral entrance, presumably as a protection against snakes and other non-avian predators that would have to approach the nest from above (Collias and Collias, 1959, 1963, 1964b; Crook, 1963). Van Someren (1956) made an interesting observation in East Africa: "I once watched a green tree-snake trying to get at the young in a spectacled weaver's nest. The brute negotiated the slender, pendant branch and reached the nest but could not manage the 12-inch tubular entrance and fell into the pond below the nest." The nest of the New World Flycatcher, *Tolmomyias flaviventris*, of Surinam, is pensile with a ventral entrance tube (Haverschmidt, 1950) that very likely serves the same function as does the entrance tube in weaverbirds.

Some protection from bird and mammal predators is aided by placement of nests in suitable cover. Nests of some species of birds are frequently placed in thorn trees and, in the case of the buffalo weavers, a thorny covering or shell to the nest has evolved. In fact, the Whiteheaded Buffalo Weaver (*Dinemellia*) even places thorny twigs along the boughs leading to its nest over a distance of several feet or more (Friedmann, 1950; Chapin, 1954; Collias and Collias, 1963).

In the warm tropics steady incubation is not so necessary nor so characteristic of birds as in colder climes, consequently nests and eggs may be left alone for prolonged periods. For example, Immelmann (1962) observed that Australian Grassfinches, which are estrildine finches, stopped incubation completely when the temperature in the nest chamber exceeded 100°F. It is evident that presence of a roof helps hide the eggs from predators during prolonged absences of the parents.

One function of the roof of domed nests must be to shed rain. Most small birds in the tropics seem to nest during the rainy season when insect food is abundant. Grasses are often arranged about the entrance of weaverbird nests in such a way as to direct away the rain. Skutch (1954) observed that during the early part of the breeding season, before the rains have begun, the nests of the Yellow-rumped Cacique (*Cacicus cela*) in Central America are all open at the top. But as the rains begin, after the eggs have been laid and even after the young have hatched, the top of the entrance is gradually roofed over and the nest entrance becomes a bent tube with the opening downward.

The roof of domed nests has an important shading effect from solar radiation. Measurements we made of the temperature inside and outside nests of the Village Weaverbird (*Textor cucullatus*) in a captive outdoor colony in Southern California showed the interior of the nests to be some 5 to 15°F cooler on a hot day. Brood nests are somewhat cooler than unlined nests, probably because of the generally thicker roof and walls of the former. The Galapagos Finches have an equatorial habitat, and unlike most other Fringillidae, members of this subfamily (Geospizinae) build roofed nests (Lack, 1947). Since predators are rare on the Galapagos, the chief function of the roof of the nest in the Galapagos Finches is probably to furnish protection from the elements, particularly from the strong tropical sun. Small naked altricial nestlings are no doubt very sensitive to direct exposure to strong sun. The danger from ultraviolet radiation is presumably greater in the tropics when compared with the temperate or colder regions than is the heating effect of the sun. However, the direct effects of ultraviolet on nestling birds seems not to have been investigated in any controlled fashion.

The roof of domed nests may be composed of very different materials in different birds—woven or thatched grasses in the weaverbirds, a mass of short heterogeneous plant materials bound together by spider silk in sunbirds and certain titmice, a leaf in the tailorbirds, and mud in the cliff swallows. The convergent evolution in these diverse instances is further evidence of the great importance of a roof in the life of nesting birds. Of interest in this connection are the few tropical families of small birds that build open cup nests. These cases are not yet fully understood, but various devices seem to substitute for a roof here. The hummingbirds frequently fasten their nests to the underside of a leaf (Ruschi, 1949), while in the bulbuls and cuckoo-shrikes both male and female incubate (Van Tyne and Berger, 1959).

The roofed nest among birds reaches its evolutionary climax of specialization in the pendulous or hanging nest and in the compound nest.

Pendulous nests are attached by their upper part while the lower part hangs free. Such nests have evolved independently in birds of different families, and these birds may use very different materials and binding techniques. For example, many sunbirds use spider silk (Chapin, 1953), while certain other sunbirds and the African Broadbills *(Smithornis)* (Chapin, 1953) may use black fungus fibers *(Marasmius)* as a binding material. The Bush Tit *(Psal-*

triparus minimus) of the western United States uses spider silk to hold together a nest made of a heterogeneous mass of moss, lichens, oak leaves, and catkins (Addicott, 1938). The Cape Penduline Tit *(Anthoscopus minutus)* of South Africa uses a combination of cobweb and felting to bind together a nest made of cottony fibers of seeds of the kapok tree, wool or hairs (Skead, 1959). Cobweb is an excellent binding material, being very strong, more or less adhesive, and easily pulled out into long strands. The tiny Black-fronted Tody-Flycatcher *(Todirostrum cinereum)* of tropical America uses for the main framework of its nest a tangled mass of strong, flexible fibers in which are entangled a great variety of short scraps of vegetable material bound together and into the nest by a liberal quantity of cobweb (Skutch, 1960).

A woven construction facilitates evolution of roofed and pendulous nests by enhancing the coherence of the nest. According to Skutch (1960, p. 544) "the pensile nests of the American Flycatchers are matted rather than woven." This statement points up the crudeness of these nests. However, these terms are both difficult to apply in any precise or objective way; the dictionary defines a mat as a piece of coarse fabric made by weaving or plaiting materials. We have designated as weaving any pattern of interlocking loops of flexible materials in the fabric of a nest (Collias and Collias, 1957). The orioles and oropendolas of the New World, like the true weaverbirds (Ploceinae) of the Old World, weave their nests of strips of flexible materials. The details of weaving techniques have been studied in few species. Herrick (1911) described the shuttle-like movements of the bill in the Baltimore Oriole *(Icterus galbula)* which repeatedly pokes in the end of one strip into its nest mass and then pulls back the end of another strip. In contrast (except in the early stages of a nest), the Village Weaverbird *(Textor cucullatus)* tends to stay with the same strip until all the strip is woven in (Collias and Collias, 1962).

Only one of the half dozen or so sub-families of the weaverbirds can be said to truly weave. The true weaverbirds normally use fresh green materials for their nests, while the rest of the weaverbirds thatch rather than weave their nests, often using dry stiff grasses. A whole series of steps can be traced from loose, crude, irregular weaving to the close, neat, regular pattern that is especially to be found among those species of weaverbirds that build pendulous nests with long entrance tubes. For example, Cassin's Malimbe (*Malimbus cassini*) of central Africa constructs the most skilfully made nest I know of in any bird (illustrated by Collias and Collias, 1963).

THE COMPOUND NEST

The compound nest refers to a common nest mass in which more than one pair of birds or more than one female of the same species occupy separate compartments. Strictly speaking there should also be some common feature of the nest which benefits all the residents. The compound nest obviously grades into instances where different pairs of birds may build their nests in physical contact with other nests; such cases may illustrate early steps in evolution of a compound nest.

The combination of abundant and concentrated food supply and relatively safe breeding sites makes possible the gregarious breeding and increased social stimulation seen, for example, in most sea birds. On the other hand, gregarious breeding is rare in small land birds (Friedmann, 1935). An example from passerine birds of highly gregarious nesting in an obviously safe nest site is seen in the cliff swallows where scores of mud nests may be in physical contact on the vertical face of a cliff. In fact, Emlen (1954) ascribes the evolution of the enclosed nest in this species to the need for neighbors to maintain a certain individual distance from one another, a need enhanced by the erection of the physical barriers provided by the contiguous walls of the nest, which in turn make possible successful nesting under the crowded

conditions and continual social strife within the colony. Emlen observed that almost without exception the entrance tunnel of completed nests was directed away from the nearest neighboring nest entrance.

An example of highly gregarious nesting in passerine birds in relation to concentrated food supply is given by the Wattled Starling (*Creatophora cinerea*) or "locust bird" of Africa, a nomadic species for which some unusually abundant insect life, as of migrating locusts, is essential for successful nesting by the flock (cf Chapin, 1954). A swarm of locusts may be followed by a flock of over 1,000 birds of this species. The young locusts usually hatch after rains during the summer, and the Wattled Starlings show a remarkable ability to turn up at the right place at the right moment. Locusts are the main food of the nestlings. Recently, Liversidge (1961) has described some observations on a colony of this species in South Africa where over 400 nests were built, with young being raised in more than half the nests all within a period of five weeks. All the nests were situated in thorny acacia trees inside the outer foliage. The domed nest is mainly of twigs with the entrance usually from one side or from above, sloping slightly downwards in single nests, but in multiple nests the entrances opened more directly downwards, some even becoming vertical. A few nests were solitary, but the majority were of two or three nests combined in one mass of twigs, and as many as eight in one mass were noted. During incubation both parents bring new nesting material to the nest and after the young hatch the males continue bringing fresh nesting twigs frequently.

A few species of birds from diverse taxonomic groups are known to build compound nests of twigs in which separate pairs or separate females occupy separate compartments in the common mass: the Palm Chat (*Dulus dominicus*) of Haiti (Wetmore and Swales, 1931); the Red-fronted Thornbill (*Phacellodomus rufifrons*) which is a South American ovenbird (cf Austin, 1961); the Monk Parakeet (*My-*

144

opsitta monachus) of Argentina (Hudson, 1920); and the Black Buffalo Weaver *(Bubalornis)* of Africa (Chapin, 1954; Crook, 1958; Collias and Collias, 1964).

The most spectacular and largest compound bird-nest known is built not of twigs but of grasses, and by a bird smaller than any of those mentioned thus far. This sparrow-like bird is the famous Sociable Weaver *(Philetairus socius)* of South Africa, whose nest masses have often been compared to haystacks in a tree. These nest masses are not woven but rather are thatched of dry grass stems. Each nest mass, and there may be half a dozen or more in one tree, is often several feet thick, of irregular extent and up to 5 meters in the longest dimension. The top of each nest mass is dome-shaped, the underside relatively flat and riddled with up to 100 or more separate nest-chambers. The common roof, on which a number of birds may build together, may be one key to the evolution of the nest of this remarkable species, since it is a communal feature that enhances protection from predation for all —like the outer thorny shell does in nests of the Black Buffalo Weaver, or the projecting eaves in nests of the Monk Parakeet (Collias and Collias, 1959, 1963, 1964b). Being domed it also helps to shed the rain.

Possible intermediate stages in the evolution of the nest of *Philetairus* are represented by the nests of its relatives, the Gray-capped Social Weaver *(Pseudonigrita arnaudi)* of East Africa, in which a number of separate nests may be built in the same tree; but when placed in ant-gall acacias many of the nests are grouped into masses. We saw up to nine nests in one mass. Colonies of *Philetairus* are frequently placed in camelthorn acacia trees, and the base of each of the many thorns may contain large swellings within which are colonies of ants. The combination of nesting birds with noxious insects and thorny trees is frequent in Africa, and may well have been important as a predisposing force to the evolution of the compound nest of the Sociable Weaver.

SPECIATION AND THE EVOLUTION OF NEST DIFFERENCES

It is evident that in analyzing variations in nest building within the species a number of subsidiary problems are involved: (1) the taxonomic status of the species may need to be reexamined; (2) the ecological factors associated with important nest variations must be investigated; (3) the genetics and physiology of nest-building, now almost untouched fields, must become better known; and (4) the role of experience in determining differences in nests between different populations of the same species needs to be investigated. These four points are discussed in sequence below.

1) Stuhlmann's Weaver *(Othyphantes baglafecht stuhlmanni)* and Reichenow's Weaver *(O. b. reichenowi)* of Africa, formerly considered separate species are now believed to be one species (Chapin, 1954). Their nests differ in some respects, Stuhlmann's Weaver builds the outer shell of pieces and strips of grass leaf, whereas Reichenow's Weaver of more arid country may use whole grass tops with leaves still attached in construction of the crudely woven outer shell of its nest. On the other hand, after considerable study Stein (1963) now believes that the two song-type races of Traill's Flycatcher *(Empidonax trailli)* represent distinct species, *E. trailli* and *E. brewsteri.* The former builds a loose bulky nest mainly of grasses as a Song Sparrow does, the latter a compact cottony nest, quite like that of a Goldfinch. Both species occur along streams and lake edges, but *trailli* prefers more wooded areas and nests closer to the ground and in smaller bushes. Actually, in avian taxonomy differences in nest form and structure may often prove to be of value in delimiting genera (Mayr and Bond, 1943; Lack, 1956; Collias and Collias, 1963).

2) As already mentioned, in the prairie country of Oklahoma where trees are few, the Mourning Doves nest by preference in trees. Those which nest in trees have a much higher nesting success than do those which nest on the ground, and those which

nest within forks of trees have a greater nesting success than those which nest farther out in the branches. The Magpie often builds a roofed nest, but where it has the protection of dense thorny bushes, it may build an open nest (Linsdale, 1937). Similarly, the New World Flycatcher, *Pitangus lictor,* makes a roofed nest in exposed situations but may build an open nest in concealed places (Smith, 1962). It would be profitable to investigate to a greater extent such instances of species in which an important characteristic, such as the presence or absence of a roof to the nest, has not yet become fixed by evolution.

3) Almost nothing seems to be known about the genetics of nest-building in birds. I suggest that a small difference in body size at certain threshold levels, or in behavior, such as nest-site selection, may lead to quite large differences in nest-types. Compare, for example, nest-sites and nests of Barn, Cliff, Bank, and Tree Swallows (cf Bent, 1942). In turn, small differences in physical traits or behavioral tendencies themselves depend on a balanced interaction of genetic, physiological, and environmental factors for their normal development.

There is as yet only a little information available on the physiology of nest-building (cf summary by Lehrman, 1959). Lehrman (1958) finds that an estrogenic hormone stimulates nest-building by female Ring Doves, and we (Collias et al. 1961) have stimulated nest building in male Village Weaverbirds *(Textor cucullatus)* outside the normal breeding season with male hormone treatment. In the latter species, the male weaves the outer shell of the nest. In Ring Doves (Whitman, 1919; Lehrman et al. 1961) there is some experimental evidence that the presence of a nest bowl and nesting material helps stimulate gonadotropic secretion and breeding behavior. It has long been known from naturalists' observations that unsuitable nesting conditions may inhibit breeding in birds.

4) The importance of experience to nest-building probably varies with the complexity of the nest built by a species. Canaries make a simple cup-nest, and Hinde (1958) found that females raised in absence of nest materials readily built normal nests. In a quite similar experiment, we found that young male Village Weaverbirds *(Textor cucullatus)* raised without nest materials needed considerable practice before the complex nest was built (Collias and Collias, 1964a). In the case of different species of parrots of the genus *Agapornis,* Dilger (1962) has recently developed some evidence showing the operation of both genetic and experiential factors on development of the ability to gather nest material.

EVOLUTION OF SPECIAL USES OF MANIPULATING ABILITY OTHER THAN NEST BUILDING

The general ability of birds to manipulate materials probably first evolved for the making of a nest. Subsequently, there has also been an evolution of additional specialized uses for this ability. For example, some birds manipulate materials or make special constructions that aid feeding, courtship and copulation, or roosting.

The use of a cactus spine or twig by the Galapagos Woodpecker-finch *(Camarhynchus pallidus)* to probe insect larvae out of crevices or holes is now well known (Lack, 1947).

"Symbolic" nest-building movements and the holding of nest-material in the bill, especially by the male, during displays preliminary to pair-formation or coition are phenomena that occur in many diverse orders of birds (Armstrong, 1942). It is as if the nest-building were projected forward in the life cycle because of its value as a signal of a given physiological state and in physiological stimulation. Immelmann (1962) has derived the courtship movements of 18 species of Australian estrildine finches mainly from nest-building activities. In some species the courting male jumps up and down on a branch while holding a piece of grass in his bill, in other species the grass is dispensed with.

The Australian Magpie Goose *(Ansera-*

nus semipalmata) moves about swamps in pairs or trios, builds platforms of rushes on which the birds stand a while to preen and court, and then moves on to build another platform elsewhere (Davies, 1962). The Red-throated Loon *(Gava stellata)* builds a special platform used for copulation, the platform being at some distance from the nest (Palmer, 1962). Special constructions used to facilitate courtship and mating activities in birds have culminated in the remarkable structures built of twigs by the male bowerbirds. These bowers in some species are decorated with flowers or various other objects. In fact, the Satin Bowerbird *(Ptilonorhynchus violaceus)* even paints its bower, using fruit-juices and charcoal. The evolution of bowers has been reviewed by Marshall (1954) and by Gilliard (1963). The latter has summarized the evolution of arena displays throughout the avian orders, and points out that there seems to be an inverse relationship between the elaborateness of the bower and the brightness and degree of ornamentation of the plumage of the male. It would appear that a large elaborate bower can substitute to some degree for conspicuous displays of bright plumage features.

Many birds sleep in their nests, and in the tropics some birds, particularly those species breeding in roofed nests, even build special sleeping nests outside the breeding season. A number of individuals of the same species may sleep in one such nest or "dormitory" together (Skutch, 1961). Many woodpeckers are solitary sleepers and in some species are known to carve special holes for sleeping in the non-breeding season (ibid).

SUMMARY OF MORE GENERAL CONCLUSIONS

The origin of nests built on land probably traces back to the origin of land life and of the land egg in reptiles. The eggs of reptiles are often buried or concealed in pits in the ground, and generally take much longer to hatch than do those of most birds. Development of eggs in ancestral birds was speeded up with the origin of direct parental incubation which in turn

probably arose in avian evolution in close association with homeothermy and ability to fly.

The primary functions of a nest are to help the parent furnish heat and safety to the developing eggs. By substituting in part for these functions, the habit of nesting in cavities tends to block evolution of elaborate increment nests, except as mere filling for the cavity. Size of bird influences the nature of the materials used in nests placed on the branches of trees and bushes. Large birds generally make twig platforms, while small birds as a rule make more compact and better enclosed nests of finer materials, often binding these firmly together and to the substrate by use of spider or insect silk.

Roofed nests are especially characteristic of small tropical passerine birds. Pensile nests with side or bottom entrances are associated with trees, and presumably enhance protection from predators. In turn, evolution of penduline nests is facilitated by a tough, woven construction of the nest as seen in New World icterids and Old World weavers, or by firm binding with spider or insect silk as seen in sunbirds, titmice, and a number of other small passerines.

Compound nests, in which a number of pairs or females occupy separate compartments in a common mass, is rare among birds, and seems to have evolved where there is an abundant food supply and under conditions where security from predation is increased by special protective devices, such as nesting close to biting or stinging social insects. The evolution of compound nests is associated with origin of some feature of the nest mass that enhances protection for all the birds, such as a communal roof.

REFERENCES

Addicott, Alice. 1938. Behavior of the Bush-tit in the breeding season. Condor 40:49-63.
Armstrong, E. A. 1942. Bird display and behaviour. Lindsay Drummond, London.
————. 1955. The Wren. Collins, St. James Place, London.
Austin, O. L. 1961. Birds of the world. Golden

Press, N. Y.

Bartholomew, G. A., T. R. Howell, and T. J. Cade. 1957. Torpidity in the White-throated Swift, Anna Hummingbird, and the Poorwill. Condor 59:145-155.

Bent, A. C. 1940. Life histories of North American Cuckoos, Goatsuckers, Hummingbirds and their allies. U. S. National Mus. Bull. 176:1-506. Smithsonian Inst., Wash., D. C.

———. 1942. Life histories of North American Flycatchers, Larks, Swallows, and their allies. U. S. National Mus. Bull. 179:1-555. Smithsonian Inst., Wash., D. C.

Chapin, J. P. 1953. The birds of the Belgian Congo. Part 3. Bull. Am. Mus. Natural Hist. 75A:1-821.

———. 1954. The birds of the Belgian Congo. Part 4. Bull. Am. Mus. Natural Hist. 75B:1-846.

Collias, Elsie C., and N. E. Collias. 1964a. The development of nest-building in a weaverbird. Auk 81:42-52.

Collias, N. E., and Elsie C. Collias. 1957. The origin and evolution of weaving by African Weaverbirds (Ploceidae). (Abstr.) Bull. Ecol. Soc. Am. 38:102.

———. 1959. Solar radiation and predation as factors determining evolution of nest form in Weaverbirds (Ploceidae) and other tropical birds. (Abstr.) Bull. Ecol. Soc. Am. 40:113-114.

———. 1963. Evolutionary trends in nest-building by the Weaverbirds (Ploceidae). p. 518-530. In Proc. XIIIth Intern. Ornith. Congr. Am. Ornithologists' Union.

———. 1964b. The evolution of nest-building in the Weaverbirds (Ploceidae). Univ. Calif. Publ. in Zoology. (In press.)

Collias, N. E., P. J. Frumkes, D. S. Brooks, and R. J. Barfield. 1961. Nest-building and breeding behavior by castrated Village Weaverbirds (Textor cucullatus). (Abstr.) Am. Zoologist 1:349.

Collias, N. E., and L. R. Jahn. 1959. Social behavior and breeding success in Canada Geese (Branta canadensis) confined under semi-natural conditions. Auk 76:478-509.

Crook, J. H. 1958. Étude sur le comportement social de Bubalornis a. albirostris (Vicillot). Alauda 26:161-195.

———. 1963. A comparative analysis of nest structure in the Weaverbirds (Ploceinae). Ibis 105 (2):238-262.

Cullen, Esther. 1957. Adaptations in the Kittiwake to cliff nesting. Ibis 99:275-302.

Davies, S. J. J. F. 1962. The nest-building behaviour of the Magpie Goose Anseranus semipalmata. Ibis 104:147-157.

Dawson, W. R., and F. E. Evans. 1960. Relation of growth and development to temperature regulation in nestling Vesper Sparrows. Condor 62:329-340.

Dilger, W. C. 1962. The behaviour of lovebirds. Scientific American 206 (1):88-99.

Dixon, C. 1902. Birds' nests. An introduction to the science of caliology. F. A. Stokes Co., New York, and G. Richards, London.

Downing, R. L. 1959. Significance of ground nesting by Mourning Doves in northwestern Oklahoma. J. Wildlife Management 23:117-118.

Emerson, A. E. 1938. Termite nests—a study of the phylogeny of behavior. Ecol. Monogr. 8:247-284.

———. 1943. Systematics and speciation. Ecology 24:412-413.

Emlen, J. T., Jr. 1954. Territory, nest building and pair formation in the Cliff Swallow. Auk 71:16-35.

Friedmann, H. 1935. Bird societies. p. 142-185. In Murchison's handbook of social psychology. Clark Univ. Press, Worcester, Mass.

———. 1950. The breeding habits of the Weaverbirds. A study in the biology of behavior patterns. p. 293-316. In Smithsonian Rept. for 1949. Smithson. Inst., Wash., D. C.

Frith, H. J. 1962. The Mallee-Fowl. The bird that builds an incubator. Angus and Robertson, Sydney and London.

Gilliard, E. T. 1963. The evolution of Bowerbirds. Scientific American 209 (2):38-46.

Haartman, Lars von. 1957. Adaptation in hole-nesting birds. Evolution 11:294-347.

Hardy, J. W. 1963. Epigamic and reproductive behavior of the Orange-fronted Parakeet. Condor 65:169-199.

Haverschmidt, F. 1949. The life of the White Stork. E. J. Brill, Leiden.

———. 1950. The nest and eggs of Tolmomyias poliocephalus. Wilson Bull. 62:214-216.

Herrick, F. H. 1911. Nests and nest-building in birds. J. Anim. Behav. 1:159-192, 244-277, 336-373.

———. 1932. Daily life of the American Eagle. Auk 49:307-323.

Hinde, R. A. 1952. The behaviour of the Great Tit (Parus major) and some other related species. Behaviour (Suppl. 2):1-201.

———. 1958. The nest-building behavior of domesticated canaries. Proc. Zool. Soc. (London) 131:1-48.

Hindwood, V. A. 1959. The nesting of birds in the nests of social insects. Emu 59:1-36.

Horváth, Otto. 1963. Contributions to nesting ecology of forest birds. Master of Forestry Thesis. Univ. Brit. Columbia.

Howell, T. R., and W. R. Dawson. 1954. Nest temperature and attentiveness in the Anna Hummingbird. Condor 56:93-97.

Howell, T. R., and G. A. Bartholomew. 1962. Temperature regulation in the Sooty Tern, Sterna fuscata. Ibis 104:98-105.

Hudson, W. H. 1920. Birds of La Plata. Vol. 2. J. M. Dent and Sons, London and Toronto.

Immelmann, K. 1962. Beiträge zu einer vergleichenden Biologie australischer Prachtfinken (Spermestidae). Zool. Jb. Sys. Bd. 90:1-196.

Irving, L., and J. Krog. 1956. Temperature during the development of birds in Arctic nests. Physiol. Zool. 29:195-205.

Kendeigh, S. C. 1941. Territorial and mating behavior of the House Wren. Illinois Biol. Monog. Vol. 18, No. 3, 120 p.

————. 1961. Energy of birds conserved by roosting in cavities. Wilson Bull. 73:140-147.

Kiel, W. H., Jr. 1955. Nesting studies of the Coot in southwestern Manitoba. J. Wildlife Mgnt. 19:189-198.

Lack, D. 1947. Darwin's Finches. University Press, Cambridge, England.

————. 1956. Swifts in a tower. Methuen, London.

Lehrman, D. S. 1958. Effect of female sex hormone on incubation behavior in the Ring Dove (*Streptopelia risoria*). J. Comp. and Physiol. Psychol. 51:142-145.

————. 1959. Hormonal responses to external stimuli in birds. Ibis 101:478-496.

Lehrman, D. S., P. N. Brody, and Rochelle P. Wortis. 1961. The presence of the mate and of nesting material as stimuli for the development of incubation behavior and for gonadotropin secretion in the Ring Dove (*Streptopelia risoria*). Endocrin. 68:507-516.

Linsdale, J. M. 1937. The natural history of Magpies. Cooper Ornithological Club, Pacific Coast Avifauna, No. 25, 234 p. Berkeley, Calif.

Liversidge, R. 1961. The Wattled Starling (*Creatophora cinerea*) (Menschen). Annals Cape Prov'l. Museum 1:71-80.

Marshall, A. J. 1954. Bower-birds, their displays and breeding cycles. Clarendon Press, Oxford.

Mayr, E., and D. Amadon. 1951. A classification of recent birds. Am. Museum Novitates, No. 1496, 42 p.

Mayr, E., and J. Bond. 1943. Notes on the generic classification of the swallows, Hirundinidae. Ibis 85:334-341.

Medway, Lord. 1960. Cave Swiftlets. p. 62-70. *In* B. E. Smythies, (ed.), The birds of Borneo. Oliver and Boyd, Edinburgh.

Moreau, R. E. 1937. The comparative breeding biology of the African hornbills (*Bucerotidae*). Proc. Zool. Soc. (London) A, 107:331-346.

Nice, Margaret M. 1922. A study of the nesting of Mourning Doves. Auk 39:457-474.

————. 1957. Nesting success of altricial birds. Auk 74:305-321.

————. 1962. Development of behavior in precocial birds. Trans. Linnaean Soc. New York 8:1-211.

Nickell, W. P. 1958. Variations in engineering features of the nests of several species of birds in relation to nest sites and nesting materials. Butler Univ. Botanical Studies 13(2):121-140.

Palmer, R. S. (editor). 1962. Handbook of North American birds. Vol. 1 Loons through flamingos. Yale Univ. Press, New Haven and London.

Pearson, O. P. 1953. Use of caves by hummingbirds and other species at high altitudes in Peru. Condor 55:17-20.

Pitman, C. R. S. 1958. Snake and lizard predators of birds. Bull. Brit. Ornith. Club 78:82-86, 99-104, 120-124.

Preston, F. W., and A. T. Norris. 1947. Nesting heights of breeding birds. Ecology 28:240-273.

Rivolier, J. 1956. Emperor Penguins. [In French,

Translated by P. Wiles]. Elek Books Ltd., London.

Ruschi, A. 1949. Classification of the nests of Trochilidae (hummingbirds). Boletin do Museu de Biologia Prof. Mello-Leitão. Santa Teresa-E. E. Santo-Brasil. No. 7. (Translated by C. H. Greenewalt.)

Serventy, D. L., and H. M. Whittell. 1962. Birds of Western Australia. Peterson Brokensha Pty., Ltd., Perth, W. Australia.

Sielmann, H. 1959. My year with the woodpeckers. [In German, translated by S. Lightman.] Barrie and Rockcliff, London.

Skead, C. J. 1959. A study of the Cape Penduline Tit, *Anthoscopus minutus minutus*. Proc. First Pan-African Ornithological Congress. Ostrich, Suppl. No. 3, pp. 274-288.

Skutch, A. F. 1948. Life history of the Olivaceous Piculet and related forms. Ibis 90:433-449.

————. 1954. Life histories of Central American birds. Cooper Ornith. Soc., Pacific Coast Avifauna, No. 31, 448 p. Berkeley, Calif.

————. 1960. Life histories of Central American birds. Cooper Ornith. Soc., Pacific Coast Avifauna, No. 34, 593 p. Berkeley, Calif.

————. 1961. The nest as a dormitory. Ibis 103a:50-70.

Sladen, W. J. L. 1958. The Pygoscelid Penguins. I. Methods of study. II. The Adelie Penguin. Sci. Rep. F. I. D. S. No. 17:1-97.

Smith, W. J. 1962. The nest of *Pitangus lictor*. Auk 79:108-111.

Smythies, B. E. 1953. The birds of Burma. 2nd ed. Oliver and Boyd, Edinburgh and London.

Stein, R. C. 1963. Isolating mechanisms between populations of Traill's Flycatcher. Proc. Am. Phil. Soc. 107:21-50.

Stresemann, E. 1927-1934. Aves. Vol. 7, part 2. *In* W. Kükenthal and T. Krumbach, (eds.), Handbuch der Zoologie. Walter de Gruyter and Co., Berlin.

Van Dobben, W. H. 1949. Nest building technique of icterine warbler and chaffinch. Ardea 37:89-97.

Van Someren, V. G. L. 1956. Days with birds. Studies of habits of some East African species. Fieldiana: Zoology 38:1-520. Chicago Natural History Museum.

Van Tyne, J., and A. J. Berger. 1959. Fundamentals of ornithology. John Wiley and Sons, New York.

Wagner, H. O. 1955. Einfluss der Poikilothermie bei Kolibris auf ihre Brutbiologie. J. für Ornith. 96:361-368.

Walkinshaw, L. H. 1941. The Prothonotary Warbler, a comparison of nesting conditions in Tennessee and Michigan. Wilson Bull. 53:3-21.

Welty, J. C. 1962. The life of birds. W. B. Saunders Co., Philadelphia and London.

Wetmore, A., and B. H. Swales. 1931. The birds of Haiti and the Dominican Republic. Smithsonian Inst., U. S. Nat. Mus. Bull. 155, 483 p.

Whitman, C. O. 1919. The behavior of pigeons. Publ. Carnegie Inst. Washington 257 (3):1-161.

Witschi, E. 1956. Development of vertebrates. W. B. Saunders Co., Phila.

ADDENDUM BY THE EDITORS

The following figures are also of interest and are reproduced from N. E. Collias and E. C. Collias, "Evolutionary trends in nest-building by the Weaverbirds (Ploceidae)," *Proc. XIIIth Intern. Ornithol. Congr.*, Ithaca, N.Y., American Ornithologists Union, 1963, pp. 520–521.

Figure 1 Variations in nests of African weaverbird species (Ploceidae) which thatch rather than weave their nests. (A) White-headed buffalo weaver (*Dinemellia dinemelli*); note thorny twigs laid on bough leading to the nest, itself covered by thorny twigs. (B) White-browed sparrow weaver (*Plocepasser mahali*); nest with side entrance and crude structure. (C) Black-capped social weaver (*Pseudonigrita cabanisi*); nest with two bottom entrances. (D) Sociable weaver (*Philetairus socius*); the huge, compound nest is viewed from beneath where the entrances to the many nest chambers can be seen.

Figure 2 Variations in nests of African true weaverbirds (Ploceinae), which weave their nests of interlocking loops of strips of fresh green materials. (A) Grosbeak weaver (*Amblyospiza albifrons*); nest with side entrance. (B) Southern masked weaver (*Ploceus velatus*); nest with bottom entrance. (C) Cassin's weaver (*Malimbus cassini*); nest with long entrance tube and finely woven structure.

6

Reprinted from *Am. Zool.*, 4(2), 191–207 (1964)

THE EVOLUTION OF SPIDER WEBS

B. J. KASTON

Central Connecticut State College, New Britain*

It was 52 years ago at a similar meeting of the American Association for the Advancement of Science that John Henry Comstock of Cornell University presented a paper on the evolution of the webs of spiders. With the advance since then in our knowledge of spiders, particularly with respect to the habits of certain species in the Palearctic, Ethiopian, and Australian Realms, my task is made immeasurably less difficult than it was for this pioneer among American araneologists.

In older classification schemes, the categories Retitelariae and Vagabondes gave an indication (as can be deduced from the latter name) that not all spiders build snares. However, although they do not all make webs, they all produce silk, the very name "spider" (*Spinne* in German and *Spindlar* in Swedish) alluding to the spinning activity. What is this material they spin, and how is it manipulated?

It has been found that silk, made of a scleroprotein, fibroin, while yet within the silk glands in liquid form, has a molecular weight of 30,000 (Braunitzer and Wolff, 1955). After emission the fibroin has a molecular weight of about 200,000 to 300,000. We can assume that polymerization occurs, forming additional peptide linkages. The four amino acids glycine, alanine, serine, and tyrosine compose over 90% of the molecule. Apparently the coagulation and hardening occur not as a result of enzyme action, but rather as a result of the mechanical stretch exercised upon the material as the spider draws the line out of the spinnerets (DeWilde, 1943). Spiders produce silk from several different glands, the physical properties varying in accordance with this, as well as from species to species. As to thickness of fiber, Hopfmann (1935) gives a range of from 12 to 0.5 μ, Witt

(1961) gives 0.1 to 0.03, and Kaestner (1956) for the individual threads of cribellate fibers gives 0.02 to 0.01 μ. Using the electron microscope, Lehmensick and Kullmann (1957) were able to show that the apparently homogeneous mass around the axis fibers of the cribellate threads, which under a light microscope has a flocculent appearance, is in reality composed of a confusion of very fine fibrils each having a thickness of 0.015 μ. For ecribellate fibers likewise they showed that any one thread may be split up into a number of fibrils, the finest having a thickness of 0.017 μ.

It is well known that the threads of spider silk show great extensibility, are very elastic, and have a high breaking stress, and the reader can find actual figures for these properties (in comparison with the silk of the silkworm, *Bombyx mori*) in the works of Benton (1907), De-Wilde (1943), and Savory (1952).

I list below (modified from Savory, 1928) some of the uses to which spiders put their silk:

I. Linear constructions
   a) drag lines
   b) parachute, or ballooning threads
II. Ribbons, or flat structures
   a) attachment discs, or anchors
   b) swathing band of orbweavers, and swathing film of Theridiids
   c) tie-band of certain hunting spiders
   d) hackled-band of cribellate spiders
   e) sperm web made by males
   f) "bridal veils" or "love nets" used by male crab spiders to fasten down females during copulation
III. Three dimensional structures
   a) snare or web
   b) egg sac
   c) nest or retreat
      i) a mere tube
      ii) a silk-lined excavation
      iii) an inverted cup near the web

---

* Author's address after Sept. 1, 1964: Dept. of Zoology, San Diego State College, San Diego, Calif.

d) hibernating chamber
e) molting chamber
f) mating chamber
g) nursery for spiderlings

For the production of this silk we know of at least seven kinds of silk glands, lying within the abdomen, and discharging through spools and spigots located on the three pairs of spinnerets (and on the cribellum of cribellate spiders). From the work of Apstein (1889), Comstock (1912), Hopfmann (1935), and Sekiguchi (1952) we know the situation best in the orbweavers. In these latter there are:

a) the aciniform glands, opening on the median and posterior spinnerets. These produce the silk of the swathing bands, and presumably also the fine fibers of the egg sacs.

b) the pyriform glands, opening on the anterior spinnerets. These produce the thin components of the frame threads, the radii, and the drag lines, as well as the attachment discs.

c) the ampullate glands, opening on the anterior and median spinnerets. These produce the drag lines, the thick components of the frame threads, and the radii.

d) the cylindrical or tubuliform glands, opening on the median and posterior spinnerets. These produce the strong fibers of the egg sacs.

e) the flagelliform glands, opening on the posterior spinnerets.[1] These were discovered relatively recently, and have been experimentally demonstrated by Sekiguchi (1952) to produce the ground line of the viscid spiral threads.

f) the aggregate glands, opening on the posterior spinnerets. These are responsible for the production of the viscid droplets. According to Richter (1956), each sticky globule consists of two kinds of substances, "a" and "b." Substance "a" can be drawn out to a very thin elastic fiber. Substance "b" covers "a" as a more glassy, more viscous, slime. The significance of this outer layer becomes clear when one carefully

washes it off. When dry the basic fiber and substance "a" lose their elasticity quickly and become brittle.

In the comb-footed spiders, the Theridiidae, the fibers of the swathing film were considered by Apstein and by Comstock to be made in what they called the "lobed" glands. But Hopfmann has shown that these are like the aggregate glands in the orbweavers, but merely much enlarged. They likewise open on the posterior spinnerets.

Finally, in the cribellate families, there are small glands opening through numerous fine tubes on the *cribellum,* a plate just in front of the spinnerets. These produce the silk which composes the woof of the hackled band, and which is combed out by the *calamistrum,* a single or double row of stiff bristles on the metatarsus of the fourth leg.

It is known that silk is produced by some mites, and by members of two lower Orders of Arachnida, the whip-scorpions and the pseudoscorpions. In all three of these groups it is used solely for the covering of eggs.[2] On this basis it has been suggested, first by Pocock (1895), then by Comstock (1912), Bristowe (1930, 1932, 1941, 1958), and Gerhardt and Kaestner (1938) that in spiders too, silk was first used for the covering of eggs. It is supposed that the web itself evolved from a mass of threads distributed around the egg sac, or from a tube constructed as a retreat in which the spider hid with its eggs. Based on the assumption that a simple tube constituted the primitive form of nest, Pocock considered that the evolution of web spinning proceeded along two main lines; one culminating in the trap-door nest, and the other in the aerial net for snaring prey. Comstock also considered two main lines (ultimately leading to the orbweavers), but these were: a) those webs of which the ground threads have cribellar silk added to them, and b) those which have viscid droplets added. Bristowe too,

---

[1] These glands were also described by Peters (1955) who was not aware of Sekiguchi's prior discovery. Peters named them *glandulae coronatae.*

[2] Young individuals of the mite, *Anystis,* were reported by Grandjean (1943) constructing silken cocoons around themselves.

considered two main lines, but he called attention to the importance of the position of the spider in its snare, with one line of evolution for those webs in which the owners stand erect, and the other for those in which they are inverted in the snare.

Taking an entirely different point of view, Savory (1928, 1934, 1952, 1960) prefers to consider that the "protection of eggs or young is not normally a primitive habit in any group of animals," that silk was originally excretory matter deposited behind as the spider ran about, and that in some way this became the silken drag line that was trailed after the spider.

Archearaneads were undoubtedly preyed upon by the early land vertebrates as early as the Carboniferous Period. Survival value became attached to the habit of hiding in holes and crevices. Possibly the most rudimentary form of snare arose from the chance spinning of a few stray threads about the mouth of the retreat (as suggested by McCook, 1889). Or possibly an irregular network was spun to interfere with the entry of such enemies as wasps. A spider would run out of its shelter to seize a passing insect; as this activity was repeated the drag lines formed a lining of silk for the shelter and radiated outwards from its mouth forming a fringe. This is similar to the primitive type of web made by many spiders even today.

These original webs were devices which merely made the spider aware that there was possible prey present to be seized. At a later stage in evolution, snaring silk was added (i.e., viscid droplets in the ecribellates and the hackled band in the cribellates). Now the web could serve to actually catch the prey.

It is generally conceded that the most highly developed type of snare is the orb, spun by most members of the Argiopidae (*sens. lat.*) and some of the Uloboridae. Pocock concluded that one factor guiding the evolution of the orb was the advantage gained by fineness of construction, which provided comparative invisibility. A large snaring area is thus provided with a minimum expenditure of silk thread. But of course, this tendency toward invisibility would have to be kept in check in order not to interfere with the strength of fiber and closeness of mesh sufficient to hold insects necessary for the animal's existence.

In constructing the snare[3] the spider first ties together the objects between which the web is to be spun. Nearly always this involves use of air currents to carry the thread. The spider tests the line emitted into the wind to determine when it has stuck. This line is then pulled taut and fastened. Then the spider crawls across on this, paying out a heavier thread which will serve as the bridge line. While some observers report that this line is simply allowed to adhere to the thin line which is already there, others report that the original line is rolled up and eaten as the new, heavier, line is laid.

A frame is then spun consisting of foundation lines which, with the bridge thread, determine the plane of the web. Next the radii are put in, and in order that they may sustain equally the stretching of the web, they are connected by means of a few turns of a spiral thread spun just outside the hub. This is the attachment, notched, or strengthening zone.

Once the hub and attachment zone are completed the spider spins a spiral thread, starting at the end of the attachment zone spiral and continuing to the periphery, with turns as far apart as the spider can stretch. This thread is not viscid, and its function is to hold the radii in place during subsequent operations. It has been called the "scaffold spiral," "helping spiral," "provisional spiral," and "spiral guyline." The construction of the actual catching spiral of viscid thread is now begun. This is always spun from periphery to center, though not all the way to the attachment zone, thus leaving an area of greater or less width, called the free zone. The original scaffold spiral is cut away turn by turn as the viscid spiral approaches it.

---

[3] For details of construction see the papers by Savory, Wiehle, Comstock, and Peters. Also the book by Tilquin (1942).

Once the viscid spiral is finished the spider has only to make a change in the hub. Some bite out the threads from the center, so that the hub becomes an open one. Others cover the hub with a dense sheet of silk. Some insert a band of silk above, or below, or both above and below, the center. This "decoration," usually called the stabilimentum, may be straight, or zigzag, or in some cases, circular.

The above account is a general one fitting most spiders in the families concerned,

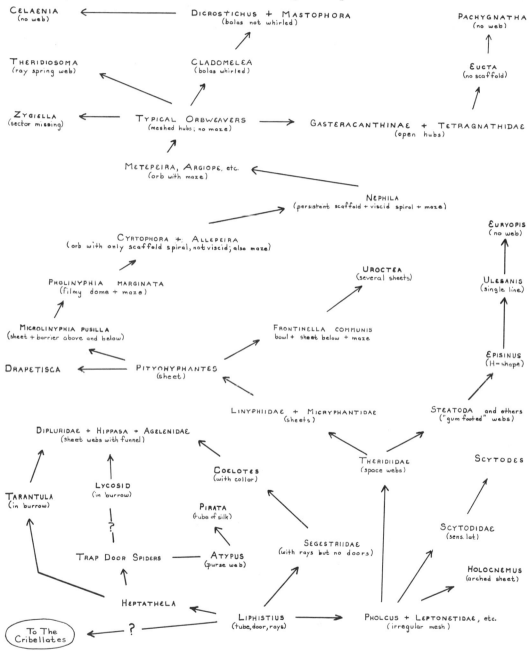

FIG. 1. A suggested phylogeny of the ecribellate web makers discussed in the text.

but there are a number of important exceptions which have a bearing on the evolution of web construction. For example, in *Zygiella* the viscid spiral threads are omitted from a definite sector between two radii, and this situation is sometimes found in other species as well. In *Nephila* the scaffold spiral is not removed, so that it can be seen in the finished snare along with the turns of the viscid spiral. Moreover, a so-called "barrier web" or "stopping maze" of irregular mesh is built to one side of the orb. In *Sybota* the scaffold spiral remains and no viscid spiral is added at all. Instead the snaring cribellar threads are fastened to the frame threads and to the radii (Wiehle, 1931). Similarly in *Cyrtophora* (Wiehle, 1928; Marples, 1949)

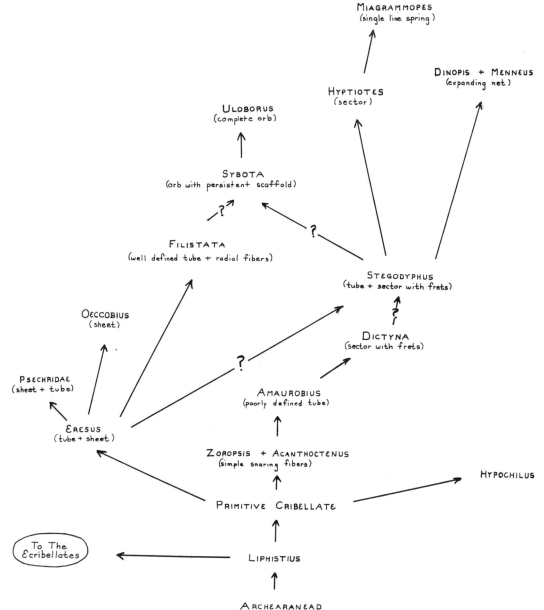

FIG. 2. A suggested phylogeny of the cribellate web makers discussed in the text.

and in *Allepeira* (McCook, 1889; Exline, 1948) the spiral thread which is present, although closely woven, is not viscid, and moreover is spun from the center outwards. It seems logical to consider it as a persistent scaffold spiral, as in *Sybota*. In *Eucta* the viscid spiral is spun without the web having any previously built scaffold spiral (Crome, 1954).

It has often been stated that young spiders build snares just as expertly as do their parents. True, they may be able to construct the snares, but it is of interest to note that the spiderlings frequently build something different from that constructed by the adults. "In some species they [the webs] are more symmetrical and less specialized, thereby indicating the inheritance of their design from some far distant ancestor" (Bristowe, 1958).

It has been reported that the webs of very young orbweavers contain fewer radii than those of the adult, but that also in aging spiders, presumably because of a loss of instinctive skill (Bristowe, 1958) the webs tend to be smaller, with fewer radii and fewer spirals. To some extent, at least, the more open mesh is due to the spider using a heavier thread, for it was found that when the spider got heavier (as it aged, or had weights added to it experimentally) it presumably had to build a thicker thread to hold its own weight (Witt and Baum, 1960; Christiansen et al., 1962). If the amount of material remains about the same, the thicker thread will have to be shorter. Hence, fewer spiral turns will be made.

The young *Zygiella* constructs a complete orb, without any sector devoid of viscid spiral threads. On the other hand the young *Araneus dalmaticus* does omit a sector. The young *Nephila* omits the stopping maze, and it is reported that the young of *Cyrtophora* (Wiehle, 1927) and of *Allepeira* (Archer, 1941) build webs that look more like those of the Linyphiidae than like those of orbweavers. No viscid spiral is put into the web by the young of *Uloborus* and the scaffold spiral is not removed. The lack of snaring threads is correlated with the fact that the spiderlings do not yet possess a cribellum or calamistrum (Wiehle, 1927; Peters, 1953; Szlep, 1961). The spiderlings molt for the first time about four weeks after emergence from the egg sac and now, possessing a cribellum and calamistrum, they make webs with the characteristic snaring spiral. It was found by Szlap that if the spiderlings were kept in small glass containers to prevent their making a primary web, and were then removed after they had molted, the web they constructed would be of the adult type. The spiderlings' past experience in web building seems to have nothing to do with the change in web structure.

Another difference in the webs of young, as compared with adult, spiders concerns the nature of the stabilimentum. These bands of silk tend to be narrow and along the radii, but have been reported to be wider, or circular, in the young of *Argiope trifasciata* (Comstock, 1912), *A. argentata* (Peters, 1953), *A. bruennichi* (Wiehle, 1928; Yaginuma, 1956), and *Uloborus* (McCook, 1889; Wiehle, 1928). In the case of *Cyclosa*, Marples (1937) found that the stabilimentum is quite variable even in the same spider from day to day, but both Wiehle (1928) and Marson (1947) indicate that the band is *always* present in the spiderling's web, and *not usually* present in that of the adult.

Then, of course, there is the situation in *Pachygnatha* (Balogh, 1934) and *Celaenia* (McKeown, 1952) in which the adults spin no webs, but where the young spiders make an orb comparable to those made by their close relatives.

Among the theridiids, there are some that construct the so-called "gum-footed" webs. These contain, attached to the substratum, a number of straight vertical threads, which have for about 3 mm from the lower end a coating of 3 to 10 viscid globules. These threads, therefore, can function in the snaring of ants and other ground insects (Wiehle, 1931; Nielsen, 1932; Nørgaard, 1948; Uyemura, 1957; Freisling, 1961). Freisling has shown that these gum-footed lines are put in last, and that they are not included in the webs of spiderlings (at least in *Theridion saxatile* never before they are six weeks of age). Even at that they are included sporadically at first, and not always vertically.

It is supposed that the ancestral spiders were wanderers, and that the eggs were carried about by the mother just as they still are today by a number of spiders including *all* Lycosidae and Pisauridae. Others, however, acquired the habit of depositing the eggs under loose bark, in a curled leaf, under a stone, or in some crevice, etc., where they could be protected to some extent from the elements, and from parasites and predators. While some females then leave their eggs, others remain, usually lining the cavity with silk or even constructing (as in the Salticidae, some Clubionidae and Anyphaenidae) a special silken bag or retreat referred to by some authors as the "cocoon."

Bristowe suggests that "At this early period, or possibly before any complete lined cell was built, the gummy silk-forming substance was probably secreted from glands which opened into the mouth or chelicerae. It is from such glands that other arachnids . . . perform their simple spinning operations today, whilst among spiders the remnants still remain to support this suggestion."[4] It has been known since the observations of Monterosso (1927) that members of the genus *Scytodes* capture prey by squirting a mucilaginous secretion from their chelicerae. From the studies of Millot (1930) we know that in *Scytodes* the gland involved is bilobed, one lobe producing gum and the other poison. In Bristowe's view the gum-producing lobe has vanished in other spiders, "and if we take the view that the two have fused we might look upon the poison glands of all spiders as being homologous, or partly homologous, to the now extinct cheliceral spinning apparatus." He further indicates that perhaps it is not quite extinct, for it is well known that in the burrow-digging

Ctenizids the particles of earth are cemented together by material from the mouth or chelicerae. We know too, of cases among the Clubionidae (*Phrurotimpus* and *Castianeira*) (Montgomery, 1903), and the Gnaphosidae (*Zelotes*) (Nielsen, 1932), where the females complete the making of their egg sacs by means of such gum, which is often responsible for the change in color of the silk, and which causes particles of dirt and grit to adhere to it. It would appear likely therefore, that ancestral spiders used such

"gum or saliva and earth to make a kind of mortar with which to cover their eggs, and build or line a retreat. Later, when silk became more abundant, the earthy mortar was retained on account of the strength it imparted to the cocoon, and finally the survival value of an inconspicuous cocoon led to many spiders independently making use of earth or other material as an external coating to make the cocoon inconspicuous. All spiders today can produce adequate supplies of silk from their spinnerets, but the liphistiids have only reached the second stage (discarded by many other spiders) where earth is woven into the cocoon and covered with silk. The third stage is met with very commonly amongst the Araneomorphs, where earth or other matter serves little or no purpose as a strengthening agent, but is plastered on the outside to disguise the contents."

This kind of activity is described in detail by Montgomery (1903) and others.

Nielsen, in addition to reporting the agglutination of grit to the cocoon in four species of *Prosthesima* (=*Zelotes*), also related how in one of them the female used her posterior spinnerets to daub small lumps of excrement all over the surface of the cocoon.[5] As indicated by Bristowe "this is just what we might expect our

---

[4] Grandjean (1948) reported finding silk glands in the pedipalps of a mite.

[5] A detail not to be overlooked in this connection, is the possibility, hinted at by Crome (1957) that *Uroctea* sometimes uses what appears to be excrement, along with silk, in wrapping its prey.

hypothetical ancestor to have done when its salivary fluid was exhausted, and in course of time the main source of covering material came not from the mouth, chelicerae, or anus, but from glands (possibly coxal glands) beside the abdominal appendages of the fourth and fifth segments, which had by that time become functionless for ambulatory purposes." Since these represent the homologs of the biramous pleopods of Crustacea there were thus originally eight spinnerets. In present day *Liphistius* they are all still there, though they are not all the same size or of the same degree of development. On each side we have a large multisegmented anterior and posterior spinneret, representing the exopodite of the appendages of the fourth and fifth somite respectively. The corresponding endopodites are represented by a pair of very small unsegmented spinnerets between each of the large pairs. Neither pair of medians appears to be functional. In *Heptathela* the posterior medians are reduced to a single unpaired tiny spinneret. In the majority of more advanced spiders the posterior medians become the definitive median spinnerets, while the anterior medians appear either as the cribellum, or as the colulus, or are lost completely. Petrunkevitch calls the small median spinneret of *Heptathela* a posterior colulus.

It is supposed by some students of the subject that the colulus is a relic of the cribellum, and by others that both the Cribellatae and the Ecribellatae have descended independently from ancestors with four pairs of spinnerets. Moreover, it would seem that in spider phylogeny the habits of snare-building, and of hunting without a snare, may have arisen in different families independently.

Bristowe considers that there have been two types of biological evolution, in accordance with whether: *a*) the spiders chose to deposit their eggs within silken cells or holes in the ground; or *b*) the egg sacs were suspended amongst vegetation. Some cell builders have continued with little change to this day, e.g., the Clubionidae, Gnapho-sidae, Salticidae, etc., while others, although still using cells, have in addition "constructed one or more little flaps or doors, from which the spider can emerge to seize insects that touch it, or pass very close. At least three families do this, and it might be said that *Liphistius* shows us how the next stage has been reached. In the Batu Caves [of Malaya] *L. batuensis* Abr. builds a cell on the walls with a thin flap or trapdoor at the top, and a smaller flap at the opposite end to serve as an emergency exit. From the rim of the aperture seven long straight threads extend."

In the beginning such threads were probably short and functioned in keeping the rim stretched, but the long threads of the present serve to inform the spider that an insect is nearby. On the cave floor some of the spiders place the silken cell so that the upper door is level with the surface and the lower door is omitted. Thus, as is the case with the other two Malayan species of *Liphistius*, the cell leads into a burrow several inches deep, though the origin of the cell is confirmed by only the first inch or so being lined with silk.

Many other spiders live all their lives in such burrows, with or without door flaps, and many live this way while carrying their eggs, or else place their eggs in such silk-lined holes. "And some of these burrowers have gone a step further than *Liphistius* . . . by extending her seven threads into a regular little sheet. In the family Lycosidae all the stages can be seen." For example, *Pardosa* is a hunter, and the female continues hunting after the eggs have been laid and the sac attached to her spinnerets. *Lycosa helluo, L. rabida,* and *Tarentula accentuata,* etc., at the time of egg laying make use of burrows which they line with silk. I have seen the burrow of *L. aspersa* provided with a turret of straw and twigs held together with silk, and others have reported that this species builds a door. Still other wolf spiders, including species of *Geolycosa,* spend all their lives in silk-lined burrows. Finally, *Hippasa* (and presumably *Sosippus* and others in the Hippasinae) spin a sheet of silk around the entrance of the burrow and run over the top of the sheet in the manner of a *Tegenaria* or *Agelena*. Bristowe (1930) reported that a close relative of the wolf spiders has a somewhat similar habit of constructing doors for burrows. This concerns a member of the trionychous family Zodariidae, which lives in an earthen cell, both ends of which are provided with doors.

Heptathela constructs a burrow with a door, but has no threads radiating from its lip. And this is more or less what is done by its numerous, wide-spread, more advanced relatives in the family Ctenizidae. Some of these, e. g., *Myrmekiaphila*, still make thin "wafer" doors at the top (as in the Liphistiids) often with another such door at the entrance to a side chamber below. Though the outer door is weak it is well camouflaged, and the side chamber provides further escape from a predatory wasp for example. As a later development the doors were strengthened so that they fitted like a cork, and other ctenizids (*Bothriocyrtum* and *Ummidia*) developed the habit of relying on a fortress guarded by a heavy cork door, which they hold shut with surprising strength.

Another line of descent led to the typical tarantulas (Theraphosidae), many of which have temporary or permanent silk-lined shallow burrows in the ground. A few have become arboreal, while the members of the closely related Dipluridae (e. g., *Euagrus*) "spin a funnel in a crevice under rocks, or in thick vegetable growth, and then continue the silk out over the ground as an expansive sheet" on top of which the spider runs (Gertsch, 1949). As with the Agelenidae and hippasine lycosids already referred to, the spider lives in the funnel waiting for insects to fall upon the sheet.

Paralleling the evolution of the theraphosids, ctenizids, and diplurids, is the evolution of the group of so-called "atypical tarantulas." These display an even closer structural relationship to the liphistiids. Some, as *Hexura*, run over the surface of sheet webs. Others, as *Antrodiaetus*, construct burrows closed with a door composed of two semicircular halves; when closed the two halves meet in a straight line over the middle of the burrow's mouth. Hence members of the Antrodiaetidae have been called the "folding-door tarantulas." But the atypical tarantulas also include the purse web spider, *Atypus*, whose web is a combination of a subterranean portion and a surface portion. The spider first constructs on top of the ground a small horizontal cell, much like that of a jumping spider, and from this it extends the tube downward. Cloudsley-Thompson (1953) has suggested that the tube represents an elaboration of the ancestral cocoon or cell. The upper portion is either extended over the surface of the ground or else up the side of a tree, and this is covered with sand and debris. Two araneomorph spiders which build somewhat similar webs are the cribellate, *Eresus niger*, and the lycosid, *Pirata piscatorius*. In the former the surface portion of the tube is expanded into a sheet, and the spider runs on the under surface of it (Nørgaard, 1941). *Pirata* builds over water in a sphagnum bog; the vertical portion is short extending through the covering of moss only to the water surface, while the horizontal portion is also short and open at the end. Still another parallel between the mygalomorphs and *Eresus* is the fact that sometimes the web of the latter has a side chamber underground.

Returning to the condition in *Liphistius* we recall that a number of threads radiate out from the lip of the burrow. Coming to the more primitive of the araneomorph spiders we find that *Segestria florentina* and *Ariadne bicolor* construct, in a crevice of rock above ground, a silken tube with a similar arrangement of threads radiating outward from the mouth, though there is no door. Gradually, by the addition of many more radial threads, a kind of funnel was perfected around the entrance of the tube. Thus we have the kind of snare such as is found in *Coelotes* today. The extension of the lower side of this funnel into a sheet gives us the typical agelenid web of *Tegenaria* and *Agelena*. On the other hand, an elaboration of the radiating fibers in all directions around the mouth of the tube produces the condition seen today in the snares of the cribellates *Amaurobius* and *Filistata*. The tube and radiating threads are not as well-defined in *Amaurobius* as they are in *Filistata*.

An added refinement to the funnel webs of many agelenids is the network of irregular threads above the sheet. This stopping maze serves to impede the flight of insects, causing them to fall on the sheet, where they are seized by the spider.

Let us now consider the ancestral spider which hunted on shrubs and herbage, and hung its cocoon amongst vegetation while remaining on guard. It seems reasonable to suppose that haphazard strands crossed one another at all angles, but radiated outward from the vicinity of the egg sac. "The logical stance for spiders in webs of this sort is an inverted one, as they would otherwise have to balance themselves on single threads like a man walking a 'tightrope'" (Bristowe, 1930). Little by little, as the spider ran about trailing its drag line, additional threads were added around the egg sac so that a simple meshwork web was formed, much like that of *Pholcus*, the leptonetids, and some theridiids of today. A continuation of this running about amongst the scaffolding threads could give rise to the formation of a rough platform or sheet, such as is found today in other pholcids (*Metagonia, Holocnemus,* and *Spermophora*), in other theridiids, and in most linyphiids. In time some of the spiders came to depend on the added protection of cracks in bark, and fissures in rocks, etc., so that the eggs would be deposited in such places where a retreat would be constructed. This is the case in several theridiids, and Bristowe (1930) reported that:

"In running out to capture insects, these spiders by reason of their drag line and inverted position, would be inclined to thicken the snare above them, and, near the entrance to their retreat, . . . a flimsy sheet would be formed automatically in the course of time. This happens in actual prac-

tice, and I recently noticed it to quite a marked extent in the webs of the southern European theridiid, *Lithyphantes paykullianus*. This represents a link between the ordinary *Theridion* web, and the sheet web such as is built by the [cribellate] psechrids of tropical countries."

These webs are quite like the sheet webs built by *Hexura, Euagrus,* and the agelenids already referred to above. Savory considers that the irregular mesh-type of web, without a sheet, as constructed by some of the modern theridiids, is not a primitive, but rather a simplified type of web derived from the linyphiid sheet-web type.

An added feature incorporated in the snares of a number of theridiids is the "gum-footed" thread. For the most part the threads used in the "space webs" or "volume webs" of theridiids are of the dry type, though viscid silk is known to be used in the swathing film wrapped around the prey. But in 1931 Wiehle reported for *Teutana* and *Steatoda* that a number of vertical threads extending from the sheet to the substratum were covered with viscid droplets for a short distance from the lower edge (as discussed above). These gum-footed threads are now also known from the webs of several species of *Theridion, Lithphantes albomaculatus, Argyria venusta,* and *Pholcomma gibbum*.

A further reduction in the theridiid web is shown by *Episinus*, reported first by Holm (1939). Here the sheet and irregular mesh are absent, and in effect the snare is reduced to two nearly vertical gum-footed threads connected transversely above so as to present an H-shaped appearance. The spider hangs from the cross piece by her hind legs and holds her front legs down along the vertical threads. A still further reduction is that exhibited by a New Zealand theridiid, *Ulesanis pukeiwa* (Marples, 1955). "In its simplest form the web consists of a single thread, usually about 10 cm. long, which may be inclined at an angle [slightly off the horizontal]. Part of the thread is covered with sticky droplets which are larger than those on an ordinary orb web, and are very easily seen with the naked eye. The resting position of the spider is on a twig . . . The web is held under tension so that when the prey is caught and the spider releases its hold on the twig it jerks forward a short distance and the thread goes slack." One can readily see that this behavior closely parallels that shown by the cribellate *Miagrammopes*, which also builds a single thread snare (see below).

Finally, at what appears to be the end of the theridiid line of evolution a few have divorced themselves completely from silk as a means of capturing prey. An example is *Euryopis*, which is reported as having no snare at all, but living under leaves and stones. It has apparently reverted to the free running hunting habits of its distant ancestors.

Another example of a spider which has given up constructing a snare to obtain prey, yet has undoubtedly descended through snare-building ancestors of the same type as *Euryopis*, is *Scytodes*. As indicated above, this spider secures its prey by ejecting a gummy substance from its chelicerae. In its morphology this spider is close to the other members of the Scytodidae *(sens. lat.)* none of which, however, show the spitting habit, and several of which *(Sicarius, Diguetia, Loxosceles, Plectreurys)* are known to construct webs that can easily be related to the irregular meshed *Pholcus*-type. Millot has shown that *Scytodes* has very small abdominal silk glands, yet when the time approaches for laying her eggs the female constructs just such an irregular *Pholcus*-type mesh web (Dabelow, 1958). This is remarkable, in view of the fact that after laying her eggs in this mesh web the female does not suspend her egg sac there, but removes it, and deserting the web carries the sac about with her under her sternum. All of this would seem to point to the probability that the immediate ancestors of *Scytodes* were web builders.

We have seen how a sheet web may arise, and how in some cases the spider constructs a stopping maze above the sheet.

In the case of those where no tube or retreat is present, the spider lives on the sheet, hanging upside down from its lower surface. The typical web of this pattern is the product of the Linyphiidae. In *Pityohyphantes phrygianus* the web is a flat sheet spun between the twigs of a shrub or tree. *Microlinyphia pusilla* makes a horizontal platform between stems of grass, and spins a stopping maze above it. *Prolinyphia marginata* makes a filmy dome, and *Frontinella communis* spins a bowl-shaped web, with an additional flat sheet below, and a stopping maze above.

There is a tendency, especially with the smaller members of the family, for the webs to show a reduction of the tangles above and below the sheet. This is especially the case in the majority of the Micryphantidae who have taken themselves and their webs down from the bushes to the ground, where they spin delicate small sheets across depressions and irregular places.

An interesting variation of the linyphiid sheet web is that made by *Drapetisca*. Here the sheet is vertical instead of horizontal and very closely appressed to the bark of the tree trunks where the spiders live. Because the delicacy of the threads makes them invisible it was thought by naturalists collecting the spiders that no web was made. However, the recent work of Kullmann (1961), who used ammonium chloride to coat the threads, has revealed the true nature of the situation.

Other sheet webs of interest are those made by *Uroctea, Oecobius,* and *Hypochilus*. In *Uroctea* there may be a single sheet or two, vaguely resembling the situation in *Frontinella;* but more usually there is a series of flat sheets one above the other, the number of sheets tending to increase as the spider ages (Gerhardt and Kaestner, 1938; Crome, 1957). In the cribellate *Oecobius* there is built a small flat sheet up against a rock (or windowsill). The spider does not station itself on the sheet, but off to one side, and captures prey by running out on the under surface of the sheet. Although a tube is usually lacking, there is a similarity between this and the cribellate psechrids already referred to, as well as to the micryphantids of the preceding section. In the cribellate *Hypochilus* the web consists of a flat sheet above, usually constructed under a rock, and with sides that slope down like a truncated cone, giving to the whole the appearance of a lampshade. This is somewhat like webs made by some of the Pholcidae, but also faintly resembles the dome of *Prolinyphia marginata*.

The evolution of orb webs presents the greatest difficulty, for this type, by general consensus, is considered the acme of the spinner's art. The question of how the elaborate method of constructing one was acquired by the ancestors of today's orbweavers is not easy to answer. On the basis of the fact that egg sacs are often placed in the center of the orb, Bristowe derives these webs from some early form of irregular mesh similar to that of a *Theridion*, which also suspended its egg sac in the center of the web. But I believe we may just as easily derive at least the ecribellate orbweavers from the Linyphiidae, to which they are very close morphologically. For example, *Cyrtophora* and *Allepeira* construct domed orbs which very closely resemble the filmy dome of *Prolinyphia marginata*. Moreover, as in this linyphiid web, and in contrast to the general situation among orbweavers, no viscid silk is present. Observations on the actual construction (Marples, 1949) indicate that the spiral is spun from the center out. This then, in effect, is really the scaffold spiral, though the turns are much more closely set than is usual for such a structure. It has already been indicated earlier in this paper, that in the webs of young *Uloborus* as well as adult *Sybota*, the scaffold spiral is the only spiral in the snare. Marples has suggested that "the web [of *Cyrtophora*] might be regarded as a specialization of the half constructed orb web, a sort of paedomorphic form, [or] the domed web might be regarded as resembling a stage in the evolution of the orb web." The

snares of *Cyrtophora* and *Allepeira* also show a stopping maze.

The adult *Uloborus* web is so similar to that of the ecribellate orb webs that a number of araneologists, including Petrunkevitch (1926) and Wiehle (1931), have considered the ecribellate Argiopidae *(sens. lat.)* as descendants of the Uloboridae. Wiehle traces the beginning of the orb from the lower cribellates. *Zoropsis* (family Zoropsidae) and *Acanthoctenus* (Acanthoctenidae) put down a few snaring fibers irregularly arranged around the entrance to the retreat. *Filistata* displays a regularity of placement around its tube such that one notices the radial pattern as in an orb web (Nørgaard, 1951). While many of the Dictynidae have an irregular arrangement of the fibers such as in the Theridiidae (of which they have been considered the cribellate homologs), some, e.g. *Dictyna volucripes* of America and *D. arundinacea* of Europe, construct snares showing lengthwise fibers with frets, such that the entire snare resembles a sector of an orb with a few radii and a number of cross fibers of the viscid spiral. A further step on the way to an orb is illustrated in the web of *Stegodyphus lineatus* (family Eresidae). This spider makes a silken tube much like that of *Filistata*, and a portion extending from this where the fretwork is more pronounced than in *D. arundinacea*, thus giving this portion a greater similarity to an orb sector (Millot and Bourgin, 1942).

Among the cribellates the orb web and its variations attain the highest degree of development in the family Uloboridae (and interesting details and comparisons are given by Marples, 1962). Wiehle (1931) studied the webs of *Sybota producta* and noted that the spider makes a regular framework, lays down radii, hub, attachment zone, and scaffold spiral, but does not finish up with a viscid spiral spun from radius to radius in the usual manner. Instead, the scaffold spiral remains, and the cribellar silk is fastened to frame threads and radii. The web is more or less vertical and the spider sits at the hub *which is always attached to a supporting twig or stalk.* This would seem to be an inter-mediate condition between *Filistata,* where the web is entirely against a solid surface, and *Uloborus,* where the web is out in space and is a complete orb with the snaring spiral bearing the cribellar silk. The *Uloborus* web is most often horizontal, with the spider stationed at the hub on the under side, and is provided in addition with a stabilimentum.

We recall that in *Stegodyphus* the web included a tubular portion and a portion looking much like a sector of an orb. If now we lift this sector away from the tube and suspend it in space between twigs, we will have something looking much like the web of the triangle-web spider, *Hyptiotes* another uloborid. It consists of four rays of non-viscid silk placed in a vertical plane, the two outermost forming an angle of from 50 to 60 degrees. Across these are laid down 10 to 20 viscid lines of cribellar silk, arranged about a centimeter apart. The outermost spiral is about 15 to 20 centimeters long. A conspicuous feature is that these viscid lines of cribellar silk are not merely attached to the radii at one point, as is the case with sticky spirals generally. Instead they extend along a radius for a short distance before extending to the next radius, and consequently they have a stepped appearance. A similar appearance is seen also in the webs of some of the species of *Dictyna,* and also to some extent in the webs of *Stegodyphus* referred to previously. The four radii of the *Hyptiotes* snare converge to a single draw thread fastened to a nearby twig.

When the web of *Hyptiotes* is to be used, the spider assumes her position at the twig upside down, with her spinnerets attached to a short line fastened to the twig and her front legs holding the draw thread, or trap-line, which she has severed from the twig. Thus the thread from the apex of the triangle to the twig is not continuous, as her body bridges the gap, suspended in mid-air between two unconnected threads. The spider gathers up a loop of thread, which she holds with her front legs, and at the same time holds the

trap-line taut. The tension is released with a spring when an insect strikes the web.

A still more remarkable spring trap is that constructed by another uloborid, the stick-spider, *Miagrammopes.* The web of a south African species was described by Akerman (1932). It consists of a single thread about a meter long, stretched across the open spaces and attached to a branch at each end. The central part for somewhat less than half its length is covered with cribellar silk. When ready for use, the spider holds one end and draws it taut exactly in the manner of *Hyptiotes,* and as with the latter, with a loop of slack thread held by the front legs. The tension is released by the spider when some insect alights on the thread: this jerks and sways the thread, thus entangling the prey. It is interesting to compare *this* single line spring web with the one operated similarly by the ecribellate *Ulesanis.*

The last group of cribellates to be discussed belongs to the family Dinopidae. This family contains but two genera, *Dinopis* from Australia and America, and *Menneus* from Africa. Their known habits, as reported by Akerman (1926), Theuer (1954) and Roberts (1955) are quite similar. These spiders construct a web consisting of a rectangle or trapezoid of dry silk across which are spun a number of bands of cribellar silk. The resulting snare, about 10 to 20 by 15 to 25 mm or slightly larger, is held with the front legs. When an insect approaches, the spider, by moving apart its front legs, expands the elastic snare to between 5 and 10 times its original length, and may even hurl the net over its victim. Did this behavior arise from a *Hyptiotes*-like ancestor, or from *Stegodyphus* through some as yet unknown type of snare builder? One can only conjecture.

I have already indicated that some authorities derive the ecribellate orbweavers from the Uloboridae. Another school of thought, to which the majority of araneologists belong, considers that the weight of evidence points to the independent evolution of orb-webs in these two groups. The phenomenon is rather a case of convergence or parallelism, just as occurs with respect to the morphology of cribellate and ecribellate groups (particularly striking in the case of the Oecobiidae and Urocteidae) and is analogous to the similarity between certain marsupial and placental groups among the Mammalia.[6]

It will be recalled that in webs of *Cyrtophora* and *Allepeira* no viscid spiral is made, and that the spiral present is actually a persistent scaffold spiral. Also, the radii are numerous, closely spaced, and branched peripherally so that the interval between adjacent radii is hardly more near the edge than near the center. Both of these characteristics are considered primitive.

In the webs of the present day silk spiders, those giants belonging to the genus *Nephila*, we still find branching radii; also the scaffold spiral is left in the finished snare, the spider placing several turns of the viscid spiral between any two turns of the scaffold. Another characteristic present in these snares, and considered primitive by Gertsch, is the tangle of threads composing the stopping maze.

With the loss of the scaffold as a permanent part of the snare, but with retention still of the stopping maze, we advance to the type of web made by very many of today's spiders, especially *Metepeira labyrinthea* and *Argiope* spp. *Metepeira* constructs the retreat within the maze, but *Argiope* is found in the center of her snare, which is generally furnished with a stabilimentum. Finally, the loss of the stopping maze, or at least its great reduction (the spider depending on the sheet alone) leads us to the typical snare made by the majority of higher orbweavers.

But even here we have modifications, and I can see at least four lines of further evolution from these typical weavers. One of the best known is that in which the spider almost always *(Zygiella)*, or very commonly *(Neosconella pegnia* and *N. thaddeus)*, or only quite rarely *(Metepeira labyrinthea)* spins an incomplete orb, en-

tirely omitting the viscid spiral threads from a sector. This is accomplished by the spider making loops and swinging back and forth many times instead of going completely around the web when laying down the viscid line. The spider takes up its station in a nearby retreat off the web, but remains connected with the latter by means of a signal thread extending from the hub and sometimes virtually bisecting the open sector.

A second modification occurs with the spider biting out the ends of the radii where they meet at the center, so that there results an open hub. This is the condition in the snares of *Meta* and *Cercidia*, of the members of the Gasteracanthinae, and of the Tetragnathidae. The gasteracanthines make webs with many radii and spirals and hence of close mesh, while the webs of the tetragnathids have few radii and spirals and are thus quite open-meshed. Moreover, they are horizontal, as are the webs of *Uloborus*, a condition considered more primitive than the vertical. It has already been stated earlier in this paper that although the spiderlings of *Pachynatha* construct snares, the adults of this tetragnathid do not.

A peculiarity of behavior in yet another member of the Tetragnathidae has been reported by Crome (1954). He found that in *Eucta kaestneri* no scaffold spiral was made! This necessitated that the spider spend much more time than is usual in the laying down of the viscid spiral.

A third modification is that seen in the Theridiosomatidae. The snare of the ray spider, *Theridiosoma*, was first described by McCook, and although considered by Comstock to be primitive, I believe, following Wiehle, that it is instead an example of a highly specialized one. As Gertsch (1949) points out "It is first spun as a reasonably typical orb snare with a meshed hub and several spiral turns in the notched [= attachment] zone, then the hub and these threads are bitten out," so that the finished web has no hub or attachment zone. Then the radii are rearranged so that they radiate from, and converge

---

[6] See Bristowe (1938) for a thought provoking discussion.

upon, a point near the center. These rays, coming from four or five main divisions, join to form a trap line attached to some nearby twig. The spider stations itself on the rays at the center and, with its greatly thickened front legs, pulls the orb into the shape of a cone or funnel. The tension is released by the spider when an insect touches the snare, and the resulting spring action serves to ensnare the prey more firmly. Thus we see here still another example of the spring web, already discussed for *Hyptiotes, Miagrammopes,* and *Ulesanis.*

The fourth and final modification concerns the loss of the orb-making habit in certain genera of the subfamily Araneinae, and the substitution of other techniques for obtaining prey. The earliest accounts of the habits of the bolas throwing spiders are those for *Mastophora* in America by Hutchinson (1903), for *Dicrostichus* in Australia by Longman (1922), for *Cladomelea* in South Africa by Akerman (1923), and an excellent summarizing account is supplied by Gertsch (1947, 1955). These spiders sit on, or suspend themselves from, a twig, and hold a thread to the lower end of which is attached a sticky globule of silk which is flung at passing insects. In *Cladomelea* the bolas is held by a third or fourth leg and whirled rapidly in a horizontal plane. In the other two genera the bolas is held by a front leg and is not whirled. Just how this remarkable behavior evolved from ancestors that undoubtedly were orbweavers I cannot guess.

Another Australian spider, *Celaenia excavata,* is known to feed solely on night-flying moths, which it catches without the aid of either orb or bolas. From the account given in McKeown's book it would appear that while the adult spiders wait in ambush and seize the moths by a lightning-quick movement of the legs, the spiderlings seem to construct snares. Details on the snares were not supplied. And so again we see, as with *Zygiella* and *Pachygnatha,* the young revert to the habit of web building abandoned completely by their adults.

Just as we find a parallelism for the structures built by certain mygalomorphs and certain araneomorphs, and again between certain Cribellatae and certain Ecribellatae, so also do we find this for members much more closely related, within the same large family, e.g., the Argiopidae *(sens. lat.).*

REFERENCES

Akerman, C. 1923. A comparison of the habits of the South African spider, *Cladomelea,* with those of an Australian *Dicrostichus.* Ann. Natal Mus. 5:83-88.

———. 1926. On the spider, *Menneus camelus* Pocock, which constructs a moth catching expanding snare. Ann. Natal Mus. 5:411-422.

———. 1932. On the spider *Miagrammopes* sp. which constructs a single line snare. Ann. Natal. Mus. 7:137-143.

Apstein, C. 1889. Bau und Funktion der Spinndrüsen der Araneida. Arch. Naturgesch. 55:29-74.

Archer, A. F. 1941. The Argiopidae or orbweaving spiders of Alabama. Alabama Mus. Nat. Hist. Mus. Pap. [for 1940] #14, 77 p.

Balogh, I. J. 1934. Vorläufige Mitteiling über radnetzbauende Pachygnathen. Folia Zool. Hydrobiol. 6:94-96.

Benton, J. R. 1907. Strength and elasticity of spider threads. Am. J. Sci. (4) 24:75-78.

Braunitzer, G., and D. Wolff. 1955. Vergleichende chemische Untersuchungen über die Fibroine von *Bombyx mori* und *Nephila madegascariensis.* Z. Naturforsch. 10b:404-408.

Bristowe, W. S. 1930. Notes on the biology of spiders. I. The evolution of spiders' snares. Ann. Mag. Nat. Hist. (10) 6:334-342.

———. 1932. The Liphistiid spiders. Proc. Zool. Soc. London 2:1015-1045.

———. 1938. The classification of spiders. Proc. Zool. Soc. London (B) 108:285-322.

———. 1941. The comity of spiders. Vol. II. Ray Society, London. p. 230-261.

———. 1958. The world of spiders. Collins, London. 304 p.

Christiansen, A., R. Baum, and P. N. Witt. 1962. Changes in spider webs brought about by mescaline, psilocybin, and an increase in body weight. J. Pharmacol. Exptl. Therap. 136:31-37.

Cloudsley-Thompson, J. L. 1953. Notes on Arachnida, no. 19. On *Atypus affinis* Eichw. Entomol. Monthly Mag. 89:88-89.

Comstock, J. H. 1912. The evolution of the webs of spiders. Ann. Entomol. Soc. Am. 5:1-10.

———. 1912. The spider book. Doubleday, Page and Co., Garden City, New York.

Crome, W. 1954. Beschreibung, Morphologie und Lebensweise der *Eucta kaestneri* sp. n. Zool. Jb. Syst. 82:425-452.

———. 1957. Bau und Funktion des Spinnapparates und Analhügels, Ernährungsbiologie und allegemeine Bemerkungen zur Lebensweise von *Uroctea durandi* (Latreille). Zool. Jb. Syst. 85: 571-606.

Dabelow, S. 1958. Zur Biologie der Leimschleuderspinne *Scytodes thoracica* (Latr.). Zool. Jb. Syst. 86:85-162.

DeWilde, J. 1943. Some physical properties of the spinning threads of *Aranea diadema* L. Arch. néerl. Physiol. 28:118-131.

Exline, H. 1948. Morphology, habits and systematic position of *Allepeira lemniscata* (Walckenaer). Ann. Entomol. Soc. Am. 41:309-325.

Freisling, J. 1961. Netz und Netzbauinstinkte bei *Theridium saxatile* Koch. Z. Wiss. Zool. 165: 396-421.

Gerhardt, U., and A. Kaestner. 1938. Araneae. p. 497-656. *In* Kükenthal's Handbuch der Zoologie, Band III, Lieferung 12.

Gertsch, W. J. 1947. Spiders that lasso their prey. Nat. Hist. 56:152-159.

———. 1949. American Spiders. Von Nostrand, New York. 285 p.

———. 1955. The North American bolas spiders of the genera *Mastophora* and *Agatostichus*. Bull. Am. Mus. Nat. Hist. 106:225-254.

Grandjean, F. 1943. Le dévelopment postlarvaire d'*Anystis*. Mem. Paris Mus. Nat. Hist. 18:46-50.

———. 1948. Quelques caractères des Tétranyques. Bull. Paris Mus. Nat. Hist. (2) 20:517-519.

Holm, A. 1939. Beiträge zur Biologie der Theridiiden. Festr. Strand 5:56-67.

Hopfmann, W. 1935. Bau und Leistung des Spinnapparates einiger Netzspinnen. Jena Z. Naturwiss 70:65-112.

Hutchinson, C. E. 1903. A bolas-throwing spider. Scientific Am. 89:172.

Kaestner, A. 1956. Araneae. p. 552-587. *In* Lehrbuch der Speziellen Zoologie, Jena. Teil I, Lief. 3.

Kullmann, E. 1961. Ueber das bisher unbekannte Netz und das Werbeverhalten von *Drapetisca socialis* (Sundevall). Decheniana 114:99-104.

Lehmensick, R., and E. Kullmann. 1957. Ueber den Feinbau der Fäden einiger Spinnen (Vergleich des Aufbaues der Fangfäden cribellater und ecribellater Spinnen). Verh. Dtsch. Zool. Ges. [for 1956], p. 123-129.

Longman, H. A. 1922. The magnificent spider: *Dicrostichus magnificus* Rainbow. Notes on cocoon spinning and method of capturing prey. Proc. Roy. Soc. Queensland 33:91-98.

Marples, B. J. 1949. An unusual type of web constructed by a Samoan spider of the family Argiopidae. Trans. Roy. Soc. New Zealand 77:232-233.

———. 1955. A new type of web spun by spiders of the genus *Ulesanis* with the description of two new species. Proc. Zool. Soc. London 125: 751-760.

———. 1962. Notes on spiders of the family Uloboridae. Ann. Zool. New Zealand 4:1-11.

Marples, M. J., and B. J. Marples. 1937. Notes on the spiders *Hyptiotes paradoxus* and *Cyclosa*

conica. Proc. Zool. Soc. London, (A) 107:213-221.

Marson, J. E. 1947. Some observations on the variations in the camouflage devices used by *Cyclosa insulana* (Costa), an Asiatic spider, in its web. Proc. Zool. Soc. London (A) 117:598-605.

McCook, H. C. 1889. American Spiders and their Spinningwork. Vol. I. Philadelphia, 373 p.

McKeown, K. C. 1952. Australian Spiders. Angus and Robertson, Sydney. 274 p.

Millot, J. 1930. Glandes venimeuses at glandes séricigènes chez les Sicariides. Bull. Soc. Zool. France 55:150-175.

Millot, J., and P. Bourgin. 1942. Sur la biologie des *Stegodyphus* solitaires. Bull. Biol. France et Belgique 76:299-314.

Monterosso, B. 1927. Osservazioni preliminari sulla biologia del genere *Scytodes*. R. C. Accad. Lincei (6) 6:171-174.

Montgomery, T. H. 1903. Studies on the habits of spiders, particularly those of the mating period. Proc. Acad. Nat. Sci. Philadelphia 55:59-149.

Nielsen, E. 1932. The biology of spiders. Levin and Munksgaard, Copenhagen.

Norgaard, F. 1948. Bidrag til danske edderkoppers biologi. I. *Lithyphantes albomaculatus* (DeGeer). Flora og Fauna Kjobenhaven 54:1-14.

———. 1941. On the biology of *Eresus niger* Pet. Entomol. Medd. 22:150-179.

———. 1951. Notes on the biology of *Filistata insidiatrix* (Forsk.). Entomol. Medd. 26:170-184.

Peters, H. 1937. Studien am Netz der Kreuzspinne. Z. Morph. Ökol. Tiere. 33:128-150.

———. 1953. Beiträge zur vergleichenden Ethologie und Oekologie Tropischer Webespinnen. Z. Morph. Ökol. Tiere. 42:278-306.

———. 1955. Ueber den Spinnapparet von *Nephila madegascariensis*. Z. Natur. 10b:395-404.

Petrunkevitch, A. 1926. The value of instinct as a taxonomic character. Biol. Bull. 50:428-432.

Pocock, R. I. 1895. Some suggestions on the origin and evolution of web spinning in spiders. Nature [London] 51:417-420.

Richter, G. 1956. Untersuchungen über Struktur und Funktion der Klebefäden in den Fanggeweben ecribellater Radnetzspinnen. Naturwissenschaften 43:23-

Roberts, N. L. 1955. The Australian netting spider, *Deinopis subrufus*. Proc. Roy. Soc. New South Wales, [for 1953-54] p. 24-33.

Savory, T. H. 1928. The biology of spiders. Sidgwick and Jackson, London. p. 137-142.

———. 1934. Possible factors in the evolution of spiders. Ann. Mag. Nat. Hist. (10) 13:392-397.

———. 1952. The spider's web. Frederick Warne and Co., London. 154 p.

———. 1960. Spider webs. Scientific Am. 202 (4): 114-124.

Sekiguchi, K. 1952. On a new spinning gland found in geometric spiders, and its functions. Annotat. Zool. Japon. 25:394-399.

Szlep, R. 1961. Developmental changes in the web spinning instinct of Uloboridae: construction of the primary type web. Behaviour 17:60-70.

Theuer, B. 1954. Contributions to the life history of *Deinopis spinosus*. M.S. Thesis, Univ. of Florida.

Tilquin, A. 1942. La toile géométrique des araignées. Press Universitaires, Paris. 536 p.

Uyemura, T. 1957. Colour changes of two species of Japanese spiders. Acta Arachnol. 15:1-10.

Wiehle, H. 1927. Beiträge zur Kenntnis des Radnetzbaues der Epeiriden, Tetragnathiden und Uloboriden. Z. Morph. Ökol. Tiere 8:468-537.

———. 1928. Beiträge zur Biologie der Araneen, insbesondere zur Kenntnis des Radnetzbaues. Z. Morph. Ökol. Tiere 11:115-151.

———. 1929. Weitere Beiträge zur Biologie der Araneen, insbesondere zur Kenntnis des Radnetzbaues. Z. Morph. Ökol. Tiere 15:262-308.

———. 1931. Neue Beiträge zur Kenntnis des Fanggewebes der Spinnen aus den Familien Argiopidae, Uloboridae und Theridiidae. Z. Morph. Ökol. Tiere 22:349-400.

Witt, P. N. 1961. Web building. In Reinhold's encyclopedia of the biological sciences.

Witt, P. N., and R. Baum. 1960. Changes in orb webs of spiders during growth (*Araneus diadematus* Clerck and *Neoscona vertebrata* McCook). Behaviour 16:309-318.

Yaginuma, T. 1956. In the world of spiders. Asahi Photo Book, Osaka. 64 p.

ADDENDUM

The following figures are also of interest and are reproduced from B. J. Kaston, *How To Know the Spiders;* copyright © 1972 by William C. Brown Co., Dubuque, Iowa.

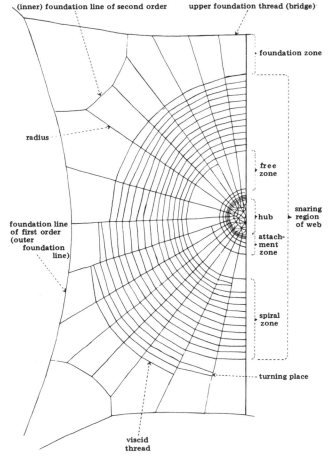

Diagram of the web of an orb-weaving spider.

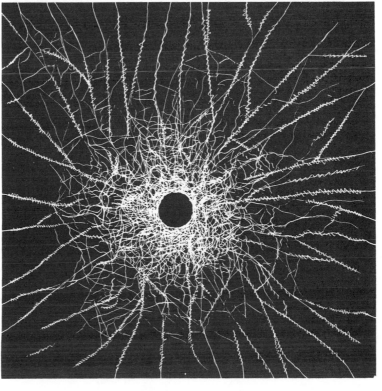

Metepeira labyrinthea snare.

Filistata hibernalis web.

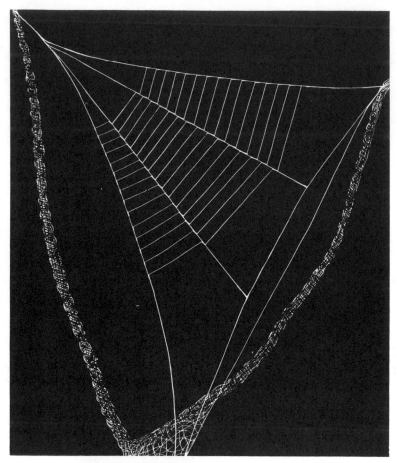

Hyptiotes snare.

Dictyna annulipes snare.

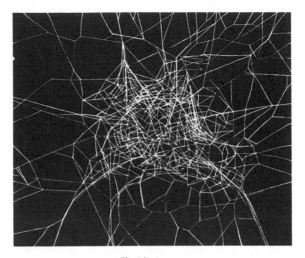

Theridion snare.

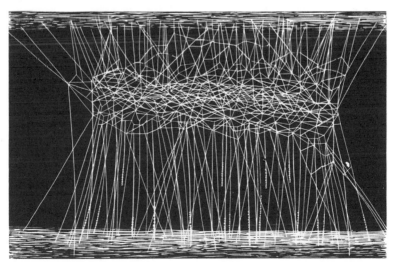

Steatoda snare.

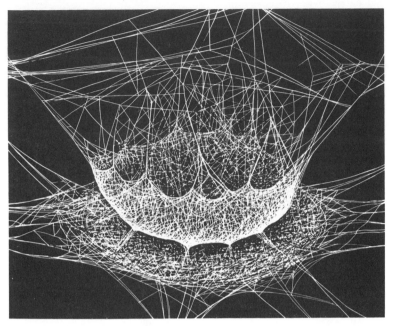

Frontinella pyramitela snare.

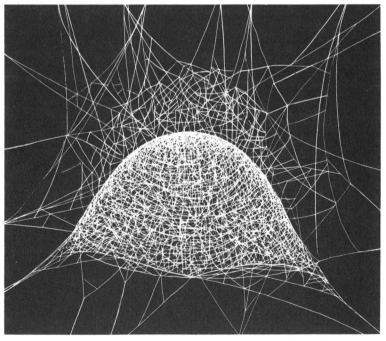

Prolinyphia marginata snare.

7

Reprinted from *Am. Zool.*, **4**(2), 209–220 (1964)

EVOLUTION OF CADDISWORM CASES AND NETS

HERBERT H. ROSS

Illinois Natural History Survey, Urbana

Probably every biologist, perhaps after being startled by a little bundle of sticks moving along the bottom of a pond, has been intrigued by caddisworms and their cases. Most of the adult caddisflies which develop from these aquatic larvae look like drab, small moths and seldom elicit much interest. The homes constructed by the caddisfly larvae, however, have been the subject of much study resulting in an extensive literature. This was reviewed in considerable detail by Betten (1934) and more recently by Botosaneanu (1963). Outstanding observations have been made by Nielsen (1948). The two major efforts to outline the evolutionary steps by which the various types of caddisworm homes arose arrived at opposite conclusions (Milne and Milne, 1939; Ross, 1956*b*). A comparison of these caddisworm constructions with findings concerning the cocoon-spinning behavior of moth larvae adds new perspectives to this evolutionary problem.

The two insect orders Trichoptera (caddisflies) and Lepidoptera (moths and butterflies) diverged from a common ancestor which must have differed from its daughter orders only in a number of small but predictable characteristics. The caddisflies have evolved into perhaps 10,000 species, few of them strikingly different from each other except in size, slenderness, and color pattern. The larvae of these same adults are also similar in a superficial way if looked at only as larvae.

ACKNOWLEDGMENTS

Figures 1, 4, 5, 7, 11, 12 and 13 are reproduced by permission of the Illinois Natural History Survey.

Figures 2, 3, and 6 are from Ross, *Evolution and classification of the mountain caddisflies*, by permission of the University of Illinois Press.

Figures 9 and 10 are reproduced by courtesy of the New York State Museum and Science Service. Other illustrations are reproduced by permission of the authors.

This research has been supported by a grant from the National Science Foundation.

The home-making habits of these caddisworms, however, are quite diverse and can be used as a basis for dividing the families comprising the order Trichoptera into five complexes. In one complex of about four families, the larva constructs a fixed retreat or net attached to the bottom of the stream or to some object in a stream or lake. The net is constructed like a fisherman's net and presumably strains foodstuffs from the flowing water or, in a lake, simply concentrates freely-moving plankton. The larva lives inside this net, eating material which has become trapped by it. In a second complex, comprising only the family Rhyacophilidae, the active larva is free-living; it spins only a ground line or anchor line of silk. In a third complex, comprising only the family Glossosomatidae, the larva makes a portable saddle-like case with a belly strap; the front and back parts of the case are open and the two ends of the larva protrude from the corresponding openings of the case. In the fourth complex, again comprising only one family, the Hydroptilidae, the larva makes a portable case, some shaped like a purse, others tubular. In the fifth complex, containing about twelve families, the larva makes a portable case, extending its thoracic legs out the front door and literally dragging its house along with it. Depending on the genus or species, these houses may be square, round, long or short, thick or thin, straight or coiled; they may be made of only silk or of various bits of material such as pieces of green aquatic leaves, bark, twigs, stones, or sand grains sewed by silken threads into the case. With many species, the case is highly diagnostic and almost unvarying; in other species, cases may have a high degree of individual variation in regard to their added materials.

The problem was to try to adduce how these different kinds of construction evolved; actually, because they result from actions of the larvae, to determine how the behavior patterns of the larvae changed so that the progeny of some common ancestor, producing only one kind of construction, gave rise to this variety of behavioral construction patterns.

MORPHOLOGICAL EVOLUTION

If we had *a priori* knowledge of the form and habits of this ancestral form, it would be simple to deduce various steps by which all the existing caddisworm behavior patterns evolved. Lacking this knowledge, the attempted reconstruction of the ancestral caddisfly and its larval

behavior pattern becomes the first pivotal point of investigation. Efforts to work out more than a highly arbitrary phylogenetic tree failed completely when only adult characters were used. Because of the relationship and similarities of the caddisflies with the moths, it was possible to postulate that the primitive caddisfly adult had three ocelli on its head, had certain sutures on the thorax, had wings with the full complement of the Comstock-Needham venation, and had two segmented claspers in the male genitalia. The difficulty is, every one of the five distinctive complexes outlined above has adults which fit this prototype almost perfectly

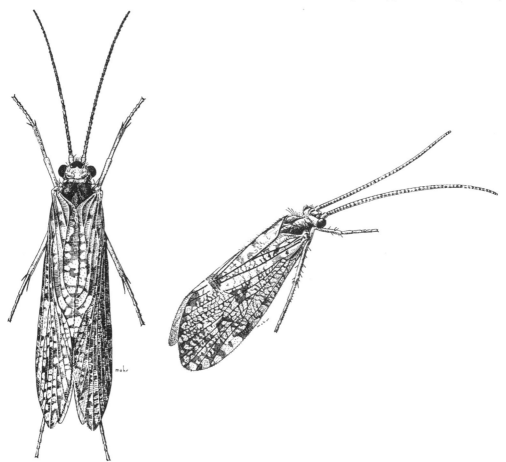

FIG. 1. Adult caddisflies of the genus *Rhyacophila* (the family Rhyacophilidae) and the genus *Banksiola* (the family Phryganeidae), primitive genera of the free living complex and the tube-case makers, respectively.

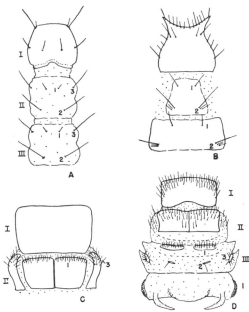

FIG. 2. Thorax of caddisfly larvae, dorsal aspect. A, *Rhyacophila*; B, *Anisocentropus*; C, *Pedomoecus*; D, *Sericostoma*. 1, 2, 3, primary dorsal setae or setal tufts.

(Fig. 1). This implies that, in the early caddisfly lineages, the different kinds of home-building behavior patterns evolved before the adults changed markedly.

The next logical step, therefore, was to examine the larvae for morphological evidence that might indicate direction of evolution. First it was noticed that in most members of the net makers and saddle-case makers, all trunk segments except the pronotum have the characteristic pattern of three pairs of hairs which are apparently homologous to the hairs designated as 1, 2, and 3 in the more primitive types of moth larvae (Fig. 2A). The homologs of these hairs are apparent in the tube-case makers, but in almost all instances they are highly modified as warts, plates, or clusters (Fig. 2B, C, D). It was next noticed that the anal legs of the larvae have a set of plates attached to the single hook at the end of each leg (Figs. 3, 4). In almost all groups, the hooks are directed downward, hanging below the level of the body. In the tube-case makers, on the other hand, the whole leg structure is shortened,

greatly consolidated, and the hooks project from about the mid-lateral axis of the body and extend primarily laterally (Figs. 3J and 5). An intermediate type was found in the purse-case makers. Here the sclerites are consolidated to a very large degree, although the hooks still projected downward. Another type somewhat intermediate occurs in the saddle-case makers, in which the legs also project downward. The net makers and the free-living types are all characterized by freely-movable legs having a freely-articulating hook capable of being brought down below the level of the body.

Because this latter is the condition and position of the anal hooks in the closely related primitive moth larvae and in forms thought to be ancestral to the caddisflies, including the dobsonflies, primitive sawfly larvae, and to some extent the primitive scorpionfly larvae, the net makers and the free-living caddisflies appear to be the two most primitive groups of the Trichoptera. Not only do their structures suggest this relationship, but unlike all the case makers, the larvae can move very rapidly both forwards and backwards (Fig. 4), as is true also of the moth larvae, dobsonfly larvae, and others. Progressions of leg morphology suggest the following evolutionary development of other members of the order: (1) from either the net makers or the free-living caddisworms arose the saddle-case makers; (2) from the saddle-case makers arose the purse-case makers; and (3) from the purse-case makers arose the tube-case makers. The three kinds of case makers, saddle, purse, and tube, exhibit progressive stages of consolidation of the anal legs and hooks associated with their use primarily to anchor the hind end of the body firmly into the case. This has culminated in an extremely efficient arrangement in the tube-case makers, with the two anal legs directed laterally and anchoring the insect in the case at the opposing mid-lateral points (Fig. 5). Small structural details of the adults strongly suggest that the case makers arose from the ancestor of the free-living caddisflies, the family Rhyacophilidae, rather than from the net makers, as indicated in the family tree (Fig. 6).

EVOLUTION OF BEHAVIOR PATTERNS

To support this family tree, up to this point based chiefly on morphology, there is needed a cohesive hypothesis of logical changes in behavior which would presumably have gone hand-in-hand with the structural changes. Experimental evidence in the Lepidoptera, the closest ally of the caddisfly, provided a foundation for such

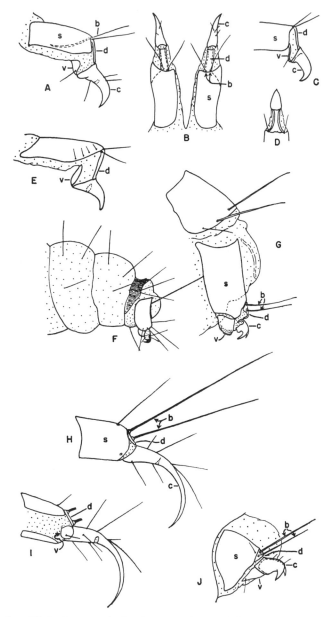

FIG. 3. Anal legs of caddisfly larvae. *A, B, Atopsyche*, a free-living larva; *C, D, Sortosa*, and *E, Stenopsyche*, both net makers; *F, Agapetus* and *G, Glossosoma*, saddle-case makers; *H, I, Agraylea*, a purse-case maker; *J, Lepidostoma*, a tube-case maker. All lateral aspect except *B* and *D*, which are the dorsal aspect, and *I* which is the mesal aspect. Abbreviations: *b*, basal tuft; *c*, anal claw; *d*, dorsal plate; *s*, lateral sclerite; *v*, ventral sole plate.

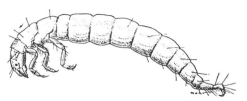

FIG. 4. Larva of the net making caddisfly *Polycentropus*.

a hypothesis. By combining information from several sources, especially the papers of Van der Kloot and Williams (1953a, b; 1954), one can outline a basic set of behavior activities followed by moth larvae in spinning larval tunnels and the cocoons in which they pupate. These activities follow this sequence of steps (Table 1):

A. An open-ended tunnel is spun by the larvae prior to each molt.

B. The larva feeds and completes its growth.

C. After feeding is completed, a wandering period occurs immediately prior to cocoon formation.

D. After the larva chooses the site of cocoon making, it spins a few anchor lines of silk which outline roughly the dimensions of the future cocoon.

E. Next, the larva spins a stout envelope

called the outer cocoon. This is normally a tough silken structure.

F. After the completion of the outer cocoon, and without a rest period, the larva spins a loose spongy intermediate or middle layer.

G. Finally, and again without a rest period, the larva spins a closely woven fine inner cocoon, in which it pupates.

The net-making caddisflies follow this

TABLE 1. *Sequence of behavioral steps in larval constructions of Lepidoptera and Trichoptera.*

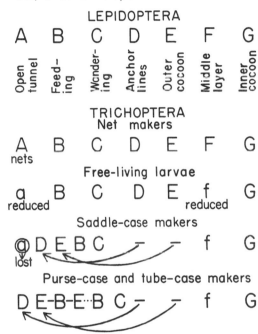

basic lepidopterous sequence of behavioral steps almost perfectly. The "fishing nets" of the active caddisworms are modifications of an open-end tunnel in which the larva can move backwards and forwards with great rapidity (representing step *A*). Figure 7 illustrates a simple type; Figure 8 a more complex type. The larva feeds and completes its growth (step *B*) in these nets. After the larva has completed its growth, it usually leaves the net and has a period of wandering (step *C*). It decides on a site for its cocoon, then starts to construct a typical lepidopterous cocoon, beginning

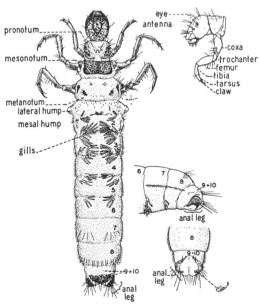

FIG. 5. Larva of the tube-case maker *Limnephilus*.

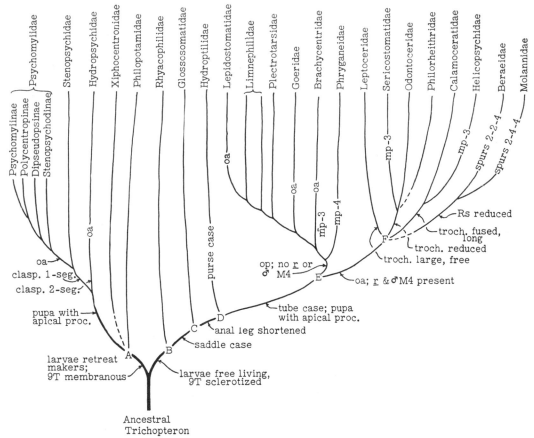

FIG. 6. Phylogenetic diagram of the Trichoptera. *mp,* maxillary palpi of male; *oa,* ocelli absent; *op,* ocelli present.

with a step of defining anchor lines (step D). The outer cocoon (step *E*) is extremely tough; the larva often incorporates materials such as sand grains, small stones, and bits of debris in it (Fig. 9). It is probable that step *F* then occurs, consisting of poorly organized threads. Within this structure the larva spins the inner cocoon (step *G*) which is extremely fine but also tough, and on all sides is united quite intimately with the outer cocoon by what seems to be the extensive intermediate layer corresponding to the products of step *F*. Once cocoon making has started, the whole operation apparently proceeds with no appreciable rest periods. In other words, steps *C* through *G* occur as an uninterrupted set of behavioral units.

In the free-living forms, the Rhyaco-

philidae, no net is made, but the larva does leave a silken strand line on the substrate as it moves about. This would indicate that step *A* is lost or greatly reduced, hence designated as step *a*. Cocoon formation is also atypical (Fig. 10); it is preceded by a period of wandering (step *C*), and

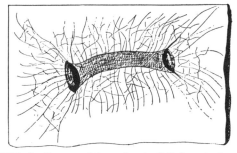

FIG. 7. Open-end tunnel net of *Polycentropus.*

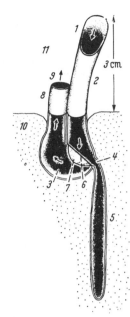

FIG. 8. Highly modified tube and net of *Macronema*. Arrows indicate direction of water flow; 7 is the net; 5 is the area occupied by the larva. (After Sattler.)

the choice of a site and spinning the anchor lines for the cocoon (step *D*). The outer cocoon (step *E*) is constructed primarily of stones, forming a hard dome shaped like a tortoise shell, glued around the edges to the substrate. Step *F* is extremely reduced. The inner cocoon, made in step *G*, forms a tough elliptical sheath which is completely free from either the substrate or the cocoon itself except for a few anchor lines at each end or at the

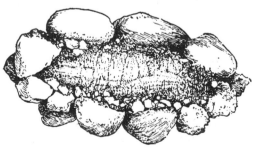

FIG. 9. Attached side of cocoon of *Hydropsyche*. The cocoon has been cut from the substrate; the outer row of stones represents the outer cocoon, the central webby area the thin layer of outer cocoon woven of the substrate; inside of this is the intermediate layer and the inner cocoon, not here visible. (After Betten.)

sides. These few anchor lines are all that remains of step *F*, henceforth designated as step *f*.

From the Rhyacophilidae line there apparently arose the first step in case making, represented by the saddle-case makers, the

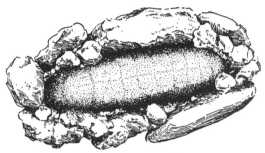

FIG. 10. Attached side of cocoon of *Rhyacophila*. This has also been cut from the substrate; the stones on the outside are the outer cocoon, the elliptic structure inside it is the inner cocoon; the intermediate layer is represented by only a few threads by which the inner cocoon is attached to the outer cocoon. (After Betten.)

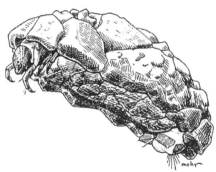

FIG. 11. Saddle case of *Glossosoma* showing the broad belly strap and the two ends of the larva protruding on each side of it.

Glossosomatidae. In this family, the larva constructs a tortoise-like case of stones with a central strap also of stones built across the under side (Fig. 11). The head and legs of the larva protrude from one side of the saddle and the end of the abdomen may protrude from the other. These larvae are extremely sluggish in their movements. When full grown, the larva cuts out the central strap, glues the dome-like part of the case to the substrate, and forms within it an inner cocoon exactly like that of the Rhyacophilidae. This larva also has

179

dispensed with step *A*, but has moved steps *D* and *E* forward to a point in the ontogeny preceding step *B*. Because step *A* has been lost, steps *D* and *E* essentially now replace step *A*. At various intervals in growth (step B), the case is enlarged, but little is known about this activity in the saddle-case makers. After the completion of growth. steps *C*, *f*, and *G* follow as in the Rhyacophilidae, with one important difference. In the Rhyacophilidae, step *f* consists in spinning anchor lines attached to the outer cocoon, these lines to end up as holdfasts keeping the inner cocoon in place within the outer cocoon. In the Glossosomatidae, the first strands of step *f* extend from the outer cocoon to the substrate, in other words, now anchor the outer cocoon securely to the substrate before the larva constructs the inner cocoon within the outer cocoon.

The next evolutionary advance in case making seems to be exemplified by the purse-case makers, the family Hydroptilidae. The circumstances in this family are peculiar. First, the species are exceedingly small, the adults attaining a length ranging only from 2 to 5 mm. The young larvae are extremely minute for the first four instars, climbing around and feeding on filamentous algae. During this period no net or case or any discernible spinning activity occurs. Thus step *A* has been lost completely. At the beginning of the fifth instar, the still minute larva constructs a case (steps *D* and *E*). Practically all of the growth of the entire larval period occurs in this one last instar after the case is made.

In primitive genera of this family, the larva makes a case shaped like a purse and usually having sand grains and bits of debris incorporated in it (Fig. 12). The

FIG. 13. Tube case of *Limnephilus*, a member of the family Limnephilidae.

two sides of the case are sewed together in a tight seam ventrally and again the front and rear of the larva project from the case. As the larva grows, it splits apart the two sides of the case on the ventral edge, enlarges the sides, then re-unites the ventral edges. In other genera the case is tubular and the larva, as it increases in size, simply spins additional material around the anterior edge. After the larva attains its full growth, it seeks a sheltered spot (step *C*), anchors the case firmly fore and aft to the substrate (step *f*), seals all the edges and spins an inner cocoon inside it (step *G*), then pupates. The sequence of steps in this family (Table 1) is exactly like that in the saddle-case makers except that no spinning activity occurs in the first four instars.

The final stage in case-making behavior is reached in the tube-making caddisflies (Fig. 13). Their homes include the great variety of cases with which limnologists are familiar. The young larvae make cases which are miniatures of those made by larger larvae, and the case is enlarged as the larva grows. The case is essentially tubular, sealed with a mesh at the posterior end, the head and thorax of the larva protruding from the anterior end. When the larva is full grown it fastens the case firmly to the substrate only at the anterior end, which it then seals with another sieve, spins an inner cocoon within it, and pupates. This behavior pattern follows the same series of steps as that of the saddle-case makers.

The evolution of the home-making behavior in Trichoptera therefore can be explained harmoniously with the morphological indications of phylogeny. The mutual support of these two sets of evidence increases greatly the probability of the phylogenetic hypothesis advanced.

FIG. 12. Purse case of *Ochrotrichia*, a member of the Hydroptilidae.

In summary, the primitive caddisflies had a behavioral pattern for making nets and cocoons comprising the same basic steps as those inherited from their Trichoptera-Lepidoptera progenitor (steps *A* to *G*). In one phylogenetic line of net makers, this behavior pattern was changed through a great reduction of step *A* and the reduction of step *F*. This resulted in a form like the free-living Rhyacophilidae. From this line with the changed behavior evolved various types of case-making caddisworms. The evolution of these case makers, stripped to its essence, may be summarized as the result of a few simple changes in the sequence of behavioral steps; step *A*, already peculiarly modified, was lost; steps *D* and *E* moved forward in the life history and became temporally removed from the other steps *C*, *F*, and *G* of cocoon making. There has also been a peculiar integration of steps *E* and *B*, in that activity in *B* (growth) results intermittently in renewed activity in *E* (enlarging the case). Surviving groups have preserved what appear to to be intermediate steps in this evolutionary progression.

Within three of the major groups of Trichoptera that make nets or cases, many types of home construction evolved. The net makers produce a total of at least ten different shapes and designs of nets. The purse-case makers construct at least half a dozen quite different-looking cases, some of them actually tubular. The tube-case makers collectively produce the greatest variety of homes, including round, elliptical, triangular, and square in cross-section; long and slender versus short and stout; straight, curved, or coiled, with or without flanges; and made of a great variety of sand grains, small stones, leaf and twig fragments, small snail shells, in fact almost anything available in the environment. Within each major group, however, these different nets and cases represent only variations on the common underlying behavior pattern with which we have been dealing. The various case makers appear to have arisen from only one line of net makers, the one line in which the primeval be-

havior pattern became disturbed. It seems that once the old pattern was broken, new possibilities arose for the natural selection of additional types. This is reminiscent of Schmalhausen's (1949) assertion that all mutations increase the rate of occurrence of mutations, but perhaps this should be paraphrased to "All changes that persist open the door for other changes."

The selection pressures involved in the origin of the case-making line are readily postulated. The protection and mobility afforded to the larva by portable cases would presumably allow the larva to crop diatoms, its chief food, from the surface of an extensive but exposed area in the stream or lake. This would free the larva from dependence on the catch of a stationary net. Once the larva had become established as a member of the exposed-site community, it and its case-making behavior would have come under strong selection pressures that would favor changes beneficial in the new environment. This would account for the rapid evolutionary expansion of the case makers.

HOMOLOGY IN BEHAVIORAL STEPS

In reviewing details of case-making behavior, certain units of different steps appear to have a common basis. The first of these is the formation of anchor lines (step *D*) prior to spinning the outer cocoon. The earliest steps made by the net maker *Hydropsyche* observed by Davis (1934) are essentially the construction of anchor lines attached to the solid substrate, preceding the weaving of a highly oriented net. In the Hydroptilidae (Nielsen, 1948), the larva first attaches anchor lines between stationary objects in the substrate prior to case formation. The tube-case maker *Molanna* observed by Copeland and Crowell (1937) must certainly begin with a few anchor lines attached to the substrate; in other genera observed by these authors the corresponding line is attached to movable pieces of the substrate gathered to itself by the larva. Anchor line formation would therefore seem to be a common initial denominator of all these home-making patterns. The continuous strand line of the Rhyacophilidae is probably a peculiar extension in time of the more typical anchor lines preceding home making. In the saddle-case makers, observations have not yet been made on this point.

A second common denominator in home making is the construction of the girdle. This structure, studied especially by Copeland and Crowell (1937), is the first ring that forms the foundation of the case. In the Hydroptilidae (Fig. 14), this ring is formed within the anchor lines attached to the stationary substrate; after this the anchor lines are broken, the ring freed, and the larva commences to build the case on it. In the tube-case makers (except *Molanna* and possibly others) the larva follows anchor-line formation by weaving a linear belt that the larva arranges around itself, then connects the free ends of the belt, and thus forms the girdle. On this point also no observations are available for the Glossosomatidae, but certain actions of some of the net makers might be construed as girdle formation. This girdle is often of a different material and texture from the case built on it, and in these cases is cut off the finished case by the larva. The formation of the girdle may be another step in the early stages of home construction common to most of the caddisflies.

Because of these early common denominators and the similarity of construction in other points discussed earlier in this paper, there is reason to believe that steps *A, E,* and *G* each represent the same basic pattern of spinning, but that in each of these steps the pattern has been modified by other controls.

Judged by the size of sand grains and pebbles used in their construction, there is evidence that each instar of the saddle-case makers constructs a new case. Similar evidence gleaned from collecting observations indicates that each instar of some of the net makers may construct a new net. Detailed observations on these two points would be of great interest. If these cursory observations prove correct, it might indicate that the primeval caddisfly construc-

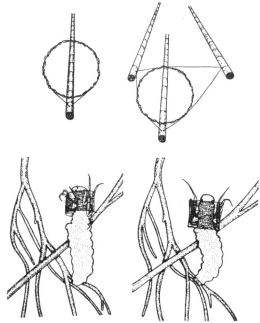

FIG. 14. Girdle and early case formation in *Orthotrichia,* a member of the Hydroptilidae. *Above,* circular girdle attached by anchor lines to three filaments of algae; *lower left,* larva starting case, its girdle still attached to alga by one thread; *lower right,* case making continuing, now free from alga. (After Nielsen.)

tion pattern embodied this basic home-making behavior pattern following each molt. This would introduce the problem that either the last larval instar of the net makers and the free-living forms (that spin both an outer and inner cocoon) would seem to have a double dose of this pattern in the one instar, or the earlier instars had only half a dose.

HORMONAL INFLUENCE ON BEHAVIOR

How did these changes of behavior come about? It has been well established (Schneiderman and Gilbert, 1964) that three glandular areas—the brain, the prothoracic gland, and the corpora allata—produce the hormones which stop and start molting and affect several other activities. The following examples cited from Van der Kloot and Williams (1954) illustrate the kinds of influences on record. The secretion of the corpora allata (the juvenile hormone) prevents cocoon spinning. When the corpora allata is removed from an early stage caterpillar, the latter spins a precocious cocoon which is normal in architecture but reduced in size corresponding to the dimensions of the young larva. It has been shown also that removal of silk glands removes the stimulus for wandering (step C). The normal change in the behavior pattern from steps E to F to G appears to be linked with the expenditure of silk by the larva, indicating that some quantitative or qualitative chemical reaction brings about a switching from one behavior pattern to another. Experimental manipulations of these organs have caused many aberrations in the behavior patterns associated with wandering and cocoon formation. Certain of these aberrations can be duplicated by the action of drugs, especially atropine and malononitrile. In certain experiments, carbon monoxide, whose principal target is cytochrome oxidase, eliminated cocoon construction temporarily, whereas exposure to carbon dioxide eventually eliminated it permanently.

The evolution of case making in the caddisflies would seem to involve the elimination, re-arrangement, or separation of inherited units in a basic synchronized set of spinning or weaving behavior patterns. The experimental evidence (of which the above examples are only a few citations) indicates conclusively that, whatever are the bases of the units of behavior we are discussing, their initiation and timing are controlled by the presence of specific chemical compounds arising from several parts of the body. Because the basic behavioral units seem to be the same throughout the entire phylogeny, and only their ontogenetic timing is different, this evolutionary sequence probably has as its basis inherited changes in the time and quantity of production of a relatively small number of chemical compounds.

TIME AND BEHAVIOR

The foregoing interpretation presents an explanation of the sequence in which behavioral units were juggled in the Trichoptera, and a possible causal mechanism. This does not answer the question to to when and how the units of the behavior pattern originally arose. Because the primeval pattern occurs in both the Lepidoptera and Trichoptera, their ancestor logically possessed it also. Members of the Hymenoptera have a superficially similar cocoon-making pattern, and members of the Coleoptera and Neuroptera also have a cocoon-making pattern, although they use anal silk instead of salivary silk. Cocoon-making occurs in the Thysanoptera and some sort of silken nets or webs are made by members of relatively primitive winged insects such as the Embioptera.

The ultimate root of the home-making behavior of the Trichoptera may therefore trace back through various ancestral forms to an early point in the evolution of winged insects. The time of occurrence of a common ancestor to the orders listed would be in the Palezoic era, at least over 300 million years ago (Ross, 1956a). The time of occurrence of the common ancestor of the Lepidoptera and Trichoptera was probably in the Triassic, perhaps 200 million years ago (Ross, 1956b). Subtracting these times, the various steps in the ancestral behavior pattern of the Trichoptera probably arose one by one over a period of some 100 or 150 million years.

Phylogenetic and biogeographic evidence indicates that the three major case-making branches of the Trichoptera had evolved by early Cretaceous, about 135 million years ago. This means that the evolution of the three types of case makers from the net makers, through the avenue of juggling steps in the behavior pattern associated with homemaking, occurred between early Triassic and the Cretaceous, a span of less than 100 million years.

Adding these time indications together, our most probable estimates are that the Trichoptera-Lepidoptera home-building pattern evolved over a span of 100 to 150 million years. This pattern was re-arranged into the major types of home-building in the Trichoptera over a span of about 100

million years, and the various modifications now known within each major caddisfly type have evolved in about the last 150
million years.

REFERENCES

Betten, Cornelius. 1934. The caddisflies or Trichoptera of New York State. Bull. N. Y. State Mus.
292:1-576.

Botosaneanu, L. 1963. Insecte . . . arhitecti si con-
structori . . . sub apa. A. Sattinger, Bucresti.

Copeland, Manton, and Sears Crowell. 1937. Observations and experiments on the case-building
instincts of two species of Trichoptera. Psyche 44:
125-131.

Davis, Marion B. 1934. Habits of the Trichoptera.
Bull. N. Y. State Mus. 292:82-106.

Milne, Margery J., and L. J. Milne. 1939. Evolutionary trends in caddisworm case construction.
Ann. Entomol. Soc. Am. 32:533-542.

Nielsen, Anker. 1948. Postembryonic development
and biology of the Hydroptilidae. Det Kgl.
Danske Vidensk. Selskab, Biol. Skrifter 5:1-200.

Ross, Herbert H. 1956a. A textbook of entomology,
2nd Ed. New York, John Wiley & Sons, Inc.

———. 1956b. Evolution and classification of the
mountain caddisflies. Univ. Ill. Press, Urbana.

Schmalhausen, I. I. 1949. Factors of evolution.
Blakiston Co., Philadelphia.

Schneiderman, H. A., and L. I. Gilbert. 1964. Control of growth and development in insects.
Science 143:325-333.

Van der Kloot, William G., and Carroll M. Williams. 1953a. Cocoon construction by the Cecropia silkworm: I. The role of the external environment. Behaviour 5:141-157.

———. 1953b. Cocoon construction by the Cecropia silkworm: II. The role of the internal environment. Behaviour 5:157-174.

———. 1954. Cocoon construction by the Cecropia
silkworm: III. The alteration of spinning behaviour by chemical and surgical techniques.
Behaviour 6:233-255.

8

Reprinted from *Am. Zool.*, **4**(2), 221–225 (1964)

APTICOTERMES NESTS

ROBERT S. SCHMIDT

Department of Surgery (Otolaryngology), University of Chicago

Nests of the approximately 200 known genera and 2000 known species of termites vary tremendously in size, location, complexity, and material of construction. Among these, the nests of *Apicotermes* are unique for the purpose of this symposium. This paper will, therefore, be restricted to a discussion of the evolution of the nests of this genus.

Apicotermes has been found only in the rainforests and savannas of Africa south of the Sahara Desert. The nests are entirely subterranean, with no surface indication of their presence. Their shape resembles that of an egg, football, or pear, and they vary in height from 4 to 42 centimeters. Their depth under the soil may be as great as 60 centimeters. These nests seem to be constructed largely of the excrement of the termites themselves.

In contrast to the constructions of some animals, the nests of *Apicotermes* are remarkably species-specific. There is little intraspecific variation, but considerable interspecific variation. In other words, each species builds a very characteristic type of nest that is distinct from the nests of even closely related species. The nests are complex, thereby providing an unusually large number of characteristics for comparison. The durability and small size of these structures greatly facilitate study. In fact, the nests of *Apicotermes* provide much better material for phylogenetic studies of this genus than do the termites themselves.

The fact that these nests can be studied by anatomical methods and described by anatomical terms should not obscure the fact that they are the products of animal behavior. Lorenz has referred to these structures as "frozen behavior."

A termite nest is not the work of a single animal, but is built by a large number of individuals of the worker caste. Since the workers are sterile and cannot pass their genes to the next generation, the evolu-

tionary significance of nest-building behavior must be in its selective value for the entire colony and, thereby, for the reproductive caste.

The fact that termite nest-building behavior is innate is shown by the method of colony founding. A new colony is started by two termites, a king and a queen, that pair and then seal themselves within a small cavity. The queen eventually lays eggs that produce workers. These workers then build a species-specific nest, although they have never experienced such a nest or previous generations of workers.

Figure 1 should provide sufficient nest illustration for the purposes of this paper. More detailed descriptions and illustrations can be found in the publications of Desneux (1952, 1956), Grassé and Noirot (1954), and Schmidt (1955a, 1955b, 1958, 1960). Emerson (1952, 1956a, 1956b) describes the termites themselves and their phylogenetic significance. Grassé and Noirot (1954) also consider the nests of some closely related genera. For discussions of termite nests in general, the reader is referred to Emerson (1938, 1956c), Grassé (1949), and Michener (1951).

NEST EVOLUTION

While reading this section, frequent reference should be made to Figure 1.

Apicotermes probably arose in the Oligocene or Miocene rainforests of central Africa (Schmidt, 1958). Initially, the nests of *Apicotermes*, or its immediate ancestors, were merely individual cells connected by tubes (Grassé and Noirot, 1954). The first major change was a coalescence or multiplication of cells to form multicellular structures of variable shape and size. At about this stage, the branch leading to *A. tragardhi* diverged.

The next step was the origin of crude, scattered perforations through the nest walls, and of a thick, friable shagreen network (a layer of sand grains cemented to the outside of the nest). The nests also became more symmetrical in shape, although their internal structure remained rather crude and cellular. The ancestors of *A. rimulifex* probably evolved at about this time.

The cellular internal structure now evolved into the more efficient storied arrangement found in all remaining nests. In *A. arquieri* and *A. occultus* these stories are separated by curved, parallel lamellae. At the center of the nest these lamellae fuse to form a complex helicoidal column

that essentially serves as a spiral staircase effecting communication between stories. In the nest of *A. lamani* and of all nests of the *porifex* branch, the stories are separated by straight, horizontal lamellae. Here, communication between stories is assured by series of small pillars with holes at their bases, called "direct ramps" by Desneux.

All of the nests with a storied internal construction also have, in addition to the shagreen network, another envelope around the outside of the nest. In the case of the *occultus* branch, nests are oriented within

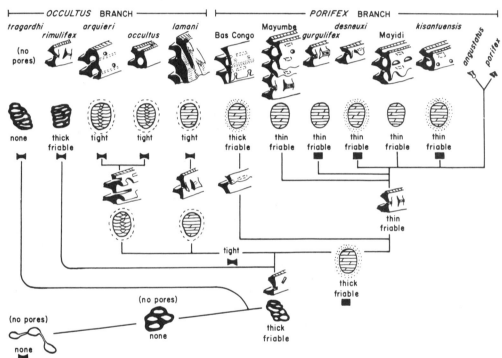

FIG. 1. Phylogeny of *Apicotermes* based mainly on nests (partly modified after Schmidt, 1955b). The top row of figures illustrates contemporary species, lower figures show various hypothetical ancestral stages. The uppermost figure of each series of drawings of a contemporary species represents a vertical section through the nest wall, with the outside of the nest facing toward the right. No attempt has been made to show the shagreen network. Below this is a schematic representation of the internal organization of the nest, i.e. cellular, storied with a central helicoidal column, or storied with direct ramps. When known, the type of orientation within the soil is shown. A broken line around a nest indicates orientation within a space, stippling around the outside of a nest indicates orientation within an envelope of loose sand. Next is indicated the type of shagreen network, i.e. "none," "thick friable," "thin friable," or "tight." Finally, the approximate shape of the soldier postmentum is illustrated. In the case of *A. angustatus* and *A. porifex*, only the shape of the conduits between the nest interior and circular galleries are shown, since this is the only important characteristic by which they differ from *A. kisantuensis*. Bas Congo, Mayumbe, and Mayidi are locality designations, since the termites constructing these nests have not yet been found.

a space. In the case of the *porifex* branch, they are oriented within an envelope of loose sand. Changes, apparently associated with these (see below), also took place in the shagreen 'network. Unfortunately, the type of orientation of the more primitive nests is not known.

The remaining feature of particular interest is the addition of some type of chamber external to the original wall perforations. Such a structure has evolved five times. In four cases it is in the form of circular galleries running horizontally within the nest wall; in the fifth case the chamber is in the form of a funnel (*A. lamani*).

The circular galleries of the *porifex* branch are best understood because of two factors. First, there is a fairly complete phylogenetic series showing the stages of gallery evolution; and second, several nests show what seem to be different stages of gallery construction within the same nest. The Mayumbe nests have pores opening through "gargoyles" at their lower stories. Most of the upper stories, however, show wall structures including complete circular galleries. Between these levels are several intermediate stages suggesting that the termites may construct the galleries by extending the upper lips of the gargoyles outward and downward until they fuse with the nest wall to enclose galleries. In the Mayumbe nest, the tips of these extended upper lips fail to fuse with the wall so that the external perforations are left opposite the internal perforations. The Mayidi nest (only a single nest and no termites known) also seems to show similar intermediate stages of gallery construction. In this case, however, the external perforations are left between the extended upper lips rather than at their tips, so that the external and internal perforations are alternate. The different perforation arrangements in the Mayumbe and Mayidi nests suggest a convergent, or at least a parallel, origin of their galleries. The fact that these stages (i.e., gargoyles, extended lips, and complete galleries) can also be found in different species of the *porifex* branch suggests that

the stages of gallery phylogeny have been similar to the stages of gallery ontogeny— that is, a type of recapitulation. Although this is admittedly an unconventional use of the term recapitulation, it is convenient and descriptive. The Bas Congo nests have almost completely enclosed galleries that seem to represent still a third independent origin of these structures.

In *A. arquieri* the circular galleries differ in at least two obvious respects from those of the *porifex* branch. The gallery is above rather than below the story with which it is connected, and the internal perforation is a continuous slit rather than a series of discrete slits. This seems, therefore, to be a fourth convergent origin of galleries. The stages of gallery evolution in this species are completely unknown. The ancestral stage illustrated is purely speculative. In the closely related species, *A. occultus*, the wall structure is similar, except for the complete absence of perforations. The presence of perforations in *A. arquieri* and the secondary addition of galleries to previously existing perforations in the *porifex* branch suggest that this is a case of regression rather than a primitive condition.

In *A. lamani*, the basic perforation system has been elaborated by the extension of the perforations into downward and outward slanting funnels. Again, there is no direct evidence on the phylogeny of these funnels.

Emerson (1952, 1956a, 1956b) finds that such information as is available from the termites themselves is consistent with the phylogeny as presented here on the basis of nests. The shape of the soldier postmentum is especially valuable in confirming the division of the genus into its two main branches.

NEST FUNCTIONS

The nests of *Apicotermes* probably have many functions in common with most termite nests (Emerson, 1938, 1956c). Of particular interest to us here are the perforation systems and their various accessories.

The wall perforations of *Apicotermes*

nests are too small to be used as passage-ways by the termites. Therefore, it seems likely that these openings serve as a type of ventilation system facilitating gas exchange between the inside and outside of the nest. Two lines of evidence are at least consistent with this hypothesis.

The known range of the genus is subjected to extended periods of heavy rainfall (Schmidt, 1960). Due to the frequent soil saturation, gas exchange difficulties would be expected. This would provide a selective advantage to some type of ventilation system.

The second line of evidence derives from the work of Brown and Escombe (1900) on diffusion through small pores. These authors found that at pore spacings of less than 7-10 diameters, the diffusion shells (clouds of diffusing molecules) of adjacent pores overlapped, thereby reducing the amount of diffusion per pore. However, at greater spacings there was little or no interference, so that each pore performed at its maximum diffusion efficiency. It is interesting that both the horizontal and vertical spacings of pores in the different species of *Apicotermes,* as well as in nests of different sizes, at least approximate this 7-10 diameter interval. This is consistent with a gas exchange function (Schmidt, 1960). Brown and Escombe (1900) also found such a spacing of the stomata of Catalpa leaves.

As noted above, the nest of *A. occultus* seems to have lost wall perforations secondarily. This species is found in an area that also has an extended dry season. It seems likely that the regression of pores in this species is an adaptation for maintenance of the high nest humidity so essential to most termites.

It was indicated above that some type of chamber, i.e. circular gallery or funnel, was independently added to the perforation system in five cases. There is a complete absence of evidence concerning the functions of these structures, although it has generally been assumed that in some way they protect the perforations from entrance of undesirable factors. The five-fold converg-

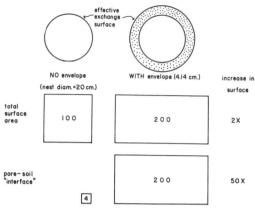

FIG. 2. Increase in effective diffusion surface due to presence of envelope. If one assumes the envelope-soil interface to be the effective exchange surface, the addition of a 4 cm envelope to a 20 cm nest will increase the total surface area by a factor of two. In a nest with perforations, however, the total area of the pores may be more pertinent. In a nest such as that of *A. angustatus*, approximately 4% of the surface is in the form of pores (Schmidt, 1955a). In the presence of the 4 cm envelope, the 4 units of pore area might be considered to be in contact with the 200 units of envelope-soil interface. This would increase the exchange surface by a factor of fifty. For the purposes of this illustration, the envelope is considered to be a space. The situation would probably be similar, however, for an envelope of loose sand or a shagreen network. The size of the nest and the thickness of the envelope (in the case of sand envelope or space) are typical. For convenience of calculation, this nest is considered to be spherical.

ent origin of such structures testifies to their high selective advantage, whatever this may be.

In all nests with perforations, one or two types of envelopes (shagreen network, envelope of loose sand, space) has been added to the outside of the nest. These probably have numerous functions—prevention of pore clogging by the soil, facilitation of drainage around the pores, and increase of effective diffusion surface (Fig. 2).

It seems likely that an envelope of loose sand and a space perform essentially the same functions as the primitive, thick, friable shagreen network—and perhaps much more effectively. It is not surprising, therefore, that all but one member of the *porifex* branch retain only a very thin, friable shagreen network. The three members of the *occultus* branch known to build nests in a space have only a thin, tight shagreen network. Therefore, it appears that the shagreen network has largely regressed in most of the more recently evolved forms.

REFERENCES

Brown, H. T., and F. Escombe. 1900. Static diffusion of gases and liquids in relation to the assimilation of carbon and translocation in plants. Phil. Trans. Roy. Soc. (B) 193:223-291.

Desneux, J. 1952. Les constructions hypogés des *Apicotermes*, termites de l'Afrique tropicale. Ann. Mus. Roy. Congo belge, sér. 8°. 17:1-98.

———. 1956. Le nid d'*Apicotermes rimulifex* Emerson. Rev. Zool. Bot. Afric. 53:92-97.

Emerson, A. E. 1938. Termite nests, a study of the phylogeny of behavior. Ecol. Monogr. 8:247-284.

———. 1952. The African genus *Apicotermes* (Isoptera; Termitidae). Ann. Mus. Roy. Congo belge, sér. 8°. 17:99-121.

———. 1956a. Ethospecies, ethotypes, and the evolution of *Apicotermes* and *Allognathotermes*

(Isoptera; Termitidae). Am. Mus. Novitates 1771: 1-31.

———. 1956b. A new species of *Apicotermes* from Katanga. Rev. Zool. Bot. Afric. 53:98-101.

———. 1956c. Regenerative behavior and social homeostasis of termites. Ecology 37:248-258.

Grassé, P. -P. 1949. Traité de zoologie, Vol. 9, Masson, Paris.

Grassé, P. -P., and C. Noirot. 1954. *Apicotermes arquiere* (Isoptère); ses constructions, sa biologie. Considérations générales sur la sous-famille des *Apicotermitinae* nov. Ann. des Sci. Nat. Zool. 11ᵉ série. 16:346-388.

Michener, C. D. 1951. American Social Insects. Van Nostrand, N. Y.

Schmidt, R. S. 1955a. The evolution of nest-building in *Apicotermes* (Isoptera). Evolution 9:157-181.

———. 1955b. Termite (*Apicotermes*) nests—important ethological material. Behaviour 8:344-356.

———. 1958. The nest of *Apicotermes tragardhi* —new evidence on the evolution of nest-building. Behaviour 12:76-94.

———. 1960. Functions of *Apicotermes* nests. Insectes Sociaux 7:357-368.

ADDENDUM

The following figures are also of interest.

Figure 1 *Apicotermes emersoni* Bouillon (formerly called "Mayumbe nest"). Height of nest, 22 cm. [Reproduced from J. Desneux, Les Constructions hypogées des *Apicotermes*, termites de l'Afrique tropicale, *Ann. Mus. Roy. Congo Belge*, 17, sér. 8, Plate XIV (1952).]

DESNEUX, J. — *Apicotermes rimulifex* EMERSON

Figure 2 *Apicotermes rimulifex* nest from the Katanga savanna in central Africa. (1) Side view of nest (12.5 cm high) after removal of all but traces of the external shagreen layer to show irregular arrangement of wall pores; entrance and exit tube is visible at upper left. (2) Vertical section of same nest; some openings for communication between cells are visible. (Reproduced from J. Desneux, Le Nid d'*Apicotermes rimulifex* Emerson," *Rev. Zool. Bot. Afric.*, 53, Fasc. 1–2, p. 97. 1956. Figures 1 and 2 by permission of Dr. P. Basilewsky, Conservateur and Chief of the Entomology Section of the Musée de Tervuren, Belgium.)

9

Reprinted from *Am. Zool.,* **4**(2), 227–239 (1964)

EVOLUTION OF THE NESTS OF BEES

CHARLES D. MICHENER

University of Kansas, Lawrence, Kansas

Nests or other objects constructed by animals provide an opportunity to see a sort of summation of results of certain of the behavior patterns of the animals as expressed over a considerable length of time. Most bees are solitary, each nest being constructed by a single adult female; but for social bees a nest permits observation of the results of behavior, not of a single individual, but of several to many (a maximum of probably 100,000).

Although 20,000 or more species of bees exist in the world, nests of relatively few have been found and described. Even among those that have been described, it often happens that characteristics which one would especially like to know about have not been recorded. Nevertheless, so much is known about bee nests that no full account can be given here.

Bees are essentially a group of sphecoid wasps that has abandoned predatory habits and makes use of pollen instead of insects and spiders as the principal protein source. The accompanying dendrogram (Fig. 1) gives some idea of the classification and probable phyletic relationships among the various groups of bees, as understood from morphological evidence.

CELLS

The only almost invariable constant of a bee or wasp nest is the presence of one or more brood cells. A cell is the cavity in which one (or rarely more) young is reared. In the sphecoid wasps, from which bees undoubtedly arose, the most common type of cell is an unlined cavity in the ground. Some sphecoid wasps construct cells in pith or make exposed nests of mud (like those of mud daubers), but these are

Contribution number 1230 from the Department of Entomology of The University of Kansas. Preparation of this paper was greatly facilitated by grant number G11967 from the National Science Foundation.

developments parallel to those that have occurred among bees, and it seems highly likely that the most primitive bees were ground nesting, like the majority of sphecoid wasps and bees. In most nonsocial forms the construction follows a strict sequence, as follows: excavation, lining (if it occurs), provisioning, oviposition, and closure. The same sequence is then repeated for the next cell. Sakagami and Michener (1962) have described the breakdown of this typical chain of operations in halictid bees, even solitary species of which usually work on several cells during the same period, and of course simultaneous construction of many cells is the rule in some thoroughly social bees such as *Apis*.

Lining: As has been indicated, sphecoid wasps do not ordinarily line their cells. Apparently ancestral bees acquired not only pollen-feeding habits and the necessary structures to permit such habits, but also the behavior patterns, structures, and glands necessary to line cells. All of the primitive (most wasplike) bees line their cells in more or less elaborate ways. The family Colletidae contains those bees in which the glossa is short and broad, like that of wasps. All species of the family apply a secreted membranous lining on the inner walls of the cells. The material is tough and waterproof, translucent to transparent, cellophane-like, sometimes with recognizable silk strands imbedded in the translucent or transparent matrix. The equivalent membrane in the other primitive families of bees (most Andrenidae and Halictidae, Fig. 4, right) as well as in some of the Anthophoridae is more delicate, often brown, more waxlike in appearance, although not or only partly soluble in wax solvents. These secreted linings are painted onto the inner walls of the cells by the mouthparts of the female bee and are presumably, at least in part, salivary secretions. It is not surprising that cell plugs, with which cells are closed after oviposition, ordinarily are not water-proofed on their inner surfaces with any such material but in the related genera *Anthophora* and *Amegilla* most of the inner wall of the plug is lined with the secreted lining. It appears that the female bee leaves a small hole in the plug as she constructs it, and finally, probably by inserting her glossa through the hole, applies the lining material to the inside of the plug surface. The final pellets of earth that close the small hole in the plug are unlined.

Presumably the secreted cell lining used by burrowing bees is related to maintenance of the proper humidity within the

cell. The texture and humidity of the pollen provided for the growing larva appears to be very important for most species; the larva dies if too much water is absorbed into the food mass or if it dries out and becomes too hard. This matter would appear to be far more critical for bees, in view of their use of a mixture of pollen and nectar or honey as larval food, than for wasps since the provisions provided for most wasp larvae already have their own epicuticular and other water-control mechanisms. Proper humidity is also important for the larval and pupal stages themselves.

The presumed primitive (for bees) secreted cell-lining has been lost in various

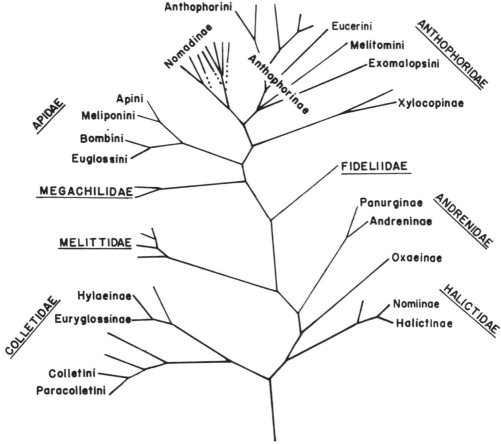

FIG. 1. Diagram of relationships among groups of bees. To facilitate reference from the text, names of some tribes and subfamilies not mentioned in the text are omitted although the branches are included. (Modified from Michener, 1944, Bull. Am. Mus. Nat. Hist.)

bees. In the Panurginae, some species (genus *Nomadopsis*) have such a lining, and in addition the subspherical food mass is covered by a very delicate secreted membrane, apparently similar to that applied to the inner walls of the cell. Rozen (1958), however, states that the cell lining is permeable to water, a condition which perhaps explains the non-permeable covering of the food mass. In other members of the same group of bees *(Perdita)* the cell lining is abandoned altogether (Michener and Ordway, 1963), and the food mass is perhaps surrounded by a thin membrane as in *Nomadopsis*. Of course the feeding of the larvae immediately breaks such a membrane and exposes the food mass to the air within the cell, but part of the food mass retains the protection for some time after the larva begins to eat.

The cells with thin secreted linings usually have these materials applied to extraordinarily smooth surfaces of the substrate (usually soil). Such smooth surfaces cannot, apparently, be constructed merely by excavation. At least it is true that with a few exceptions, such as *Colletes*, these bees excavate cells that are considerably larger than the ultimate cell size. They bring earth from elsewhere in the nest and form an earthen lining often as much as one or two mm thick. This lining of worked substrate (usually earth) is frequently recognizable because, consisting of earth from another part of the nest, it differs in color from that surrounding it. Also, it is usually of very fine texture and is firmly tamped by the pygidial area of the bee. Only after the inner surface is made extraordinarily smooth, much smoother than one would think possible, is the secreted lining applied. The earthen lining is sometimes permeated by liquid applied at about the same time as, and perhaps as part of, the secreted lining (Batra, 1964). As a result, the earthen lining is sometimes quite firm, and in forms such as *Melitoma (Anthophoridae)* intact earthen cells can be separated from the substrate.

It is interesting that the behavior patterns involved in the construction of such cell walls and their secreted linings—patterns apparently essential for the survival of the more primitive bees such as most of the short-tongued forms of the families Colletidae, Andrenidae, and Halictidae as well as some of the long-tongued Anthophoridae—should have been abandoned independently in various specialized, long-tongued bees.

A great many bees, particularly those in the families Megachilidae and Apidae (that is, in the most specialized families of long-tongued bees), have abandoned the thin secreted cell-lining and do not make the smooth lining of worked substrate or the secreted lining. In Megachilidae the cell lining may be partly the substrate in which the cell is constructed but it is always partly, and usually wholly, made of other materials brought into the nest from the outside, for example leaf pieces, chewed leaf material, resin, pebbles, etc. In the Apidae the relatively thick cell walls are made of resin (Euglossini) or of wax *(Apis, Bombus)* or of a mixture of the two (Meliponini).

Cocoons: In nearly all of the forms in which the thin secreted cell-lining of the more primitive families of bees has been abandoned, it appears to have been replaced by a cocoon spun by the mature larva. This cocoon is in many ways similar to the thin cell-lining. It consists of a layer of silk fibers to which is applied a liquid which hardens as a matrix surrounding the fibers, making the cocoon impervious except frequently for a small area, perhaps serving for ventilation. Those bees in which the thin cell-lining is applied by the adults virtually never construct cocoons as larvae, except for some anthophorids (e.g., Eucerini). Perhaps the lining and cocoon-making materials are produced from the same glands which mature in the late larval stage in some forms, and in the adult in others, and function in both stages in some anthophorids. Probably the importance of the cocoon is maintenance of proper humidity for the late larval stage and the pupa. Many bees live

193

for months or even for several years in the late larval (prepupal) stage.

Sphecoid wasps, like most Hymenoptera, usually spin cocoons as mature larvae. It seems that the line leading to the bees lost this behavior, and probably used the same materials in the adult stage for the secreted cell-linings. However, most of the highly evolved long-tongued bees reverted to cocoons and no longer make secreted cell-linings. The salivary spinnerets of bee larvae are associated with this picture; they are well developed only in the cocoon-spinning specialized families of bees (Michener, 1953a).

Food masses: It seems reasonable to consider in a discussion of constructs by bees the food mass with which each cell is provisioned. In all the members of the family Colletidae (those with the cellophane-like cell-linings), the food mass is liquid and even watery. It fills the bottom part of the waterproof cell. The relatively small amount of pollen in such material is responsible for its liquid state and is presumably a primitive feature. In two groups of colletids (Euryglossinae and Hylaeinae) the polled is carried from the flowers back to the nest along with the liquid nectar in the crop instead of on the outside of the body of the bee. This is quite possibly the primitive manner of collecting pollen for bees, although in all other groups it is carried in a scopa made up of hairs on the hind legs, the abdomen, or sometimes in part on the propodeum.

In bees other than Colletidae, the provisions contain a higher percentage of pollen so that they are firmer. In many bees (Halictidae, Andrenidae, Xylocopinae, and others) the provisions are quite firm, obviously consisting largely of pollen. In such forms, there is an interesting series of shapes for the pollen mass, all apparently related to the curious matter of allowing this mass to contact the inner wall of the cell in the minimum possible area. The reason for this is not obvious but may have something to do with the control of humidity in the food mass or the inhibition of mold which may tend to start its growth

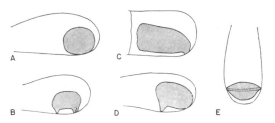

FIG. 2. Diagrams of cells of certain bees which make firm pollen masses (shaded) showing ways in which contact of these masses with the cell walls is minimized. a, *Lasioglossum inconspicuum*; b, *Dasypoda plumipes* (after Malyshev, 1935); c, *Xylocopa valga* (after Malyshev, 1935); d, *Exomalopsis chionura* (after Rozen and MacNeill, 1957; relation of cell size to pollen mass was not illustrated by those authors and is possibly incorrect); e, *Nomia triangulifera* (After Cross and Bohart, 1960).

at points where the cell wall and the pollen mass are in contact. In any event, pollen masses have the shapes described as follows (Fig. 2):

1) The most frequent shape is spherical or subspherical. Obviously such a mass will have a relatively limited area in contact with the cell wall. This shape is found in most Andrenidae and Halictidae. (2) In some species of *Nomia* (Halictidae) the pollen mass is flattened and provided with a projecting rim around the circumference so that it is supported in the lower part of the vertical cell by this rim, the bottom of the pollen mass thus not reaching the bottom of the cell. (3) In the Xylocopinae the food mass is loaf-like; the upper end (in a vertical cell) of the loaf is in contact with and attached to the cell wall but the remainder is not, the posterior end being free or supported by the egg (*Ceratina australensis,* see Michener, 1962) or the posterior end may have a small area of contact with the cell wall. (4) In *Exomalopsis* (Anthophoridae) the pollen mass is in contact with the cell at its rounded posterior end; the anterior end is supported by a median projection (Rozen and MacNeill, 1957). (5) The most remarkable shape is that made by *Dasypoda* (Melittidae), in which the pollen mass is supported by three short legs projecting from it, thus constituting a tripod (Malyshev, 1935).

(6) The activities of the larva also contribute to reduction of the contact of provisions with the surface of the cell, for in various bees, especially Panurgini, the larva works under the food mass and feeds with the food supported on its venter. (7) The provisions made by some long-tongued bees (e.g., *Anthophora*, Meliponini, some Megachilidae) are soft, semi-liquid, and flow; therefore they fill the lower part of a cell. In their texture such provisions are presumably independent reversions toward the liquid colletid type.

Of course in those few bees in which food to the larvae is provided gradually instead of by mass provisioning, there is no preservation problem. However, the only bees which exhibit this behavior are *Apis, Bombus* (in part and sometimes only to a limited degree), and *Allodape* (Xylocopinae) and its relatives; these probably constitute not over 400 of the 20,000 species of bees.

Cell shape: The cells of bees do vary in shape, but in general, particularly in the bees that are thought to be most primitive (Paracolletini and Euryglossinae), the cells are homomorphic—that is, all are more or less of the same shape. This is the commonest situation. However, various groups have acquired a degree of plasticity in cell shape, which then conforms more or less to the shape of the cavity in which the cell is located. Some members of the primitive family Colletidae, including especially those that nest in pith (e.g. *Hylaeus*) and also the ground nesting bees of the genus *Colletes*, make burrows into the substrate and then divide the burrows by means of transverse partitions made of the cellophane-like material already described as characteristic of cell linings in this family. The result is that the innermost cell, that is, the cell farthest from the surface or nest entrance, has its distal end rounded to fit the rounded end of the burrow, whereas the other cells have both ends truncated. In other colletids, e.g. some forms of *Hylaeus* and its relatives such as *Meroglossa,* the cells are sometimes made in the same way but using the burrow of a beetle or other insect in twigs or in wood. The volume of the various cells is more or less constant, hence they are shorter in portions of the burrow having large diameters than in portions having small diameters. (By no means do all bees which construct cells in series, end to end, have heteromorphic cells. For example, the Xylocopinae regularly construct cells in series, but the partitions between the cells are so shaped that the distal end of each cell is rounded just like the distal end of the apical cell of the series.)

Many primitive bees (most Paracolletini, Halictidae, Andrenidae) have homomorphic cells which are flatter on one side (the lower side in the case of horizontal cells) than on the other (Fig. 2a). Such cells are called bilaterally symmetrical by Malyshev (1935). Radially symmetrical cells (round in cross-section), however, occur in some Colletidae and in almost all of the long-tongued families, and are modified as hexagonal in *Apis*. At least in the latter families and possibly also in Colletidae, radial symmetry must be derived from bilateral.

Certain bees, principally or exclusively in the family Megachilidae (e.g., *Osmia lignaria*), make heteromorphic cells whose shape corresponds to cavities that the bees happen to have discovered and used. Thus, in at least two families (Colletidae and Megachilidae), the generally homomorphic form of bee cells has been abandoned, obviously independently.

The wax cells of *Bombus* are a most extraordinary development. They are somewhat variable in size and shape, and therefore heteromorphic, but are unique among the bees in that in most species they contain a cluster of eggs or immature stages instead of only one, and in that they grow (i.e. wax is added to them) as their contents grow. At first such a cell need only enclose the eggs, because no large quantity of provisions is placed in the cell; thus the larvae are fed progressively instead of by mass provisioning.

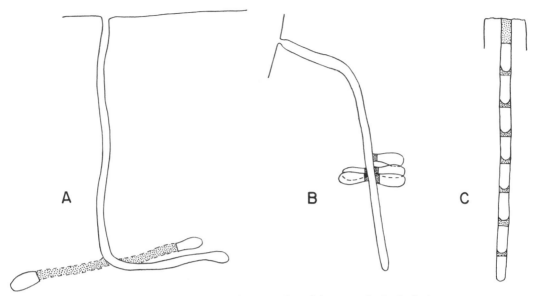

FIG. 3. Diagrams of nests of bees. a, *Andrena erythronii* in ground; b, *Lasioglossum guaruvae* in earth bank; c, *Hoplitis pilosifrons* in pithy stem.

NESTS

Because of its widespread occurrence through the more primitive families of bees, I believe that the primitive type of bee nest consists of a burrow entering the soil, and giving rise at its lower end to laterals, each ending in a single cell (Fig. 3A) (See Michener and Lange, 1957). This sort of nest is also widely distributed among sphecoid wasps. The lateral burrows are built one at a time so that at first the nests will have only the main burrow continued as one lateral ending in a cell. When that cell is provisioned and sealed, a second lateral is started, the soil from which is used to fill the first. The result is that there is rarely more than one lateral open at a time, although if one considers all the earth-filled laterals, there may be several radiating from the lower end of the main burrow. Such nests are well known in the ground-nesting Colletidae, Andrenidae, and in some of the Anthophoridae.

Many Halictidae have such nests, differing in that the main burrow extends downward below the level of the laterals. In other Halictidae, however, there is a tendency to concentrate the cells or save work by shortening the laterals until the cells arise without laterals from the walls of the main burrow. Further developments in the Halictidae, as detailed by Sakagami and Michener (1962), include: (1) concentration of the cells in limited spaces (Fig. 3B); (2) excavation of burrows around the cells or around portions of them so that there are only thin walls between the burrows and the interiors of the cells; and (3) expansion of these burrows to form a space around the cluster of aggregated cells, the cell cluster often being supported only by a few earthen pillars or only by rootlets which pass through the cell cluster. Such clusters are small combs of cells, usually made of earth. Photographs of such nests are shown in Figure 4.

It is interesting that this air space or vault surrounding the cell cluster has arisen independently in several groups of the subfamily Halictinae, as well as in the genus *Nomia* in the Nomiinae, and even in the subterranean nests of the distantly related genus *Proxylocopa* (Xylocopinae). Such a space surrounding the cell cluster occurs also in the halictine, *Augochlora*, which nests in rotting wood. The function of this space is unknown, but it seems that it must play some role related to the

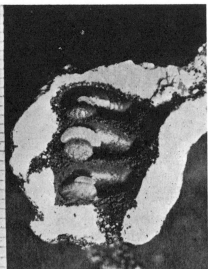

FIG. 4. Nests of *Augochlorella striata* in soil. Left, nest opened to show cell cluster and burrow leading down to it (photo by Ellen Ordway); center, nest pored with plaster of Paris, then excavated to show form of whole nest (photo by C. W. Rettenmeyer); right, cell cluster opened and showing smooth linings of three adjacent cells, very thin earthen cell walls, and earth pillars supporting cell cluster in air space here filled with white plaster of Paris. (Immature stages shown are an egg in the upper cell and larvae in the others.) (Photo by C. W. Rettenmeyer.) Scales are in millimeters.

environmental conditions within the cells. It certainly is not a labor-saving device since it obviously requires considerable work to construct it. The thinness of the earthen walls of the cells is sometimes noteworthy; in some genera they are consistently less than one-half a millimeter thick, and one can see the details of the shapes of all cells from the outside of the cell cluster.

In burrowing halictids which do not make such cell clusters, the cells are excavated into the substrate and lined with worked earth and a secreted lining, as described above. Some of those which make cell clusters construct the cells in the same way and then excavate the earth from around the group of cells. Others, however, excavate the space and then build the cells within it. The constructed wall in this case appears to be homologous to the earthen lining added to excavated cells. Intermediate conditions are known, as described especially by Stockhammer (1964) for *Augochlora*, in which the bee sometimes excavates part of a cell into the substrate and then continues the walls outward into the nest cavity with independent walls as a constructed cell. In spite of such intermediates, however, the diversity of the methods of producing the cells in those bees which isolate a cell cluster from the surrounding earth (or rotten wood) emphasizes the independent origin of such cell clusters in various phyletic lines.

The shortening of lateral burrows and clustering of cells as seen in many halictids may originally have been a matter of economy of labor. An entirely different development of a similar sort which arose repeatedly was the habit of placing cells in series end to end. Sometimes such an arrangement can be seen in lateral burrows, each of which thus contains two to several cells [*Pseudagapostemon* (Halictidae), see Sakagami and Michener, 1962; some *Andrena*, see Michener and Rettenmeyer, 1956]. In other cases no lateral burrows are constructed, but cells are placed end to end in the main burrow or its branches (e.g., *Hylaeus*, some *Colletes*, Xylocopinae, many Megachilini, Fig. 3C).

Still another interesting labor-saving device is re-use of cells after refurbishing.

Curiously, this seemingly simple device seems to have arisen as a regular arrangement only in some of the Halictidae that make clusters of cells (Sakagami and Michener, 1962) and in the genus *Apis*.

Many bees tend to start their nests in pre-existing holes in the substrate, and as has been indicated, some always use such cavities instead of making their own. These are labor-saving arrangements of a different sort. The family Megachilidae includes many which consistently use pre-existing holes and rarely or never dig or gnaw their own. The family Megachilidae also includes nearly all the bees (except the social ones) in which the cells are no longer constructed in the substrate but are placed on the surfaces of materials in fairly exposed situations. Presumably this development was possible only among bees using more or less weather-resistant foreign materials (resin, pebbles, mud) in cell construction. Examples include the genus *Dianthidium* which builds nests of pebbles and resin on boulders or branches of trees or bushes, and *Chalicodoma* which constructs mud nests on cliffs or buildings. For such an exposed nest, the cells and the nest are synonymous; there is no other part of the nest. For the majority of bees, however, the nest is much more than merely the cells.

In some nonsocial bees, all of the cells that a female produces are located in a single nest. From the standpoint of economy of labor, this is the best arrangement. In others, however, the female may disperse her cells in several nests. For example, in *Megachile brevis* the female moves from place to place, sometimes over distances of several miles, making nests with one to several cells in different places (Michener, 1953b). In *Andrena erythronii* the female remains in the same vicinity but nonetheless constructs two or three nests during her life, each containing a few cells (Michener and Rettenmeyer, 1956). In *Anthocopa papaveris* (Megachilidae) (Malyshev, 1935) and *Perdita maculigera* (Andrenidae) (Michener and Ordway, 1963) each nest burrow ends in only a single cell (rarely two in the first mentioned species). The number of nests made by each such bee is unknown, but must be several. Since the bees listed are quite unrelated, the tendency for multiplicity and simplification of nests must have arisen independently in various groups.

Reduction of nest structure reaches its extreme in the primitively social xylocopine genus *Allodape* and its relatives. Here the nest itself is nothing but a burrow, usually in the pith of a stem. The young are reared together in it, there being no separate cells for them (Sakagami, 1960).

By contrast, in some bees the number of cells per nest is very large. This is particularly true in the social forms. Sociability has arisen independently in various bees. Even in the Halictinae, social behavior including a worker caste and an increased number of cells has arisen several times. In this group, neither nest nor cell structure is influenced by the social behavior, except for the more extensive burrow system and larger number of cells in nests containing more bees. Thus there are both solitary and social forms with cells scattered through the earth, with cells gathered into clusters, and with cells in clusters or combs surrounded by air spaces.

Even within a single species of facultatively or temporarily social halictine, the cells and nests are essentially the same whether made by a single bee or by a colony. Batra (1964) has seen that the cells in the nests of a social halictine, *Lasioglossum zephyrum*, are excavated by several bees and that several individuals may cooperate in lining and provisioning. Yet a single bee of the same species can do the same job. The bees in such nests have little contact with one another and it is probable that the stimuli that result in appropriate construction are those from the nest itself. There is no evidence for integration of activities through contacts among members of the colony, such as is obvious in socially more specialized insects including the Meliponini and *Apis*. Communication is therefore not direct, but by way of the construct. In a nest occupied by a single bee, constructed features must also serve as stimuli for further constructing, i.e., it is as though the bee communicates with itself. Similar stimuli evidently result in similar construction activity whether there is only one bee or a colony.

Among the most interesting bees are those of the family Apidae. Some Euglossini are solitary, constructing their cells of resin, sometimes mixed with bits of bark, in pre-existing cavities (Euglossa, Euplusia). Some species of Euglossa (e.g. E. dodsoni) construct an exposed nest of resin in which they place their cells; in other words, they make their own cavity for the cells instead of finding one. Among the close relatives of Euglossa is the genus Eulaema, in at least some species of which there is a certain amount of social organization since several individuals which are probably workers are involved in the construction of each mud cell and in its provisioning. (C. Dodson, personal communication.) Such nests may contain numerous individuals and occupy large cavities in the soil, in wood, in termite nests, or in other situations.

The related bumblebees (Bombus) make use of similar cavities or abandoned nests of small mammals, and secrete wax for construction. In Bombus we encounter for the first time pots made of wax (and old cocoons) for storage of food material for the consumption of adults. Food supplies discussed above were in brood cells and for the use of larvae (although limited autumnal storage of food for adult consumption is known in the nest burrows of Xylocopinae).

Such storage pots are also characteristic of the stingless bees (Meliponini) which exist in large permanent colonies with a social organization equivalent in complexity to that of the common honeybee. The nests of most stingless bees are found in large cavities. One of the architectural features which seems to be of evolutionary interest is the arrangement of the brood cells. In the Australian species of the subgenus Plebeia of the genus Trigona, the cells are spherical and clustered together

in a mass which is surrounded by a thin wax and resin involucrum. It seems quite possible that this is the primitive cell arrangement (see Michener, 1961). In another species of the same subgenus found in northern Australia and New Guinea, the cells are arranged in concentric layers with spaces between them. It is possible that from such an arrangement the usual meliponine nest organization with a series of horizontal combs of cells (Fig. 5, left) may have arisen. In all such nests there is a wax and resin involucrum surrounding the brood chamber which contains the combs. Independently, in several different subgenera of Trigona and the derived genus Lestrimelitta, the neat arrangement of cells into combs appears to have broken down. Derivation from ancestors with cells in combs is suggested by the vertically elongate cells like those crowded into combs, in contrast to cells of the Australian Plebeia thought to have the primitive cell arrangement for the group. In those forms in which the combs are partly or entirely disorganized to form clusters of cells, the involucrum is absent. The significance of the disappearance of the involucrum and the breakdown of the combs appears to be adaptation to small spaces for nesting. Nests of these species with clusters of cells are to be found in large hollow stems, in crevices in logs, in small spaces between boards in the walls of houses (Fig. 5, right), and in similar situations where a nest organized into combs surrounded by an involucrum could not possibly be accommodated. It is interesting that in some species whose involucrum is virtually absent, small patches of it are sometimes constructed. These would appear to be vestigial structures (resulting from "vestigial behavior") having no functional significance in the nest as it is used by the bees today.

From the standpoint of cell arrangement, the most remarkable of the Meliponini is the African Dactylurina. In this genus the brood cells, instead of forming horizontal combs or clusters as in other members of the group, and instead of be-

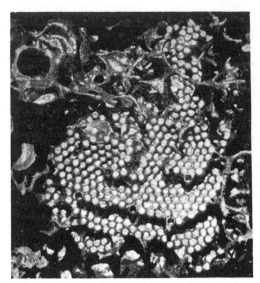

FIG. 5. Brood cells of *Trigona*. Left, horizontal combs of *Trigona carbonaria*, with large honey pots (one open) at upper left; right, sheet of cells of *Trigona wybenica* filling space between two boards, showing how most species which do not make combs can crowd their nests into small spaces.

ing vertically elongate as in most members of the group, are horizontal and form vertical combs with cells facing in both directions just as in the genus *Apis*. In other features the nests of these bees are similar to those of other Meliponini.

The brood cells of *Apis* are arranged as in *Dactylurina* but there are two sizes, one for males, the other for workers. The occasional very different brood cells for queens are comparable to those that occur in the genus *Trigona* of the stingless bees.

Honeybees are remarkable for the fact that, unlike the bumblebees and the stingless bees, they do not have honey pots but store provisions for adult consumption in cells similar to the brood cells. Perhaps this is related to the fact that the presumably primitive species of the genus *Apis* do not nest in cavities as do their relatives but construct nests in the open, hanging from a large branch or an overhanging rock or ledge. In such situations, large and rather delicate honey pots might be easily destroyed by natural forces. It is true that the more specialized species of *Apis* such as the common honeybee do nest in cavities, but this is probably a derived rather than a primitive nesting site for the genus, as judged from behavioral evidence relating to communication.

A comparable evolutionary development in stingless bee nests is the progressive independence of the nests from the cavities usually used as nesting sites by members of that group. Nests in cavities are surrounded by batumen layers. The thick walls of cerumen (a mixture of resin, wax, and sometimes mud or other materials) cut off the hollow (for example, in the trunk of a tree) and form a cavity of proper size for the nest; thin walls of lining batumen cover the inside of that portion of the hollow used for the nest. Some nests (e.g., *Trigona cupira*) are partially exposed, being constructed in hollows in banks or in trees, one side of which is open. In such cases the batumen is much elaborated on the exposed side of the nest, being several layers thick with airspaces between. Finally, in certain species such as *Trigona corvina,* the nest is constructed in the open, surrounding the branch of a tree. In such cases the batumen layers are numerous, sometimes forming a multilayered insulating and protective barrier as much as 20 cm thick surrounding the whole nest except for the entrance hole,

BEHAVIOR PATTERNS, HOMOLOGIES, AND CONSTRUCTS

Although relatively few of the nests of the 20,000 species of bees in the world are known, it is very obvious that nest structures have some taxonomic value. It is often easy to recognize the tribe or genus of a bee by its nest structure. Such characteristics, however, are useful at all levels, from the species to the family. Indeed, Kerfoot (1964) has described what appear to be subspecific differences in the shape of the pollen mass in *Nomia nevadensis*.

Comparative studies of nests designed to add to our knowledge of affinities among groups of bees are primarily useful if one can recognize homologous structures among the various nests.

The structures of bee cells or nests result from the interactions of the behavioral patterns of the bees with the environment in which the nest is constructed. It is legitimate to ask to what extent the nest structure can be considered a reflection of species specific behavior patterns of the insect, and to what extent is it a response to the environment in which the insect finds itself. It is at least possible to point out that, with exceptions in the Colletidae and Megachilidae and of course cells producing different castes in social bees, the shape and structure of cells are very uniform within each species. The arrangement of the cells and the formation of the whole nest may vary greatly with the local variations in the substrate or the nest environment. Indeed, Stockhammer (1964) was able to find virtually all possible arrangements of cells and of the surrounding vault in nests of *Augochlora* in different situations in logs and in artificial nests. The cells themselves, at least internally, varied but little, in spite of the impressive flexibility in nest architecture.

Furthermore, among different kinds of bees, cells are often quite similar although the nest structure may be very different. Batra (1964) and Stockhammer (1964) have observed the behavior patterns involved in cell construction and cell lining in species of different genera (*Lasioglos-*

sum and *Augochlora*) of the family Halictidae in nests constructed in substrates between two sheets of glass in a bee room where the insects have been maintained under laboratory conditions. Nests of these two bees are extremely different, although the cells are similar. It is quite obvious that the behavior patterns involved in the construction of the cells can be recognized and homologized in bees of the two genera. It therefore seems entirely legitimate to homologize the parts of the cells. For example, the smooth, worked lining of substrate material; the thin secreted lining; the inner end; the cell opening; the lower rather flat side of the cell as distinguished from the other, more concave sides; and the cell plug all seem to be homologizable between the two genera of Halictidae. The flat side of the cell is particularly significant in this case because in the cells which have such a surface, it is typically the lower one. However, in various Halictidae, the cells become slanting or even vertical (see Sakagami and Michener, 1962). Even when the cells are vertical, one of the sides is flatter than the rest. It is to this side that the pollen mass is attached in a vertical cell, being somehow stuck to the smooth, waxy surface. In horizontal cells it is held in position on this surface by gravity alone. The evidence of homology is strengthened by this sort of observation, as well as by the common behavior patterns involved in the construction of the various features of the cells.

On the other hand, cells of more distantly related bees such as *Megachile* and *Apis* are constructed by means of largely different movements on the part of the insect and by use of entirely different materials. The cells, moreover, have different shapes, and in *Apis* they are not plugged in the same way after oviposition, for there is progressive feeding of the larvae. It is difficult to speak of homology of the details of the structure, as one can do when comparing the two species of Halictidae. However, it would be unreasonable to feel that the cells as wholes are not homologous among all bees (exceptions are *Allodape*

238 CHARLES D. MICHENER

and its relatives, see Sakagami, 1960).

The problem in homologizing these constructs of bees is exactly comparable to that in homologizing structures. The difficulty or impossibility of defining the idea of homology in a satisfactory way is the same for constructs as for morphological features. For example, it is possible to homologize fine details of the fore limbs of two mammals, but on comparing one of them with the pectoral fin of a fish, only general statements about homologies can be made. Clearly, however, it seems as reasonable to speak of homologous constructs as it is to speak of the homologous structures of the organism itself.

One of the noteworthy features of the nests of many bees, especially of the Halictidae and the social Xylocopinae related to *Allodape,* is the constriction of the entrance. The nest burrows are originally dug with a constricted entrance, and the bees replace the constriction in the event that it is destroyed. It is obvious that the long-lived bees such as these would have special needs for nest defense; similar constrictions are weak or absent in the majority of the burrowing bees which are solitary and whose nests are open only for brief periods in each generation. Some such bees have other protective devices such as tumuli of loose dirt completely closing the nest entrances. The noteworthy feature of the entrance constrictions is the exactly parallel way in which they are used for protection in the unrelated Halictidae and Xylocopinae. In both groups, guard bees commonly occupy the entrances, consistently if more than one bee is in the nest at the same time. The guard normally has her head in the entrance, her body down in the burrow. At any disturbance, she bites at the disturbing object, often without touching it. If the disturbance continues, she will turn and block the entrance with the dorsal apical portion of her abdomen; this action is practical only in a nest having a constriction at the entrance but a relatively large burrow diameter just below the entrance. In this position the bee is difficult to dislodge and we

have seen her successfully prevent the entrance of ants, mutillid parasites, etc. The constricted entrance and the associated behavior patterns seem to have arisen independently in at least the two unrelated groups of bees mentioned above. Superficial examination would suggest that they are homologous, but the great taxonomic differences and absence of comparable behavior among other, and more or less intermediate, groups suggest that this is not the case.

SUMMARY AND CONCLUSIONS

Bee nests include or consist of cells for rearing young. Nests also serve as resting places and often as wintering places for adults (mostly for females). Among thoroughly social bees, nests include storage places for food for adults as well as for larvae. Primitive nests are branching burrows, each branch ending in a cell lined with impervious material and containing liquid food for the larva. Labor-saving grouping of cells ultimately leads to series of cells end to end along a burrow, or to clusters or comblike arrangements. Adaptation to pre-existing cavities leads to cells of various shapes rather than of the same shape, as in most bees. Many other features of the cells, the food mass, and the nest as a whole have evolved, and it is easy to cite instances which can be described as convergence, parallelism, reversion to a structure made by an ancestral type, simplification of structure, vestigial structures, and the like. Most features have arisen more than once, if our understanding of bee phylogeny is at all correct. In general, influence of the substrate is minimal on cells but may be very great on gross nest form or the manner of cell clustering. Flexibility of behavior serving to take advantage of local differences in the substrate is so great that in nests of a single species virtually all possible arrangements of cells can be found in rotten wood, but species that nest in more homogeneous substrates probably possess less potential for flexibility.

There is no obvious relationship between the development of social organization and the nest architecture, except that the larger the social groups, the more cells are present. The organization of the nest and the structure of the cells is similar in social groups and their close solitary relatives (if extant). This means that there are distinctive features of nest structure in the various taxa of bees, and that these features are more conservative than the degree of social organization, a conclusion at first surprising considering the complex of interactions among individuals that would seem necessary to make a society effective. However, communication among individuals of primitively social bees seems largely based upon stimuli of the construct, without contact among individuals of the colony. The same stimuli must guide construction in nests of related noncolonial bees, the bee making features which later cause it to do something else. In that case, it is as though the bee is communicating with itself; in a colony, features made by one bee may have their effect on another. In this light, similar constructs would be expected whether made by a lone individual or by a colony.

Comparative studies of nests show not only distinctive features of the taxa from subspecies to superfamily, but also homologies among nest structures. Such homologies can be supported in detail for related forms by ontogeny, behavior patterns involved in construction, as well as relative position, appearance, and function of nest parts but suffer from the same problems of definition as homologies of structural parts of organisms themselves.

REFERENCES

(Review or recent works which can serve as guides to other literature, rather than historically significant works, are cited in most cases.)

Batra, S. W. T. 1964. Behavior of the social bee, *Lasioglossum zephyrum*, in the nest (Hymenoptera:Halictidae). Insectes Sociaux, in press.

Cross, E. A., and G. E. Bohart. 1960. The biology of *Nomia (Epinomia) triangulifera* with notes on other species of *Nomia*. Univ. Kansas Sci. Bull.

41:761-792.

Grandi, G. 1961. Studi di un entomologo sugli Imenotteri superiori. Boll. Instituto Ent. Univ. Bologna 25:1-659.

Kerfoot, W. B. 1964. Observations on the nests of *Nomia nevadensis bakeri* with comparative notes on *Nomia nevadensis arizonensis* (Hymenoptera: Halictidae). J. Kansas Entomol. Soc., in press.

Malyshev, S. I. 1935. The nesting habits of solitary bees. Eos 11:201-309.

Michener, C. D. 1944. Comparative external morphology, phylogeny and a classification of bees. Bull. Am. Mus. Nat. Hist. 82:157-326.

———. 1953a. Comparative morphological and systematic studies of bee larvae with a key to the families of hymenopterous larvae. Univ. Kansas Sci. Bull. 35:987-1102.

———. 1953b. The biology of a leaf-cutter bee *(Megachile brevis)* and its associates. Univ. Kansas Sci. Bull. 35:1659-1748.

———. 1961. Observations on the nests and behavior of *Trigona* in Australia and New Guinea (Hymenoptera, Apidae). Am. Mus. Novitates 2060:1-46.

———. 1962. The genus *Ceratina* in Australia, with notes on its nests (Hymenoptera:Apoidea). J. Kansas Entomol. Soc. 35:414-421.

Michener, C. D., and R. B. Lange. 1957. Observations on the ethology of some Brazilian colletid bees (Hymenoptera, Apoidea). J. Kansas Entomol. Soc. 30:71-80.

———. 1958. Observations on the ethology of neotropical anthophorine bees (Hymenoptera: Apoidea). Univ. Kansas Sci. Bull. 39:69-96.

Michener, C. D., and E. Ordway. 1963. The life history of *Perdita maculigera maculipennis* (Hymenoptera:Andrenidae). J. Kansas Entomol. Soc. 36:34-45.

Michener, C. D., and C. W. Rettenmeyer. 1956. The ethology of *Andrena erythronii* with comparative data on other species (Hymenoptera, Andrenidae). Univ. Kansas Sci. Bull. 37:645-684.

Rozen, J. G. 1958. Monographic study of the genus *Nomadopsis* Ashmead (Hymenoptera, Andrenidae). Univ. California Publ. Entomol. 15:1-202.

Rozen, J. G., and C. D. MacNeill. 1957. Biological observations on *Exomalopsis (Anthophorula) chionura* Cockerell, including a comparison of the biology of *Exomalopsis* with that of other anthophorid groups (Hymenoptera : Apoidea). Ann. Entomol. Soc. Am. 50:522-529.

Sakagami, S. F. 1960. Ethological peculiarities of the primitive social bees, *Allodape* Lepeltier (*sic*) and allied genera. Insectes Sociaux 7:231-249.

Sakagami, S. F., and C. D. Michener. 1962. The nest architecture of the sweat bees (Halictinae). Univ. Kansas Press, Lawrence. 135 p.

Stockhammer, K. A. 1964. Manuscript in preparation on *Augochlora pura*.

10

Reprinted from *Anim. Behav. Monogr.*, **1**, Pt. 3, 165, 194–210, 298–300 (1968)

THE BEHAVIOUR OF FREE-LIVING CHIMPANZEES IN THE GOMBE STREAM RESERVE

Jane Van Lawick-Goodall

[*Editors' Note:* In the original, material precedes this excerpt.]

The Study Area

The Gombe Stream Reserve supports a semi-isolated population of *P. troglodytes schweinfurthi*, the eastern or long-haired chimpanzee (see Plate). This area was selected for a field study because the country offers favourable opportunities for observation. It consists of a narrow mountainous strip stretching for some 10 miles along the east shore of Lake Tanganyika, between Kigoma and Burundi, and running inland about 3 miles to the peaks of the mountains of the Rift escarpment, which rise steeply from the lake (2334 ft) to heights of about 5000 ft. Numerous steep-sided valleys and ravines intersect the mountains, many of which support permanent streams. The dense gallery rain forests of the valleys and lower slopes give place to more open deciduous woodland on the upper slopes and many of the peaks and ridges are covered only by grass). In the valleys stands of trees reach a height of up to 80 ft, but the trees of the open woodlands seldom exceed 40 ft. During the rainy season, between October and May, the grass grows as high as 14 ft, but during the dry season grass fires often sweep through the reserve, started by African farmers outside the boundary.

Temperatures in this area may reach 100°F in the sun and drop below 60°F at night. In normal years this area has about 50 in. of rain. During the wet season, from October to May, at least an hour's heavy rain falls on most days (though not in 1966/67 when rainfall was abnormally low), and sometimes it rains for 10 hr without stopping.

[*Editors' Note:* Material has been omitted at this point.]

Nest-making Behaviour

In its natural habitat the chimpanzee, like the gorilla (Schaller, 1963) and the orangutan (Schaller, 1961), constructs a sleeping platform or nest on which to rest at night. There is evidence that the construction of a nest is characteristic of the chimpanzee throughout its range: Yerkes & Yerkes (1929) summarize the early reports, and there is more recent information from French Guinea (Nissen, 1931; de Bournonville, 1967), the Congo (Kortlandt, 1962), Uganda (Bolwig, 1959; Reynolds & Reynolds, 1965), and Tanzania (Azuma & Toyoshima, 1961–62, 1965; Goodall, 1962; Izawa & Itani, 1966). The information from these reports is summarized by Izawa & Itani (1966).

I observed chimpanzees in the Gombe Stream area construct night nests on well over 500 occasions. Each individual made its own nest, with the exception of infants which slept with their mothers. New nests were normally made each night in trees at heights between 15 and 80 ft from the ground; only on rare occasions were chimpanzees observed to sleep at night on the ground.

I found nests in nearly all parts of the reserve; in the thick gallery forests of the valleys, in the trees fringing the lake, and in the more open *Brachystegia* woodlands of the upper slopes. However, I found few nests above 4500 ft where the vegetation is sparse and provides little food, since chimpanzees normally nest close to the trees in which they are feeding at night. There are some 'nesting sites' where the presence of old nests, in varying stages of decay, shows that the tree or group of trees has been used on several occasions. (Today, for example, there are several such nesting sites near the artificial feeding area.) When the chimpanzees were seen feeding near such a place in the evening, they invariably moved to the 'site' at dusk to make their nests.

Trees overhanging a gulley or stream were often chosen, as were trees at the edge of a forested area overlooking a more open stretch of country. Normally chimpanzees nested where there were well defined tracks below and where the undergrowth was relatively sparse.

[*Editors' Note:* Certain plates and parts of plates have been omitted owing to limitations of space.]

(a) Type of Tree Utilized

Almost any type of tree may be used for nesting, providing it is 20 ft or more in height and is fairly well foliaged. Trees with branches lower than 10 ft above ground level are usually avoided. *Brachystegia busei*, one of the common trees in the area, fulfills these conditions and is frequently utilized.

Another common tree is the oil nut palm. In Queen Elizabeth Park, Uganda, about one-third of the chimpanzee nests observed were in the tops of palms (Schaller, pers. comm.). In French Guinea, Nissen (1931) saw no nests in palm trees, and I observed none in my area between June 1960 and August 1961. At the end of August, however, I saw one such nest, and by October of the same year they were conspicuous in palm trees in many parts of the reserve. The making of such nests apparently became a 'fashion' amongst the chimpanzee population.

When a chimpanzee is feeding at dusk it generally leaves the food tree and makes its nest in a non-food tree close by. I observed this behaviour on almost every occasion when I watched chimpanzees making their nests, and I seldom found a nest in a tree bearing ripe food. Sometimes this was due to the unsuitability of the tree, since when the food was shoots or blossoms, the branches were bare of leaves; at other times there was no apparent reason for this behaviour.

(b) Height of Nest

Nests are normally constructed at any height above 15 ft from the ground. In the valley forests, the trees form stands up to 80 ft, but in most places they seldom exceed 50 ft. In a tree that does not exceed 30 ft, nests are usually constructed near the top, but in a higher tree they can be found near the top or in the lower branches. The position of a nest in a tree cannot be related to the size of the chimpanzee that made it, except insofar as a small animal is unable to use the thicker branches and can make its nest where the slender boughs would not support a mature animal.

Figure 11 shows the approximate heights of 384 nests. The 33 per cent between 30 and 40 ft from the ground represents the nests found at the tops of the lower trees together with many constructed in the lower branches of taller trees. The 25 per cent between 60 ft and 80 ft represents nests found in the thick valley forests along the streams where branches often occur only near the tops of tall trees. Nests found low in the trees were constructed almost always in branches overhanging a gully or stream so that the actual height from the ground was increased considerably.

Only on one occasion was a healthy chimpanzee seen to sleep on a ground nest at night. He was travelling with a female whom he was 'forcing' to follow (see p. 219) and she nested in the only tree available to him in her immediate vicinity—a palm tree which could only support one nest (McGinnis, pers. comm.).

(c) Nesting Groups

Direct observations of nesting chimpanzees, together with data gathered from examination of groups of nests, show that there is no rigid social pattern controlling the number of apes sleeping in a group. Usually between two and six

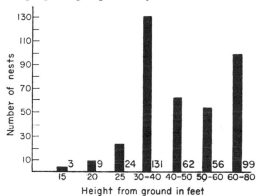

Fig. 11. Approximate heights from the ground of 384 nests. (The number of nests found at each height is shown in the figures at the base of the relevant columns.) (From Goodall, 1964).

animals slept together, building their nests in a single tree or in closely adjoining trees. Often a female with her infant and her 3- to 5-year-old child formed a small nesting group. It was not uncommon for a single male to nest alone.

Sometimes two or four of these groups joined up at dusk and made their nests within a small area, but at other times a large number of chimpanzees feeding together in the evening split up into two or three groups which nested some distance apart. The largest group of nests I found, all of which had been slept in at the same time, was seventeen; the largest number in one tree was ten. Such large groups occurred when the ripening of certain fruits or the presence of a receptive female caused a concentration of the chimpanzee units in the area.

Mature or adolescent males sometimes made their nests 100 yd or more away from the rest of the group, but females and juveniles usually slept within a few yards or even a few feet of the others. On three separate occasions a young juvenile, after constructing a nest between 40 and 60 yd away from her mother, left it as dusk fell, in order to make a new one closer to her mother. One adolescent female who made her nest early abandoned it and made another amongst the rest of her group when she found the others were nesting some 60 yd away. By contrast, I observed one mature male make a careful nest near the observation area which he left when a small group arrived in the late evening to feed on bananas. These individuals moved off and made their nests some 500 yd distant, but the mature male climbed back into his nest and so slept by himself.

(d) Time of Making and Leaving the Nest

The time at which nests were made varied slightly, depending on the season of the year, on whether or not the chimpanzees were in large groups and also on the individuals which composed the nesting group.

During the rainy season nests were often constructed as early as 6.00 p.m. local East African time (about $1\frac{1}{2}$ hr before darkness) and sometimes even earlier. During the same season, however, when a number of animals were moving about together, nests were occasionally made after sunset when it was almost too dark for observation. In the dry season the normal time was between 6.45 and 7.45 p.m.

During 1965 an adolescent male (Pepe) visited the feeding area at 7.45 p.m. By the time he had finished his bananas the only light came from the quarter moon; suddenly he looked round and at once got up and hurried to a nearby palm tree, whimpering to himself as he went. He continued to whimper whilst making his nest and for a few moments as he lay in the completed structure. He then gave some rather wavering 'pant hoots' and was quiet. On another occasion a whole group of chimpanzees arrived after our lamps were lit. They stayed only long enough to seize some bananas and left carrying these. There was no moon but they nevertheless travelled some 500 yd before stopping for the night. It was too dark to watch them making nests but that they did so was evident from the sounds of breaking branches. Other evidence of chimpanzees travelling at night will be discussed later (pp. 200-201).

The fact that the sun sets some 16 min later during the wet season is offset by the low cloud banks that give rise to a 'false dusk' on many evenings. That early nesting is induced by rain rather than the time of year is suggested by the fact that one of the earliest times I saw nests made was 5.30 p.m., after a severe rainstorm during September, when nests were normally made after sunset.

Within a group of chimpanzees there may be as much as 30 min difference between the making of the first and last nests. When a group was feeding at dusk, individuals moved from the food tree at irregular intervals to make nests nearby. Sometimes the entire group left the food tree simultaneously, but while some apes at once made their nests on reaching the nesting tree, others sat quite motionless for 10 to 15 min, and it was sometimes almost dark before they moved to a suitable nesting place.

The time of leaving the nest also varied. During the dry season, and particularly when a number of chimpanzees were sleeping together, they left the nests early, but only on ten occasions did I hear them moving about in the tree while it was still dark. Usually there was some movement at dawn, when defaecation and urination commonly occurred. After that, the chimpanzees often sat quietly beside their nests or even lay down again for a further 10 to 15 min.

In the rainy season chimpanzees tended to leave their nests much later, and one group, in March, did not get up until 8.10 a.m.—1 hour after sunrise.

Normally the individuals in a group leave their nests within 5 to 10 min of each other, but there may be a time lag of 30 minutes between the first and the last to leave. In particular one female was always the last to get up of any group she happended to be with. Her infant was restless in the mornings and moved off on its own for longer and longer periods as it grew older.

(e) Building Techniques

The time taken for the construction of a normal nest varies between 1 and 5 min. A nest consists basically of a main branch or branches forming the 'foundation', over which smaller branches or 'crosspieces' are bent. A typical foundation is a horizontal fork from which several leafty branches fan out (Plate 3 b), a shallow crotch at the top of an upright branch, (e.g. Plate 3 a), or two adjoining parallel branches.

The chimpanzee, standing on the foundation, takes hold of a fairly stout branch and bends it down over the foundation to form the first crosspiece. Holding this in place with his feet he bends a second branch across it. From four to six main pieces are bent over in this way, interspersed with a number of smaller branches (usually not more than ten). Each one is held in place by the weight of the animal. Branches forming crosspieces may be supple enough to bend without breaking, but usually at least half the fibres snap.

The chimpanzee finishes the nest by bending in the small leafy twigs that project from the larger branches. Each one is held down with a hand or foot, the next bent over it, and so on.

Some chimpanzees worked methodically, turning in a complete circle during the making of the nest. Others bent branches first from one side

Plate 3b A mother and infant asleep in the nest. (*Reproduced by permission of the National Geographic Society.*)

Plate 4 (a) Male brandishing a stick prior to throwing it toward his mirror image. (f) Juvenile female termite "fishing." (*Reproduced by permission of the National Geographic Society.*)

and then another, turning round several times during the process.

Fig. 12. Diagram of rough interweaving of cross-pieces—plan view. Examination of nests revealed that this was sometimes quite complex—one branch was often bent backwards and forwards over the foundation several times. (Sketched from observations)

In addition to bending in the crosspieces, a chimpanzee usually breaks off smaller branches and leafy twigs which are then laid on top of the structure. An ape may lay down and then after moving restlessly for a few minutes reach out, pick a few twigs, put them under its head or other part of the body and then lie still.

On many occasions the nest is positioned where the branches of two trees meet, so that material from both goes into the making of the nest. The fronds of adjacent palms are often used in this way. Three times chimpanzees left the nest site to fetch branches from a few yards away when they could not reach them from the nest.

(f) Atypical Nests

Sometimes a nest differs in construction or position from the typical nest described above. One such was only 15 ft from the ground in the *centre* of a low bushy tree. Another, which had been slept in, consisted simply of three small branches bent over a horizontal fork. In this case there were three more branches that could have been used. On other occasions when I found skimpy nests of this sort, all available branches had been used. One female went to the other extreme and continued to bend in branches until she had amassed a heap of some twenty-two boughs, on which she finally lay. Twice I found nests suspended between two small trees, so that the nests had no main supporting branch. The foundation was made by bending down a small bough from each tree and resting it on a branch of the opposite one. One juvenile (Fifi) by looping palm fronds *over* herself and standing

on the ends (as in Fig. 13 a) constructed a nest with a 'roof' during the 1965 rainy season.

A seemingly healthy chimpanzee in Uganda apparently failed to make a nest and spent the night in a crotch of a tree (Reynolds & Reynolds, 1965). One chimpanzee in the Gombe Stream area also failed to make a nest, but this was a sick individual (see p. 199).

(g) Abandoning Nests: Errors in Technique

On eleven occasions I saw chimpanzees make complete nests which they then abandoned. Four of these instances have been described above and occurred when an adolescent female (once) and a juvenile (three times) left their nests in order to sleep closer to other individuals. On two occasions a female and her juvenile offspring left their nests when other chimpanzees arrived to feed in the late evening. Both times mother and child went off with the group afterwards and made new nests elsewhere. Twice a chimpanzee abandoned its nest after suddenly noticing my presence. In the final case the nest was abandoned because of faulty construction. Occasionally a nest is maintained in the normal horizontal position only by the weight of the chimpanzee itself. When the ape leaves, the branches spring up and the nest appears to have been constructed at an angle. A female made a nest of this sort, lay down and was joined by her infant. The added weight of the child upset the balance and tilted the nest. After moving about, presumably trying to adjust the level, the mother and child left the nest and the female made another nearby.

Sometimes a chimpanzee starts to make a nest which is then discontinued because there is insufficient material. I watched three individuals bend over the first few branches of a nest, look around and then, since no other branches were easily available, move away and start to build in another place. On four occasions I found nests which had been abandoned—presumably because the wood was too brittle, since large branches (that would have formed the main supports) had snapped off and fallen to the ground.

It has been noted that chimpanzees nests first appeared in palm trees towards the end of my first period in the field. Soon after this 'fashion' began, I watched an adolescent female trying to make such a nest. She tried to bend a frond over, but it looped over her head (Fig. 13 a), and after struggling for about a minute she gave up and made a nest in another tree.

Apparently the making of nests in palm trees was a new development in these chimpanzees, and called for a slight modification of the basic technique which was acquired only by practice.

(h) Use of an Old Nest

Normally a nest is used for one night only, but on twenty occasions between 1960 and 1964 a chimpanzee re-used a nest that had been previously slept in. Twice the nest concerned had been made by the same individual, six times by a different one, for the remainder, the original owner was not known. In no instance was the nest more than two weeks old, and on each occasion the chimpanzee bent new branches down over the existing structure before settling down to sleep.

This pattern of nest use has been modified near the observation area now that the animals resort to a limited number of 'nesting sites', and chimpanzees sleeping in these places tend to use old nests on a good many occasions. Possibly all the 'best' foundations have already been used.

(i) Body Position in Nest

After making its nest, a chimpanzee may sit on it for a while, or lie down at once. Initially many individuals lie on their backs, holding a higher branch with one hand or foot (Plate 3 b); later they turn over onto their sides, sometimes

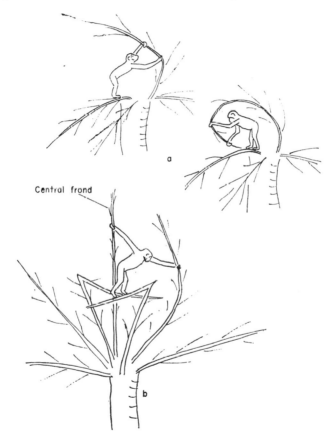

Central frond

Fig. 13. (a) adolescent female attempting to make a nest in a palm tree in 1961. The juvenile female who constructed a nest with a 'roof' during the rains of 1965 used this method.
(b) Normal method of constructing palm nest. The ape first climbs half-way or more up a central frond. Holding on to this it reaches out, pulls another frond towards it, bends this down and holds it in place with its feet. It then bends in up to ten other fronds in a similar manner,

still holding the higher branch. When nests were made early, so that it was possible to watch the chimpanzees during the period before sleep, there was generally little movement save for occasional arm movements, scratching, or turning over.

During the night a chimpanzee often moved round in its nest, so that in the morning its head faced in the opposite direction. Thus an infant which started the night cradled by its mother's arms, its head resting on her shoulder, was seen at dawn in the reverse position, its head resting on her groin.

(j) Nest Hygiene

The chimpanzee, unlike the gorilla, does not foul its nest. I never found excrement in a nest, (except in the case of an old male paralysed in both legs described below) and observation showed that these apes are careful to excrete and urinate over the edge of the nest, even during the night. There was invariably fresh excrement and traces of urine beneath a nest that had been slept in deposited either during the night or, more commonly, at dawn.

(k) Nest-Making Behaviour of Sick Chimpanzees

On four occasions when mature chimpanzees were suffering from bad 'colds' they made nests between 4 and 4.30 p.m. and lay there quietly until darkness. On the mornings following they got up later than normal—between 8.45 and 9.30 a.m. Two of them again made nests during the late morning where they lay for 3 and 4 hr respectively. Two other chimpanzees suffering from a form of respiratory illness remained in elaborate day nests for 4 and 5 hr respectively.

When the old male chimpanzee lost the use of both legs he spent a major part of each day in a nest. During the 7 days that he was under observation (before we had to destroy him) he spent an average of 7 hr 12 min in his nest between 6.30 a.m. and 7.30 p.m. During these days he slept in only two different trees, in each of which he made more than one nest (two in one and three in the other). Typically he pulled himself up to the lowest after climbing into the tree from the ground. Then, towards dusk he pulled himself up to a higher nest. In the morning after spending an hour or so in the 'night nest' he usually swung down to a lower one. On the last evening of his life, after he had dislocated one arm and was no longer able to move, he repeatedly gazed up at the trees above him as night fell. When I broke off a pile of leafy branches he managed to work them under him,

using one arm and his mouth, to form a crude ground nest.

For 3 successive nights an old female suffering from a type of bad 'cold' was apparently too weak to climb into trees. She was not observed to construct a nest before nightfall nor could any trace of a nest be found the following mornings.

Orangutans have been observed to make nests when wounded (Brooke, 1843; Wallace, 1869).

(l) Day Nests

In the rainy season nests are frequently constructed during a rest period in the day time. Sometimes the chimpanzee simply bent a few leafy twigs across the branch on which it was resting, to form a 'cushion'. At other times an elaborate nest was constructed exactly similar to that made for the night. Females with small infants invariably made the latter type of structure (Plate 3a). Reynolds (1965) saw day nests in Uganda and Nissen (1931) found traces of them in French Guinea. Gorillas (Schaller, 1963) and orangutans (Harrisson, 1962) also make nests during the day.

Day nests were constructed at any time during the morning or early afternoon when chimpanzees were resting. They were frequently made at the end of rain, and on several occasions I saw a chimpanzee construct a nest during a lull in heavy rain. When the nest was completed he alternatively lay or sat in a 'hunched over' position as the rain eased off or fell more heavily.

Adults and adolescents of both sexes probably construct day nests on most days throughout the rainy season. Youngsters between the ages of $2\frac{1}{2}$ and 5 years, however, were observed to construct nests more frequently. For instance, one juvenile was seen to make forty-five nests during a 5-month period as compared with her mother's fifteen and her elder sibling's thirteen. She frequently made elaborate nests on which she lay still for only a few moments before swinging off to make another. Sometimes she played on her nests, bouncing up and down or turning somersaults. At other times she constructed crude cushions on which she sat whilst feeding. Similar behaviour was observed in other youngsters. The early attempts at nest-making by small infants also occur during the day (see below).

During the dry season, when chimpanzees rest on the ground, they frequently bend over a few handfuls of vegetation to make crude cushions of the same type as those made in the trees. The material used ranged from a few blades of grass to a small sapling. On six occasions elaborate ground nests were made which involved the bending of up to ten branches from nearby bushes and the breaking off of leafy twigs for a lining.

(m) Development of Nest-making Ability

The extent to which learning is necessary for the development of the nest-making patterns is not known yet, but experiments at the Yerkes Laboratories (Bernstein, 1962) suggest that experience does play some part. A wild-born chimpanzee, even when taken from its natural habitat at a few years of age and kept up to 30 years in conditions where it has no opportunity for building nests, will, nonetheless, build a nest at once when given suitable material. But a number of laboratory chimpanzees which were separated from their mothers after a few days, never built nests when tested in situations where wild-born ones did so. This suggests that some kind of experience obtained under natural conditions during the first few years, and not obtained by laboratory animals (for example, the opportunity to watch others or constantly to manipulate branches and twigs) is necessary for the development of nest-making behaviour.

Certainly in the wild there is much opportunity for the chimpanzee to learn the nest-making patterns, initially by watching and subsequently by imitation and practice. For the first 3 or 4 years the infant sleeps with its mother at night and therefore has no need to construct a sleeping nest for itself. It has, however, ample opportunity for watching the mother while she makes their communal nest and, on two occasions, infants of about one year old 'helped' their mothers by bending down one or two twigs at the edge of the structure. In addition (as has been said above), infants frequently construct small nests during the day, often, apparently, as a form of play activity.

One infant was observed to make his first crude attempts at nest-making when he was 8 months old, whilst crawling around near his mother on the ground. First he pulled down one twig and sat on it. Next he bent down a few blades of grass but, instead of pushing them under him, folded them carefully *onto* his lap. A month later he was observed to bend a few twigs under him whilst dangling from one hand in mid-air! Another infant of approximately 9 months was observed to bend one twig under him successfully, but when he reached out for a second the first sprang up, and when he reached

for that one again, the second sprang up—and so on.

Gradually, however, the infant increases in proficiency, partly due to maturitional development and partly no doubt as a result of constant practice. (One infant, for instance, made six nests in 1 hour, sitting in each for a few minutes only.) Thus by the time the infant has to make its own sleeping nest, it has already acquired a fully developed technique. Schaller's observation on mountain gorillas suggests that infant gorillas also make little nests as a form of play activity before they have to do so as a necessary part of the daily routine (Schaller, 1963).

Sleeping and Resting

Data obtained during 30 nights spent within 20 yd or so of chimpanzees suggested that they sleep fairly soundly. When they were observed in the moonlight, in small groups, there was little movement. Occasionally, however, when several groups were sleeping within earshot there was inter-group calling during the night, especially when there was a moon. Even when I stayed by a group which called eight times during one night, I was unable to ascertain the reason.

Prior to 1966 there was no evidence that chimpanzees ever left their nests or travelled during the night: every time I returned at dawn to a place where I had watched chimpanzees construct nests the previous evening they were still in, or close by, the same nests. However, in May 1966 the top-ranking male chimpanzee (Mike) arrived at the feeding area at 3.00 a.m. on a bright moonlit night. He found a few bananas, moved about, seemingly quite unafraid, for about 30 min, and then moved back up the slope to where his group was sleeping. Prior to, during and after this visit there was a good deal of calling from this group and another group that was nesting just across the valley. It seemed as though the latter moved its position during this time, but the observers were not certain.

Later, during the same month, an adolescent male, (Pepe, see also p. 196) made four visits to the feeding area during the night. It was not known whether or not he was with a group, or sleeping nearby on his own. The first two visits were made during moonlit nights: he arrived at 2.00 a.m. and 1.00 a.m. and stayed on each occasion between 15 and 20 minutes. Each time he found some bananas. The next two visits were made at midnight and 2.40 a.m. on very dark nights when the only light came from the stars. He found no bananas on either visit, stayed between 20 and 30 min, and then moved back up the slope.

After this there were no more night visits for two months. Then in July Mike arrived at 11.50 p.m. on a bright moonlit night. Prior to his arrival there had been a great deal of calling (pant hoots, screams etc) as from a large group of chimpanzees. Judging from these sounds the group had moved over the top of the opposite mountain ridge, travelled down into the valley and finally moved up towards the feeding area. When Mike arrived he gave a number of pant hoots. Shapes as of other chimpanzees were descerned moving about further up the slope, but it was not possible to identify them. Mike stayed for about 30 min, and then moved off up the slope. The calling, now very close at hand, continued on and off for the rest of the night and at 4.45 a.m. Mike returned for another short visit.

Thus it appears that, despite popular belief that most primates are too afraid to leave their sleeping places during the night, these chimpanzees sometimes do move about at night even when there is no moon. Izawa & Itani (1966) also conclude, from evidence of calling during the night, that the chimpanzees in their area, south of Kigoma, also travel occasionally during the night.

Mature chimpanzees normally rested for at least two hours every day, generally at some time between 9.30 a.m. and 3.0 p.m. During the dry season they sprawled out along comfortable branches or (as we have seen) in day nests. During rest periods the apes often slept for 30 min or so; for the rest of the time they simply sat or reclined, idly grooming themselves or each other. Some examples of body positions during resting are illustrated in Fig. 14 a–g. A chimpanzee always lies down in the same way: first by lowering the forebody, holding an overhead bough if possible, and then, when the shoulder is on the ground (or branch), the rump is carefully lowered. On many occasions they lie with their heads at a lower level than their feet, whether on the ground or in nests.

Self-grooming

In captivity the frequency of self-grooming behaviour increases with age (Riopelle, 1963) and this is also true of individuals in the wild. Infants were not seen to groom themselves before their seventh month and only occasionally (and for brief periods) during their second year. Older

Fig. 14. Some typical sitting and resting positions. (e) is a posture commonly adopted by an old female and may be the result of cradling successive infants in a similar postion (compare Fig. 18, p. 224); (g) mature male in 'huddled' position during rain.

infants and juveniles sometimes groomed themselves for up to 5 min at a time, usually when they were themselves being groomed by others. Adolescent and mature animals often groomed themselves for 15 min or so at a time, sometimes very intently.

The areas most frequently groomed are the thighs, arms, chest and abdomen, and close visual inspection normally accompanies grooming behaviour. Chimpanzees sometimes use both hands for grooming, pushing the hair back with the thumb or index finger of one hand and holding it back while picking at the exposed skin with the nail of the thumb or index finger of the other. Or they use one hand, parting hair in the same way and holding it back with the lower lip. Flakes of dried skin etc. are scratched loose and then removed either with the lips or between thumb and index finger. Self-grooming is frequently interspersed with slow, deliberate scratching movements against the direction of hair growth; the area scratched is then groomed. Parts of the body which are not accessible for self-grooming are typically scratched in this way and the chimpanzee then carefully inspects the nails of its fingers. These inaccessible areas (such as the rump, back and neck) are, significantly, the parts of the body most frequently presented to other individuals for grooming (see p. 263).

After heavy rain, when the chimpanzees are wet; they frequently scratch downwards over all accessible areas of their bodies, pulling the hair between their fingers. They also help to dry their coats by shaking, by rubbing their backs and shoulders against a tree (Fig. 15), or (occasionally) by rubbing themselves with handfuls of leaves or by licking off drops of water.

Fig. 15. Male rubbing against a tree trunk after rain.

Sticky or unpleasant substances were usually wiped off, either against the ground or a branch, or with leaves (p. 208). Blood was sometimes removed in the same way   , sometimes licked off, and sometimes the chimpanzee would dab the wound with its fingers, which it then licked. After sneezing, chimpanzees usually removed mucus from their noses by a downward movement from nostril to mouth with the index finger; the mucus was then eaten.

Self-grooming probably functions primarily to keep the body free from dirt etc. However, it also occurs frequently in anxiety or conflict situations when it may be regarded as a displacement activity (pp. 269, 273).

One other behaviour pattern which may also be a displacement activity will be described here since it was most frequently observed in a grooming context. Sometimes when a chimpanzee was grooming himself or socially grooming (and occasionally when he was just sitting resting) he suddenly reached out (often with scarcely a glance in that direction), seized a handful of

leaves, and started to 'groom' them. Peering closely at the leaves and often lip-smacking, he held them with both hands, palms turned slightly upwards and facing inwards, and made grooming movements with one or both thumbs. Occasionally the leaves were brought close to the mouth and minute specks picked off with the lips. After a few moments the leaves were dropped and the chimpanzee resumed whatever he was doing before. Usually other individuals nearby gathered round and also put their faces close to the leaves while being 'groomed'.

When I first saw this I was under the impression that it was related to feeding behaviour: many close examinations of leaves (whilst they were being 'groomed' and afterwards), together with the facts that the chimpanzee often reached out for the leaves at random and took them from any plant, and that the behaviour was usually observed in a grooming context, suggests that it is some form of displacement grooming.

Infant chimpanzees have been observed to make grooming movements at the base of palm trees where remains of old fronds have become frayed. Three instances when chimpanzees groomed dead animals have been cited. One juvenile groomed a dead rat, a mature male groomed the hair on the head of a baboon he was eating   , and one adolescent female groomed a baboon that had just died

Aimed Throwing and Tool-using

The use of natural objects as tools in free-living non-human primates is of interest, not only as shedding light on the capacities of the species itself, but also in connection with theories concerning tool-using and the development of tool cultures in prehistoric man. There is abundant evidence in the literature that captive apes, and some monkeys, can use objects both as aimed missiles and as tools (e.g. Köhler, 1925; Yerkes & Yerkes, 1929; Klüver, 1937). Reports of missile-hurling in wild primates are not uncommon in the literature, but accurate descriptions of apparent aiming and throwing are rare, and there are few references to the use of objects as tools. The evidence in the literature has been summarized and discussed by Hall (1963) and Kortlandt & Kooij (1963). Recently rhesus monkeys in Singapore have been seen using leaves to rub dirt from food or other objects (Chiang, 1967).

The only reliable reports referring to the chimpanzee in the wild appear to be the following: Kortlandt (1963) observed female chimpanzees in the Congo throwing sticks (with poor aim) in the direction of a stuffed leopard, and using saplings as 'whips' in response to the same stimulus. They swayed the saplings to and fro, frequently hitting the leopard with the distal branches. More recently the same author has reported accurate throwing of sticks at the leopard. Beatty (1951) observed a chimpanzee pick up a rock and break open a dried palm nut. Savage & Wyman (1843–44) refer to a chimpanzee, observed in the wild, that used a rock to crack open a small fruit like a walnut. Merfield & Miller (1956) watched a group of chimpanzees poking sticks into an underground bees' nest and then licking the honey, Izawa & Itani (1966) observed similar behaviour in one female. Kortlandt & Kooij (1963) refer briefly to the use of a stick or fruit as a 'toilet aid'.

(a) Aimed Throwing

At the Gombe Stream area the throwing of sticks, stones or handfuls of vegetation, apparently at random, frequently formed part of the branch-waving and frustration displays of mature and adolescent males (chapter IV 6 (c)). Ten of the eleven mature males and three of the six adolescent males listed in Table 9 were observed to throw in this way. Schaller (1963) noted the same non-directional throwing during the male gorilla's chest-beating display.

Chimpanzees sometimes threw underarm with a swinging-forward movement either from a tripedal or bipedal stance. At other times they threw overarm either swinging the arm in an arc through the air, from a bipedal position or hurling the object forward from a tripedal or bipedal position, with the action of a javelin thrower.

In addition to the throwing seen during displays, I observed prior to 1965 ten different individuals throw towards an objective with what appeared to be definite aim*. Two of these instances occurred when an adolescent (once) and a juvenile (once) broke off sticks, retained them whilst peering intently at another individual below and finally threw them carefully with an overarm movement in the direction of the other animal. On these occasions the action was apparently playful. However, at other times objects

*Data on aimed throwing collected since 1965 have not yet been analysed, but there is no doubt that the chimpanzees throw objects at baboons at the feeding area far more frequently than before. This is undoubtedly due to the abnormal competition for bananas between the two species,

were thrown at humans, baboons, other chimpanzees etc. when the context indicated that the actions was aggressive. Table 12 (Appendix) summarizes the observations of aggressive throwing. Five of the six occasions when chimpanzees threw objects at humans occurred when our presence prevented nervous individuals from approaching to take bananas or, in one case, prevented an adolescent male from approaching to investigate a toy chimpanzee. The sixth happened when I walked past a group which had just tried, and failed, to break into the banana store. All cases of throwing at baboons occurred during aggressive encounters at the artificial feeding area, when the baboons approached chimpanzees too closely (Plate 4 b, 5 d). Objects were thrown at other chimpanzees during chasing or bluff charging. The table shows that only three individuals were observed to throw more than four times in this context. It also shows that of the forty-four objects thrown only five hit their objectives: each time these were less than 6 ft away from the chimpanzees concerned. In most cases the aim was good, but the objects fell short.

My earlier observations on throwing (Goodall, 1964) suggested that the chimpanzees threw anything that was to hand. This may be true, but more recent data revealed that 51 per cent of the 44 objects thrown (prior to 1965) were large enough to intimidate baboons and certainly humans. Nevertheless, the rarity of aimed throwing as compared with non-aimed throwing, together with the fact that it was recorded in only eight out of seventeen males suggests that, as a form of threat or defence, throwing, at least prior to the setting up of a feeding area, was not highly developed in this chimpanzee community.

(b) Development of Throwing Behaviour

Play-throwing was seen in infants and occasionally in juveniles and adolescents. One infant threw a palm nut about 2 ft when he was 9 months old. Older infants threw a variety of objects, including leaves, twigs and stones. The youngsters threw overarm with a downward, hitting movement or (frequently) with an upward movement of the arm so that the object went up into the air and landed either in front or behind of them.

The behavioural context from which aimed throwing has derived, however, is probably different from this play-throwing. Hall (1963) has suggested that the behaviour is a simple form of operant conditioning imposed upon the threat-

gesture repertoire of the species. That this may
be the case is suggested by the fact that male
chimpanzees when threatening baboons or
humans often made movements exactly similar
to those seen during aimed throwing (Plate 5 c),
and occasionally let fly an object which they
happened to be holding—such as a banana skin
or even a handful of bananas! In such a case there
was apparently no deliberate throwing movement
but the object often went in the right direction.
This, together with the fact that aimed throwing
is well developed in so few individuals, suggests
that individual experience and learning may
play a role in the development of throwing in
an aggressive context.

(c) 'Whipping' and Hitting with Branches and Sticks

During aggressive interactions with other
chimpanzees, baboons and humans chimpanzees
at the Gombe Stream sometimes swayed saplings
or branches to and fro so that the ends hit,
or nearly hit, the objects which had aroused
their aggression. On two occasions this was the
reaction to my close proximity: once a mature
male, in the early days of the study, climbed
half-way up a sapling when he suddenly saw me
and then proceeded to sway the whole sapling
violently back and forth (by alternately flexing
and extending his arms and legs) until the end
was hitting my head. Another time a male
accustomed to me suddenly jumped down to a
branch just over my head and swayed it in the
same way. Once an adolescent male 'whipped'
a baboon with a low branch, and on another
occasion a high-ranking mature male 'whipped'
another mature male and female with a thin
sapling whilst they were copulating. Other
examples have been observed but not yet extracted from the records.

At the Gombe Stream Reserve chimpanzees
have also been seen brandishing sticks and
hitting towards conspecifics, their own reflections
in a mirror (Plate 4 a), baboons and, once, some
sort of insect on the ground. In most cases the
stick, held firmly in one hand, was raised high
in the air and brought down forcefully in the
direction of the 'opponent'. Usually the weapon
was released prior to the moment of contact.
One of the exceptions was when a 3-year-old
infant 'clubbed' an insect on the ground several
times in succession (Fig. 16).

(d) The Use of Objects as Tools

I saw chimpanzees using natural objects as
tools for a variety of purposes. As we have

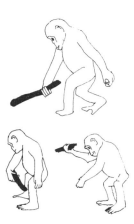

Fig. 16. Sequence showing infant using stick to hit towards an insect on the ground.

seen, sticks, stalks, stems and twigs were used
in connection with insect-eating (chapter III
2 (b)); leaves were used as 'drinking tools'
and for wiping various parts of the body. In
addition, chimpanzees used sticks or grasses as
'investigation probes' and used sticks, at the
artificial feeding area, to try and prise open boxes
containing bananas.

(i) Use of sticks, twigs and grasses for ant and
termite feeding. Chimpanzees use sticks when
feeding on two species of ants. Most of the sticks
which had been used in this way were estimated
as between $1\frac{1}{2}$ and $2\frac{1}{2}$ ft long (although I measured one which was $3\frac{1}{2}$ ft), and they were either
broken from nearby branches or picked up from
the ground. Once an infant picked up a stick
and pushed it into some loose soil where Anomma
were present. Her behaviour was the same as
that of an adult but her tool did not penetrate
the nest.

Stalks, stems and small twigs were used
when the chimpanzees fed on termites (Plate 4 f).
The material used was generally between 6 and
12 in. in length. Some animals inspected several
clumps of grass etc. before selecting their tools;
sometimes they picked several to carry back to
the termite heap and then used them one at a
time. Other individuals used more or less anything within reach, including tools left by others.
It was not uncommon for an individual to
develop a temporary preference for a particular
type of tool. Once, for example, when two
mature females were working at the same termite
heap, one repeatedly climbed a nearby tree and
selected long thin pieces of liana, whilst the

215

other consistently chose green blades of grass. Sometimes tools were carefully prepared: leaves were stripped from stems or twigs with the hand or lips, and long strips were sometimes pulled from a piece of grass that was too wide. Occasionally the leaf blades were stripped from the mid-vein of a palm frond leaflet, and on two occasions I saw chimpanzees pull off strips of bark fibre from which they detached thin lengths for use as tools. When the end of a tool became bent the chimpanzee usually bit off the bent part, or turned it round and used the other end, or else selected a new one. When a chimpanzee has been working for some time at the same nest the termites gradually stop biting onto his tools. When this happens the chimpanzee may do one or more of four things: (i) give up and move away; (ii) move back and forth between two or more holes which he has been using; (iii) search for a new hole; (iv) try a number of different tools in quick succession, discarding each one after one or two pokes if no termites are extracted. I watched one male who had used each tool for an average of about 10 min during an hour's 'fishing', pick up and discard thirteen grasses in 6 min when the termites stopped biting. No chimpanzee was seen to move further than 10 yd from the site to select a new tool, but individuals were seen to pick a tool for subsequent use on a heap that was out of sight and as far as 100 yd away. One male twice carried a tool for over half a mile whilst inspecting a series of termite heaps, none of which were ready for working.

Infants under 2 years of age were not observed to poke grasses etc. into holes in termite heaps although I frequently saw them playing with discarded tools whilst their mothers fed there. In addition, from about 9 months of age they sometimes watched their mothers or other individuals closely and picked up and ate an occasional termite. Slightly older infants between 1 and 2 years of age often 'prepared' grasses etc., stripping them lengthwise or biting pieces from the ends, apparently as a form of play. One infant of 1 year and 7 months once picked up a small length of vine and, holding it with the 'power grip' (Napier, 1961) jabbed it twice at the surface of the termite mound (there was no sign of a hole) and then dropped it.

The youngest chimpanzee of known age that I watched using a tool in a termite hole was 2.2 years old. From this age to about 3 years old, the tool-using behaviour of an infant at a termite heap was characterized by the selection of inappropriate materials and clumsy technique. In addition, no infant under 3 years was seen to persist at a tool-using bout for longer than 5 min—as compared with 15 min during the fourth year and several hours during maturity.

On six occasions three different infants between 2 and 2½ years were observed using, or attempting to use, objects as tools in the termite feeding context, i.e. Goblin aged 2.2 years (twice), Flint aged 2.5 years (twice) and Gilka aged approximately 2 years (twice).

The types of material selected were normally the same as those used by adults, but the infants generally tried to use tools that were far too short to be effective. Thus of the thirty tools which I saw being used by this age group twenty were less than 2 in. Usually these tiny tools were bitten or broken by hand from a tool discarded by an adult, although sometimes an infant picked its own material. Thus one infant broke off a 1½-in. long bit of stick: it was of similar diameter to the hole into which he pushed it and where it became firmly stuck. Of the other ten tools, three were too thin and flexible, two were bent and broken at the end and two were too thick: thus only three were suitable for the purpose. However, that selection occurs in some cases is suggested by the fact that Flint, at 2.5 years, picked up and discarded, *before* trying to use, three tiny bits of green grass and one dead stick about twice as thick in diameter as a normal hole.

The infants held their tools in more or less the same way as an adult (i.e. between thumb and side of index finger) but their techniques were clumsy. Thus typically, after prodding a tool into a hole (often one just vacated by another chimpanzee) the infant pulled it out immediately with a quick jerking movement (unlike the slow careful withdrawal made by an adult). Only once was one of these infants seen to push more than about an inch of its tool into a hole. This was when Flint, aged 2.5 years, pushed a 4-in. straw in with one hand and then used his other hand to push it in further. On this occasion he did withdraw it slowly, but the attempt was unsuccessful. Only on two occasions were infants of this age group observed to 'catch' a termite from a total of twenty-two bouts, ranging from a few seconds to 5 min.

Three infants were observed using tools for termite fishing between the ages of 2½ and 3 years old. (Flint at 2.7 years (twice); Goblin at 2.10 years (once) and Merlin at about 2½ years (eleven times)). They showed only slight improvement in their selective and manipulative ability. Thus whilst they normally used longer tools these were frequently too flexible—and one was about twice the diameter of the hole into which the infant tried twice to push it. Also the infants usually held the tools so that only 1 or 2 in. projected beyond their fingers, and only this end was usually pushed into the hole. However, each of the infants occasionally inserted his tool in more or less the adult manner, pushing it in with one hand and then sliding that hand back up the tool and pushing it again. Usually this age group pulled out the tool slowly, although one of them several times did this in such a way that the grass bent as it was withdrawn: any termites that had bitten on would undoubtedly have been scraped off. He used both hands on these occasions.

One infant (Flint), when he was 2.8 years old, was twice observed using grass tools out of context. Both instances occurred during the termite-fishing season: once he pushed a grass carefully through the hair on his own leg, touched the end of the tool to his lips, repeated the movement and then cast the grass aside. On the other occasion he pushed a dry stem carefully into his elder sibling's groin three times in succession. Schiller (1949) working with captive chimpanzees found that once a complex manipulative pattern had been mastered, an individual frequently performed the action when it was not necessary, as a form of play. It seems likely that Flint's behaviour was of a similar nature. The same may be true of the repeated picking up or making of new tools by infants.

Four-year-olds, with the exception of one individual, showed an adult technique although they did not normally persist for more than 15 min at a stretch. The exception was the infant Merlin, who, during his third year, lost his mother. During the termite season immediately following her death his tool-using behaviour was not observed. The year after, however, when he was about $4\frac{1}{2}$ years old, his tool-using behaviour showed no improvement at all from that which I had watched when he was about 2 years younger—if anything it had regressed. Thus nine tools of the total of nineteen used during two sessions were under 2 in. long, one was bent and the others were always held within an inch or so of their ends. He used the same fast insertion and jerky withdrawal shown by the 2 to $2\frac{1}{2}$-year-olds, and he only caught two termites during the total 40 min. (Four others working at the same time were getting plenty

of insects.) Only in one respect had he advanced: he persisted, during one of the sessions, for 33 min without interruption.

A juvenile female, estimated as about $5\frac{1}{2}$ years of age, fished for termites for consecutive periods of up to $1\frac{1}{2}$ hr, and her technique was exactly comparable to that of an adult.

(ii) Sticks, twigs and grasses used as 'olfactory aids'. On a number of occasions chimpanzees used grasses or sticks as 'olfactory aids' or 'probes'. This was seen frequently during termite-fishing behaviour: after scratching away a small amount of the earth sealing a termite passage a chimpanzee often poked a thin grass some way into the hole, withdrew it and carefully sniffed the end. Subsequently it either enlarged the opening of the passage and worked there, or else left that hole and tried elsewhere.

On four occasions juvenile chimpanzees (two females and one male) poked thin sticks or twigs into small holes in dead branches (probably entrances to wasps' nests). Twice the same female (Fifi) after withdrawing her probe and intently sniffing the end, proceeded to break the wood apart with both hands and teeth and eat some sort of grubs from within. The other, younger female, broke the end of her probe in the hole, effectively sealing it. She tried to poke it out, gave up and moved away. When the male withdrew his probe a large wasp flew out. He broke the branch in half but apparently found the nest empty.

Another time I saw an adolescent female push a long stick three times into a hole in a tree trunk. She sniffed the end each time she withdrew it, then dropped it and moved away.

A juvenile female (Fifi) on three occasions pushed a long grass stalk right into my trouser pocket, subsequently sniffing the end, when I prevented her from feeling there with her hand for a banana. Each time there was, in fact, a banana there, and she followed me, whimpering until I gave it to her. A 2-year-old male infant once used a long twig to inspect the genital area of a female (see p. 220), touching her vulva and then sniffing the end of his probe.

A final example of this type of tool-using occurred when a group of females and youngsters was staring at a dead python (chapter II 4 (d)). A female of about 8 years of age (Fifi), after first sniffing an 8-ft palm frond on which the python had lain previously, three times pushed it, hand over hand, until the tip touched the reptile's head (covered in blood). She then withdrew the frond, again hand over hand, and very intently sniffed the end.

(iii) Other uses of sticks. Some 4 or 5 months after the setting up of the artificial feeding area, three adolescents used sticks to try to open boxes containing bananas. After pulling and pushing at the boxes for up to 5 min, each one broke off a stick and stripped it of leaves. Two of them then tried to push their sticks under the box lids. The third pushed his into the bananas through a hole in the bottom of the box. None of the three had seen either of the others trying to solve the problem in this way.

Throughout the following year, more and more chimpanzees were seen using sticks for this purpose. Some persisted for more than 10 min at a time and broke off anything up to ten different sticks or twigs, trimmed them and poked vainly at the boxes. Such persistence may be explained by the fact that we sometimes opened the boxes when chimpanzees happened to be working at them in this way—the tool-using behaviour was thus, occasionally, rewarded.

Another instance of tool use was as follows: a mature male, who was afraid to take a banana held out to him by hand, shook a clump of tall grass as a mild threat directed towards the human. When there was no response he shook the grasses more violently so that one actually touched the banana. He stared at the fruit and suddenly released the grasses, pulled a long soft plant from the ground, dropped it, turned to break off a thicker stick and then hit the banana to the ground and ate it. When a second banana was offered in this way he hit it from the hand with no hesitation.

(iv) Use of leaves as a drinking tool. In the Gombe Stream area chimpanzees, when unable to reach water which had collected in hollows of trees, normally used leaves like sponges as drinking tools (Goodall, 1964). I saw individuals drinking in this way from natural hollows, but in order to study the behaviour in more detail bowls were scooped out of tree trunks at the feeding area on two occasions. A mature female, two adolescent males, two adolescent females, a juvenile and four infants drank from these bowls at different times; all used leaves, after first drinking as much as they could reach with their lips, with the exception of two infants aged 9 months and 2 years respectively.

Three of the five types of leaves which were used were of known food plants, but it seems likely that any non-poisonous leaves would be used—one chimpanzee used a food wad which it had been chewing prior to drinking, one used a handful of dry dead blossoms and two used pieces of grass. Leaves (from one to eight or more depending in size) were stripped by hand or with the lips from a nearby twig and then usually chewed briefly so that the surface was crumpled and the water-carrying capacity thereby increased. The leaf mass, held between the index and second fingers, was then pushed into the bowl, withdrawn and the water sucked. The process was repeated until the bowl was empty or the chimpanzee was no longer interested. On the one occasion when leaves were not crumpled prior to use (they were tiny leaflets from a big compound leaf) the chimpanzee stuffed them into her mouth *after* wetting them for the first time and chewed as well as sucked.

Only in two cases did infants fail to use leaves when drinking from water bowls: one of 9 months (who did, in fact, use leaves some 3 years later) and one of 2 years. Both drank by licking water from the backs of their fingers*. Another 2-year-old (previously (Goodall, 1964) estimated as $1\frac{1}{2}$ years old.

picked, crumpled and used leaves for drinking in an adult manner, but twice selected very small ones. Once, after using one tiny leaf, she tucked it into her groin and then picked and used another leaf of similar size. Another time she picked and crumpled two leaves when the bowl was completely dry: she pushed them in and left them there. A $3\frac{1}{2}$-year-old, however, was even less efficient, despite the fact that at 9 months he was seen to watch his sibling using leaves for drinking. After drinking some water with his lips he then dipped his hand into the water and licked off the drops. Next he picked up a dry grass, pushed it down into the bowl using similar movements to those seen in termite fishing (Plate 4 e), withdrew and licked it. He then pushed it again into the water, but this time after withdrawal he not only licked the drops but also chewed the end of the grass. He continued using the tool in the same manner for another 3 min but each time he chewed a little bit more until the entire grass was a tiny crumpled mass   . He then dropped this into the water, hooked it out with his fingers,

*Two of the infants seen using leaves also, occasionally, dipped their hands into water bowls and licked off drops; once an adolescent female did the same. A chimpanzee drank thus from a water bowl in Uganda (Reynolds & Reynolds, 1965).

sucked it and dropped it in again. He continued in this way for approximately 10 min after which he moved to pick up a short narrow piece of dead leaf, which, once more, he poked into the water as though fishing for termites. When he withdrew it, however, he immediately crumpled it in his mouth. Two min later he left the bowl. This infant was seen at the bowl on three subsequent occasions: once he used a similar piece of dry leaf which he poked in twice and dropped, another time his hand, the third time, a leaf mass left in the bowl by another individual.

A 4-year-old* (Fifi) used a completely adult technique when she used leaves for drinking. Once when the bowl was nearly empty she let go of the leaves which she had been using and apparently had difficulty in reaching them again. After a few moments she stripped the leaves from a nearby twig, picked it, poked it into the bowl and then licked the end. She next picked and stripped a second twig and repeated the process. Finally she tried to drink with her lips and then moved away. It was not clear whether the twig-using was a tool-using pattern appearing out of context or whether, in fact, she was trying to use a second tool to reach the first.

On one occasion a young juvenile and a 3-year-old infant took it in turns (two or three times each) to push a few small crumpled leaves down into a hole (about ¼ in. in diameter, and 4 in. deep) in a tree trunk. Once the leaves had been pushed down, neither child was able to extract more than a few of them, and soon the hole was packed tightly with foliage, and the youngsters moved away. When I removed the leaves later, I found that there were a few drops of water at the bottom.

A 2½-year-old infant once crumpled leaves in his mouth and then dabbed at tiny specks of banana that were adhering to the branch of a tree.

(v) Leaves used for body-wiping. I saw chimpanzees remove sticky or unpleasant substances from their bodies on eighty-six occasions during 1963 and 1964. Twenty-six times they merely wiped their feet, hands or mouths against the ground or a tree, but on all other occasions and many other times since 1964 they cleaned themselves with leaves. A 3-year-old, dangling above a visiting scientist, Professor R. A. Hinde, wiped her foot vigorously with leaves after stamping on his hair. Sometimes they used

*Previously (Goodall, 1964) estimated as 3½ years old

leaves that were still attached to the stem; at other times they stripped a handful from a twig or plant. Faeces were wiped off with large handfuls of leaves on twenty occasions, seven times when the animals concerned were suffering from diarrhoea and thirteen times when they stepped or sat in dung, or when a mother was dirtied by her infant. Other substances removed with leaves were sticky foods, such as over-ripe, bananas (six times), blood, from a cut or sore which was dabbed at rather than wiped (five times   ); urine (twice), mud (once) and ejaculate (once). On four occasions chimpanzees pulled great sprays of vegetation towards them and rubbed themselves vigorously during heavy rain. On the other occasions it was not possible to identify the substance which was wiped off.

On only one occasion was a chimpanzee seen to wipe another individual: a juvenile removed sticky banana from the head of her infant sibling by gently rubbing with leaves.

I saw wiping behaviour in all age–sex classes except in infants under 10 months. A 9-month-old infant who fell into some diarrhoea made no attempt to wipe himself, either with leaves or against an object. He looked at the dung, whimpered and ran to his mother who ignored the mess. A month later he was seen to wipe himself with leaves for the first time—when he got sticky banana on his chest. Another infant was also first seen to wipe himself with leaves when he was 10 months old.

(e) Development and Evolution of Tool-use

Two questions present themselves in connection with tool-using in the wild: firstly, to what extent, if at all, does individual experience and learning play a part in the development of the behaviour; secondly, from what fundamental behavioural patterns or situations has tool-using behaviour been derived?

(i) Experience and learning. Schiller (1952) suggests that certain inborn manipulative patterns are available to the chimpanzee from which adaptive behaviour such as tool-using may be derived. He found that a maturitional gradient was involved in the development of these patterns: thus chimpanzees of under 2 years of age normally touched or held objects passively without relating them to their environment, whilst older individuals manipulated them actively and used them to poke or hit at other objects. This change in manipulatory patterns, in chimpanzees of different ages could be related to a gradual improvement in the same individuals in the solving of problems which entailed the use of objects as tools. Mason found a similar maturational gradient in the development of object manipulation in infant rhesus monkeys (Mason, Harlow & Rueping, 1959).

In wild chimpanzees a similar development of ability was apparent: infants were clumsy and inefficient in their first attempts at manipulation, and completely sophisticated behaviour (including powers of perseverance as well as selection and manipulation of materials) was not seen until the third or fourth year. This applied to feeding, nest-making and grooming techniques, in addition to the use of objects as tools.

Schiller (1949) also showed that a chimpanzee must have the opportunity to become familiar with various types of objects before he can use them successfully as tools in the test situation. Thus 2-year-old chimpanzees, deprived for 1 year of all opportunity of playing with sticks, were unable to solve food-getting problems involving the use of sticks as tools with a speed comparable to that of individuals of the same age which had been playing freely with sticks during the preceding year. Furthermore, Köhler (1925) found that even when his chimpanzees were familiar with the use of sticks as tools, they were slow in associating a branch growing on a tree with a potential tool; his chimpanzees were not very familiar with trees.

In the wild, from infancy onwards, the chimpanzee has ample opportunity to become familiar with the objects of his environment. Small infants reached out to leaves and branches from the sixth week. Older infants continually investigated their surroundings with their eyes, lips, tongues, noses and hands. Objects were frequently picked up for closer inspection: leaves were plucked from twigs, sticks were lifted up and waved around. Finally, from about 8 months onwards, infants gained further familiarity with twigs and branches whilst feeding, nest-making and moving about through branches. Thus during the gradual maturation of the manipulatory patterns, the infant not only becomes familiar with the objects used as tools, but also acquires practice in handling them.

At this point we should consider whether imitation plays some part in the appearance of tool-using behaviour, over and above the normal development of manipulatory behaviour. That some primates can learn by watching the actions of others has been shown experimentally. Viki, a female infant chimpanzee who was raised like a human child, was able to learn to perform a number of actions as a result of watching the behaviour of the experimenters (Hayes & Hayes, 1952). Darby & Riopelle (1959) showed that a rhesus monkey which observed another individual performing a long series of discrimination problems did indeed benefit from this experience.

In the wild I saw infant chimpanzees, on many occasions, not only watching adults as they worked, but also picking up and using the same tools when the adults moved away. On two occasions a 3-year-old infant watched intently as his mother wiped dung from her bottom: he then picked leaves and wiped his own rump although, almost certainly, it was not dirty.

Infants under 2 years were not seen to use tools for termite eating; nevertheless, one 8-month-old infant, during the first month of the termite season, started to pick and play with grass stems as opposed to the sticks, fruit clusters etc. he had more often played with before. In addition (although this is not specifically related to tool-using) he suddenly began to 'mop' objects (such as stones or his mother) with great vigour and frequency. This gesture is normally seen only during termite-fishing behaviour (p. 188). Other small infants were also observed to play frequently with grasses during termite seasons; 'preparing' of tools such as shredding the edges off wide grass blades, biting the ends of long stems etc. was also incorporated into their play repertoire at this season.

The use of sticks on banana boxes represented a new behaviour pattern, and it is possible that its acquisition was precipitated in some cases by imitation. For instance, a female new to the feeding area remained in the safety of the bushes during her first three visits. On her fourth visit she ventured into the open and immediately picked up a stick and began to poke at the box. It is scarcely probable that such a response would have appeared during her first encounter with the boxes and bananas unless it had some reference to the fact that she had been watching other animals using sticks in a similar manner. The behaviour may be comparable to the spread of 'potato washing' in a group of Japanese macaques (Kawamura, 1959).

Thus, although the evidence is somewhat slender it seems possible that much of the tool-using behaviour described on the preceding pages can be regarded as a series of primitive cultural traditions, passed down from one generation to another by processes of learning and imitation. One observation, however, suggests that this may not always be the case: when the $3\frac{1}{2}$-year-old male poked a grass stem into a water bowl and gradually transformed it into a crumpled water-absorbent mass. This behaviour may have been an example of spontaneous modification of a known behaviour pattern (that of termite-fishing) to suit a different purpose.

(ii) **Behaviour from which tool-using has been derived.** Hall (1963) has suggested that all primate tool-using behaviour has been derived from a fundamental behavioural situation which has 'evolved primarily in the context of agonistic tendencies towards opponents that inhibit direct attack'. He argues that when a primate is frustrated in a food-getting situation, displacement activities and redirection of aggression may occur and that tool-using attempts in the laboratory often consist of throwing movements barely distinguishable from threat gestures.

So far as the Gombe Stream chimpanzees are concerned only one instance of tool use (as opposed to weapon use) appears to comply with this theory. This was when the mature male used a stick to hit a banana from a human hand: it occurred under abnormal conditions which may to some extent be compared to those operating in many laboratory situations.

Hall, however, was considering primarily the use of objects as weapons and was unaware of the frequency with which objects are used for other purposes by the chimpanzees of the Gombe Stream area. Indeed, in this population, tool using in feeding contexts is highly developed, both in selection and manipulation of objects, as compared with aimed throwing. In view of this it may be helpful to discuss another aspect of behaviour from which the use of sticks, twigs and leaves for insect feeding and drinking might have been derived.

In the laboratory, chimpanzees (particularly young ones) show highly developed investigatory behaviour (Butler, 1965). New objects are approached, inspected and often manipulated (Menzel, 1964), and in addtion, chimpanzees which have had an opportunity to handle sticks usually learn to apply them to other objects: they poke at things or hit them prior to

manual investigation (Butler, 1965). Finally, chimpanzees and other primates such as rhesus monkeys will manipulate objects for no other reward than the performance of the activity (Schiller, 1949; Harlow, 1950).

The same characteristics, as we have seen, apply also to wild chimpanzees. The latter, in addition, are constantly practising all the major behaviour patterns involved in the use of tools in the course of their normal daily activities. During nesting and feeding they bend over and break off branches, sticks and twigs, strip leaves from their stems, peel and strip off the bark from twigs. As they move through the trees they frequently break off dead branches and drop them to the ground (behaviour which is not easily understood, but which is relevant to our discussion). Finally, as Andrew (1963c) has pointed out, during grooming behaviour much of the close inspection and detailed manipulation seen in, say, termite-fishing, is also practised.

Keeping in mind the above facts, it requires little imagination to envisage a situation in which a young chimpanzee playing or merely passing by a termite heap might first scratch enquiringly at a spot of wet earth sealing a termite passage, and secondly poke a grass down into the revealed hole. If the youngster was already familiar with termites (and all other species of primates in the reserve do eat winged termites) the insects clinging to the grass might well have proved reward enough to reinforce the behaviour. Similar situations might well have led to ant-eating. The manner in which the use of leaves for drinking may have evolved from a more typical type of tool use, the poking of grasses etc. into holes, was demonstrated by the $3\frac{1}{2}$-year-old infant (p. 207 and Plate 4 d, e).

Thus, it is suggested that whereas aimed throwing has almost certainly been derived from an aggressive behavioural context, as postulated by Hall, the use of other objects, in connection with feeding, may have evolved from a combination of investigatory and manipulatory behaviour which, in certain contexts, was reinforced by food rewards

The use of leaves for wiping the body presents an entirely different problem. No evidence regarding its origin has so far been obtained.

REFERENCES

Altmann, S. A. (1962). A field study of the sociobiology of rhesus monkeys, *Macaca mulatta*. *Ann. N.Y. Acad. Sci.*, **102**, 338–435.

Andrew, R. J. (1963a). Origin and evolution of the calls and facial expressions of the primates. *Behaviour*, **20**, 1–109.

Andrew, R. J. (1963b). Evolution of facial expression. *Science*, **142**, 609–649.

Andrew, R. J. (1963c). *Comment* on Hall, 1963 (q.v.)

Asdell, S. A. (1946). *Patterns of Mammalian Reproduction.* N.Y.: Comstock.

Avis, V. (1962). Brachiation: the crucial issue for man's ancestry. *Southwestern J. Anthrop.*, **18**, 119–148.

Azuma, S. & Toyoshima, A. (1961–62). Progress report of the survey of chimpanzees in their natural habitat, Kabogo Point area, Tanganyika. *Primates*, **3**, No. 2.

Azuma, S. & Toyoshima, A. (1965). *Chimpanzees at Kabogo Point; Sociological studies of monkeys and apes* (ed. S. Kawamura & J. Itani). Tokyo: Chuokoron.

Bally, G. (1945). *Vom Ursprung und von den Grenzen der Freiheit: eine Deutung des Spiels bei Tier und Mensch.* Basel: Karger.

Bartlett, A. D. (1885). On a female chimpanzee now living in the Society's Gardens. *Proc. zool. Soc. Lond.*, 673–676.

Beatty, H. (1951). A note on the behaviour of the Chimpanzee. *J. Mammal.*, **32**, 118.

Berghe, van den (1959). Naissance d'un gorille de montagne à la station de zoologie experimentale de Tshibati. *Folia Scientifica Africae Centralis* **4**, 81.

Berkson, G., Mason, W. A. & Saxon, S. V. (1963). Situations and stimulus effects on stereotyped behaviours of chimpanzees. *J. comp. physiol. Psychol.*, **56**, 786–792.

Bernstein, I. S. (1962). Response to nesting materials of wild born and captive born chimpanzees. *Anim. Behav.*, **10**, 1–6.

Bernstein, I. S. & Schusterman, R. J. (1964). The activity of gibbons in a social group. *Folia primatol.*, **2**, No. 3.

Bingham, H. C. (1928). Sex development in apes. *Comp. Psychol. Monogr.*, **5**, 1–161.

Birch, H. G. (1945). The relation of previous experience to insightful problem-solving. *J. comp. Psychol.*, **38**, 367–383.

Bolwig, N. (1959). A study of the nests built by mountain gorilla and chimpanzee. *S. Afr. J. Sci.*, **55**, No. 11.

Brooke, J. (1843). *Narrative of Events in Borneo and Celebes.* Vol. 1., London. (Cited in Wallace, 1869, q.v.)

Budd, A., Smith, L. G. & Shelley, F. W. (1943). On the birth and upbringing of the female chimpanzee Jaqueline. *Proc. zool. Soc. Lond.*, **113A**, 1–20.

Butler, R. A. (1965). Investigative behaviour. In *Behavior of Non-human Primates* (ed. Schrier, Harlow & Stollnitz). New York/London: Academic Press.

Carpenter, C. R. (1935). Behaviour of red spider monkeys in Panama. *J. Mammal.*, **16**, 171–180.

Carpenter, C. R. (1940). A field study in Siam of the behaviour and social relations of the gibbon. *Comp. Psychol. Monogr.*, **16**, 38–206.

Carpenter, C. R. (1958). Soziologie und Verhalten freilebender nichtmenschlicher Primaten. *Hb. Zool.*, **10**, 1–32.

Carpenter, C. R. (1965). The howlers of Barro Colorado Island. In *Primate Behavior* (ed. DeVore). New York: Holt, Rinehart & Winston.

Chiang, M. (1967). Use of tools by wild macaque monkeys in Singapore. *Nature*, **214**, 1258–1259.

Collias, N. E. (1944). Aggressive behavior among vertebrate animals. *Physiol. Zool.*, **17**, 83–123.

Cross, H. F. & Harlow, H. F. (1963). Observation of infant monkeys by female monkeys. *Percept. mot. skills*, **16**, 11–15.

Darby, C. L. & Riopelle, A. J. (1959). Observational learning in the rhesus monkey. *J. comp. physiol. Psychol.*, **52**, 94–98.

de Bournonville, D. (1967). Contribution a l'étude du Chimpanze en Republique de Guinée. *Bull. Inst. fondamental Afr. noire*, XXIX A, No. 3.

DeVore, I. (1963). Mother–infant relations in free-ranging baboons. In *Maternal Behavior in Mammals* (ed. by H. L. Rheingold). New York/London: Wiley.

DeVore, I. & Hall, K. R. L. (1965). Baboon Ecology. In *Primate Behavior* (ed. I. DeVore). New York: Holt, Rinehart & Winston.

Frank, L. K. (1959). Tactile communication. In *ETC: a Review of General Semantics* 1953–1958 (ed. S. I. Hayakawa). New York: Harper.

Gillman, J. & Gilbert, C. (1946). The reproductive cycle of the chacma baboon (*Papio ursinus*) with special reference to the problems of menstrual irregularities as assessed by the behaviour of the sex skin. *S. Afr. J. med. Sci.*, **11**, Biol. Suppl.

Goodall, J. (1962). Nest-building behaviour in the free-ranging chimpanzee. *Ann. N.Y. Acad. Sci.*, **102**, 455–467.

Goodall, J. (1963). Feeding behaviour of wild chimpanzees: a preliminary report. *Symp. zool. Soc. Lond.*, **10**, 39–48.

Goodall, J. (1964). Tool-using and aimed throwing in a community of free-living chimpanzees. *Nature*, **201**, 1264–1266.

Goodall, J. (1965). Chimpanzees of the Gombe Stream Reserve. In *Primate Behavior* (ed. I. DeVore). New York: Holt, Rinehart & Winston.

Haddow, A. J. (1952). Field and laboratory studies on an African monkey, *Cercopithecus ascanius schmidti* Matschie. *Proc. zool. Soc. Lond.*, **122**, 297–394.

Hall, K. R. L. (1962). The sexual, agonistic and derived social behaviour patterns of the wild chacma baboon, *Papio ursinus. Proc. zool. Soc. Lond.*, **139**, 283–327.

Hall, K. R. L. (1963). Tool-using performances as indicators of behavioural adaptability. *Current Anthropol.*, **4**, 479–494.

Hall, K. R. L. (1964). In *The Natural History of Aggression* (ed. by J. D. Carthy & F. J. Ebling). London/New York: Academic Press.

Hall, K. R. L. & DeVore, I. (1965). Baboon social behaviour. In *Primate Behaviour* (ed. I. DeVore). New York: Holt, Rinehart & Winston.

Hall, K. R. L. & Gartlan, J. S. (1965). Ecology and behaviour of the vervet monkey, *Cercopithecus aethiops*, Lolui Island, Lake Victoria. *Proc. zool. Soc. Lond.*, **145**, 37–56.

Harlow, H. F. (1950). Learning and satiation of response in intrinsically motivated complex puzzle performance by monkeys. *J. comp. physiol. Psychol.*, **43**, 289–294.

Harlow, H. F. (1960). Primary affectional patterns in primates. *Am. J. Orthopsychiat.*, **30**, 676–684.

Harlow, H. F. & Harlow, M. K. (1962). Social deprivation in monkeys. *Sci. American*, **207**, 137–146.

Harlow, H. F., Harlow, M. K. & Hansen, E. W. (1963). The maternal affectional system of rhesus monkeys. In *Maternal Behavior in Mammals* (ed. H. L. Rheingold). New York/London: Wiley.

Harlow, H. F. & Zimmermann, R. R. (1958). The development of affectional responses in infant monkeys. *Proc. Am. phil. Soc.*, **102**, 501.

Harlow, M. K. & Harlow, H. F. (1966). Affection in primates. *Discovery*, **27**, 11–17.

Harrisson, B. (1962). *Orang-Utan*. London: Collins.

Hayes, C. (1951). *The Ape in our House*. New York: Harper & Row.

Hayes, K. J. & Hayes, C. (1952). Imitation in a home-raised chimpanzee. *J. comp. physiol. Psychol.*, **45**, 450–459.

Herscher, L., Richmond, J. B. & Moore, A. U. (1963). Maternal behavior in sheep and goats. In *Maternal Behavior in Mammals* (ed. H. L. Rheingold). New York/London: Wiley.

Hinde, R. A. & Rowell, T. E. (1962). Communication by postures, gestures and facial expressions in the rhesus monkey (*Macaca mulatta*). *Proc. roy. Soc. Lond.*, **138**, 1–21.

Hinde, R. A., Rowell, T. E. & Spencer-Booth, Y. (1964). Behaviour of socially living rhesus monkeys in their first six months. *Proc. roy. Soc. Lond.*, **143**, 609–649.

Hinde, R. A. & Tinbergen, N. (1958). The comparative study of species-specific behaviour. In *Behavior and Evolution* (ed. A. Roe & G. G. Simpson). New Haven: Yale University Press.

Hoof, J. A. R. A. M. van (1967). The facial displays of the catarrhine monkeys and apes. In *Primate Ethology* (ed. D. Morris). London: Weidenfeld & Nicholson.

Imanishi, K. (1965). The origin of the human family—a primatological approach. In *Japanese Monkeys— a collection of translations* (ed. and publisher S. A. Altmann).

Imanishi, K. (1960). Social organisation of subhuman primates in their natural habitats. *Current Anthropology*, **1**, 393–404.

Izawa, K. & Itani, J. (1966). Chimpanzees in Kasakati Basin, Tanganyika. *Kyoto Univ. Afr. Studies*, **1**.

Jay, P. (1963). Mother–infant relations in langurs. In *Maternal Behavior in Mammals* (ed. H. L. Rheingold). New York/London: Wiley.

Jay, P. (1965a). The common langur of North India. In *Primate Behavior* (ed. I. DeVore). New York: Holt, Rinehart & Winston.

Jay, P. (1965b). Field studies. In *Behavior of Non-human Primates* (ed. Schrier, Harlow & Stollnitz). New York/London: Academic Press.

Kaufmann, J. H. (1967). Social relations of adult males in a free ranging band of rhesus monkeys. In *Social Communication Among Primates* (ed. S. A. Altmann). University of Chicago Press.

Kawamura, S. (1959). The process of sub-culture propagation among Japanese macaques. *Primates*, **2**(1), 43–60.

Klüver, H. (1937). Re-examination of implement-using behaviour in a cebus monkey after an interval of three years. *Acta psychol.*, **2**, 347.

Koford, C. B. (1963). Group relations in an island colony of rhesus monkeys. In *Primate Social Behavior* (ed. C. H. Southwick). Princeton, N.J.: van Nostrand.

Köhler, W. (1925). *The Mentality of Apes*. New York: Kegan Paul, Trench, Trubner.

Kohts, N. (1935). Infant ape and human child. *Sci. Mem. Mus. Darwin, Moscow*, **3**, 1–596.

Kortlandt, A. (1962). Chimpanzees in the wild. *Sci. American*, **206**, (5), 128–138.

Kortlandt, A. (1965). How do chimpanzees use weapons when fighting leopards? *Yb. Am. philosoph. Soc.*, 327–332.

Kortlandt, A. (1966). On tool-use among primates. *Current Anthropol.*, **7**, 215–216.

Kortlandt, A. & Kooij, M. (1963). Protohominid behaviour in primates. *Symp. zool. Soc. Lond.*, **10**, 61–88.

Kummer, H. & Kurt, F. (1963). Social units of a free-living population of hamadryas baboons. *Folia primat.*, 14–19.

Kummer, H. (1967). Tripartite relations in hamadryas baboons. In *Social Communication Among Primates* (ed. S. A. Altmann). University of Chicago Press.

Lawick-Goodall, J. van (1967). Mother-offspring relationships in chimpanzees. In *Primate Ethology* (ed. D. Morris). London: Weidenfeld & Nicolson.

Lawick-Goodall, J. van (1968). Expressive movements and communication in free-ranging chimpanzees: A preliminary report. In *Primates: Studies in Adaptation and Variability* (ed. P. Jay). New York: Holt, Rinehart & Winston.

Loizos, C. (1967). Play behaviour in higher primates: a review. In *Primate Ethology* (ed. D. Morris). London: Weidenfeld & Nicolson.

Lorenz, K. Z. (1956). Plays and vacuum activities in animals. In *L'Instinct dans le Comportement des Animaux et de l'Homme* (no ed.). Paris: Masson.

Marler, P. (1965). Communication in monkeys and apes. In *Primate Behavior* (ed. I. DeVore). New York: Holt, Rinehart & Winston.

Mason, W. A. (1964). Sociability and social organisation in monkeys and apes. *Adv. Exp. soc. Psychol.*, **1**, 277–305.

Mason, W. A. (1965a). The social development of monkeys and apes. In *Primate Behavior* (ed. I. DeVore). New York: Holt, Rinehart & Winston.

Mason, W. A. (1965b). Determinants of social behavior in young chimpanzees. In *Behavior of Non-human Primates* (ed. Schrier, Harlow & Stollnitz). London/New York: Academic Press.

Mason, W. A. (1967). Motivational aspects of social responsiveness in young chimpanzees. In *Early Behaviour: Comparative and Developmental Approaches* (ed. H. W. Stevenson). New York: John Wiley & Sons.

Mason, W. A. & Berkson, G. (1962). Conditions influencing vocal responsiveness of infant chimpanzees. *Science*, **137**, 127–128.

Mason, W. A., Harlow, H. F. & Rueping, R. R. (1959). The development of manipulatory responses in the infant rhesus monkey. *J. comp. physiol. Psychol.*, **52**, 555–558.

Mason, W. A., Hollis, J. H. & Sharpe, L. G. (1962). Differential responses of chimpanzees to social stimulation. *J. comp. physiol. Psychol.*, **55**, 1105–1110.

Menzel, E. W. (1964). Patterns of responsiveness in chimpanzees reared through infancy under conditions of environmental restriction. *Psychol. Forsch.*, **27**, 337–365.

Menzel, E. W. (1966). Responsiveness of objects in free-ranging Japanese monkeys. *Behaviour*, **26**, 130–150.

Merfield, F. G. & Miller, H. (1956). *Gorillas Were My Neighbours.* London: Longmans.

Napier, J. R. (1961). Prehinsility and opposability in the hands of primates. *Symp. zool. Soc. Lond.*, **5**, 115–132.

Nissen, H. W. (1931). A field study of the chimpanzee. *Comp. Psychol. Monogr.*, **8**, 1–22.

Nissen, H. W. (1954). The development of sexual behaviour in chimpanzees. *Symp. genet. psychol. Maintenance Patterns of Sexual Behaviour in Mammals.* Cited in *Sex and Internal Secretions* (ed. W. C. Young).

Osborn, R. (1963). Some observations on the behaviour of the mountain gorilla. *Symp. zool. Soc. Lond.*, **10**, 29–38.

Reynolds, V. & Reynolds, F. (1965). Chimpanzees of the Budongo Forest. In *Primate Behavior* (ed. I. DeVore). New York: Holt, Rinehart & Winston.

Riopelle, A. J. (1963). Growth and behavioural changes in chimpanzees. *Z. Morph. Anthropol.*, **53**, 53–61.

Riopelle, A. J. & Rogers, C. M. (1965). Age changes in chimpanzees. In *Behavior of Non-human Primates* (ed. Schrier, Harlow & Stollnitz). New York/London: Academic Press.

Rowell, T. E. (1962). Agonistic noises of the rhesus monkey (*Macaca mulatta*). *Symp. zool. Soc. Lond.*, **8**, 91–96.

Rowell, T. E. & Hinde, R. A. (1962). Vocal communication by the rhesus monkey (*Macaca mulatta*). *Proc. zool. Soc. Lond.*, **138**, 279–294.

Sade, D. S. (1965). Some aspects of parent-offspring and sibling relations in a group of rhesus monkeys, with a discussion of grooming. *Am. J. Phys. Anthropol.*, **23**, 1–17.

Savage, T. S. & Wyman, J. (1843-44). *Boston J. nat. Hist.*, **4**, 383.

Schaller, C. B. (1961). The orang-utan in Sarawak. *Zoologica*, **46**, 73–82.

Schaller, C. B. (1963). *The Mountain Gorilla: Ecology and Behavior.* Chicago: University of Chicago Press.

Schiller, P. H. (1949). Innate motor action as a basis of learning. In *Instinctive Behavior* (ed. C. H. Schiller). New York: International Universities Press.

Schiller, P. H. (1952). Innate constituents of complex responses in primates. *Psychol. Rev.*, **59**, 177–191.

Scott: J. P. (1958). *Aggression.* Chicago: University of Chicago Press.

Southwick, C. H., Beg, M. A. & Siddiqi, M. R. (1965). Rhesus monkeys in North India. In *Primate Behavior* (ed. I. DeVore). New York: Holt, Rinehart & Winston.

Simonds, P. E. (1965). The bonnet macaque in South India. In *Primate Behavior* (ed. I. De-Vore). New York: Holt, Rinehart & Winston.

Sparks, J. (1967). Allogrooming in primates: a review. In *Primate Ethology* (ed. D. Morris). London: Weidenfeld & Nicolson.

Thorpe, W. H. (1956). *Learning and Instinct in Animals.* London: Methuen.

Thorpe, W. H. (1963). Antiphonal singing in birds as evidence for avian auditory reaction time. *Nature*, **197**, 774–776.

Wallace, A. (1869). *The Malay Archipelago.* London: Macmillan.

Washburn, S. L. & DeVore, I. (1961). The social life of baboons. *Scient. Am.*, **204**, (6), 62–71.

Yamada, M. (1963). A study of blood-relationship in the natural society of the Japanese macaque. *Primates* **4**, 43.

Yerkes, R. M. (1943). *Chimpanzees, a Laboratory Colony.* New Haven: Yale University Press.

Yerkes, R. M. & Elder, J. H. (1936). Oestrus receptivity and mating in chimpanzee. *Comp. Psychol. Monogr.*, **13**, No. 5.

Yerkes, R. M. & Yerkes, A. W. (1929). *The Great Apes.* New Haven: Yale University Press.

Zuckerman, S. (1932). *The Social Life of Monkeys and Apes.* London: Routledge.

Part III
HABITAT CONTROL AND
BEHAVIORAL ENERGETICS

Editors' Comments
on Papers 11 Through 15

11   WELLINGTON
     Tents and Tactics of Caterpillars

12   PEARSON
     *The Oxygen Consumption and Bioenergetics of Harvest
     Mice*

13   FRITH
     *Temperature Regulation in the Nesting Mounds of the
     Mallee-Fowl,* Leipoa ocellata *Gould*

14   WHITE and KINNEY
     Avian Incubation

15   WHITE, BARTHOLOMEW, and HOWELL
     *The Thermal Significance of the Nest of the Sociable Weaver
     Philetairus socius: Winter Observations*

In Part III we discuss the function of external construction by animals. Further advancement in the study of other problems, such as the evolution or mechanisms of external construction, demands a closer understanding of the uses of external constructions in the lives of animals.

However, a comprehensive consideration of all the diverse functions of animal architecture would probably require another volume. Instead, we have decided to illustrate the general area of function and its experimental analysis by selecting one general topic for presentation, habitat control and energetics, with special reference to temperature regulation in relation to external construction. There are several reasons for this choice. Temperature control is well suited for quantitative study, and a good deal of research has been done in this area. Temperature control is particularly important in the life of animals that live on land, where temperature variations are generally much greater than in water. Temperature regulation provides an excellent example of dynamic and social homeostasis.

Temperature regulation by very small animals, such as insects, is often most conveniently studied in groups of animals, and such study exemplifies what W. C. Allee (1931) called "mass physiology," one aspect of sociobiology. Such work has been pioneered by W. G. Wellington on tent caterpillars, by many early apiculturists, and by M. Lüscher on termite nests. Vertebrate zoologists have more often measured the body temperature of individual animals. During the last thirty years a new breed of naturalist and physiologist has arisen, interested in the physiology of whole animals in nature. Leading investigators have included S. C. Kendeigh and E. P. Odum, as well as G. A. Bartholomew, H. F. Frith, L. Irving, O. P. Pearson, and K. Schmidt-Nielsen.

These physiological ecologists soon discovered that the animal often determines the physiological conditions to which it is exposed by means of its behavior. Thus Schmidt-Nielsen and his wife (1953), after careful environmental and physiological measurements and calculations based on these, determined that life in the desert for the kangaroo rat, *Dipodyomys spectabilis*, would be physiologically impossible, were it not that this creature avoids the full impact of desert conditions by living in a burrow by day and by being active above ground only at night.

Problems of temperature regulation are closely related to those of energetics, since energy requirements and expenditures of animals have often been measured in terms of calories. In recent years there has been a shift in emphasis from a study of temperature control per se to a study of the interrelations between temperature control and energetics in the natural life of animals. Much the same investigators are involved as before, as well as many younger investigators, and interest in the field is developing almost explosively, for example in birds, where the first general symposium on the subject of avian energetics was published recently (see Paynter, 1974).

Interest in animal energetics is of course a very old subject but one that is of great current interest and of potentially great significance. The problem of animal energetics attracted attention long ago for honeybees. Darwin (1859, pp. 233–234) remarks: "I am informed by Mr. Tegetmeier that it has been experimentally found that no less than from 12 to 15 pounds of dry sugar are consumed by a hive of bees for the secretion of each pound of wax; so that a prodigious quantity of fluid nectar must be collected and consumed by the bees in a hive for the secretion of the wax necessary for the construction of their combs." Wilson (1963), in his review of the social biology of ants, suggested the

227

term "ergonomics," borrowed from human sociology, for the quantitative study of the distribution of work, performance, and efficiency in insect societies.

The energetics approach can be applied to the study of evolution as well as to the ecology and behavior. When one of us (N. E. C.) was a graduate student at the University of Chicago, Alfred Emerson, then editor of *Ecology*, invited him to review a book for that journal; the theme of this book was the comparison in terms of evolutionary success of a bird and a mammal. The faculty was duly consulted, and Professor Sewall Wright suggested that, although there is no absolute standard for judging success in evolution, for purposes of measurement and objective comparison one could study the amount of energy transformed by any two groups of animals being compared. Collias (1939) then extended the idea to various ecological concepts in the belief that eventually the study of evolution, population ecology, and behavior could be synthesized from the viewpoint of energetics.

Measurement, or at least estimation, of the amount of energy required by the various activities of a species of animal can be combined with various measures of the amount of these activities, for example, time budgets, and makes possible a quantitative analysis of behavior and a relatively new field of biology that could be termed "behavioral energetics." For example, we have found in comparing colonies of different races of African village weaverbirds (*Ploceus cucullatus*) from cool and hot localities, respectively, that birds from the former locality had a thicker nest and had to do more flying, that is, expend more energy, in building the nest. In turn the better insulated nest helps protect these birds from the cool mountain climate they inhabit (Collias and Collias, 1967, 1971).

In our selection of papers for this part we have attempted to gather together articles of key significance that illustrate some problems and principles of temperature regulation and of behavioral energetics in relation to various types of external construction in different habitats.

The article by W. G. Wellington (Paper 11) on the western tent caterpillar, *Malacosoma pluviale* (Dyar), on the west coast of Canada, lays stress on the importance of population quality and of individual differences in relation to the tent and survival. The silken tent made by a cluster of caterpillars functions as a sort of greenhouse, collecting heat in sunny periods and slowing its loss during cloudy intervals. The tiny larvae are then able to feed on the enclosed leaves, digest their food, and grow in body size.

After these leaves are consumed, the group must forage farther afield, and for this the colony needs to have some active leaders. A colony composed of all sluggish individuals, even though they weave a warmer tent with closer mesh, may starve with fresh leaves only a foot away.

Wellington is director of the Institute of Animal Resource Ecology at the University of British Columbia in Vancouver. When he began his studies of insect populations, he chose tent caterpillars because their conspicuous tents made them easy to survey. He then found that individual differences in behavior could affect the welfare of whole colonies and began studying the interactions between individual activity patterns and group survival. The article reproduced here is actually a brief synthesis of work that he and his co-workers have been carrying on for over twenty years and that they have published as a series of many technical articles. Perhaps the really key article, which space limitations have prevented us from using, was published by Wellington in 1957 and is entitled "Individual Differences as a Factor in Population Dynamics: The Development of a Problem." It outlines a good part of the research program he subsequently followed (see also Wellington, 1960, 1964).

Some of the best examples of control of the microclimate within the nest come from studies of the highly social insects, particularly honeybees and termites. We lack sufficient space to reprint any article in this part on these insects, but two key examples are well worthy of a brief description.

Many apiculturists have studied the classic case of temperature control within honeybee hives. The earlier literature is reviewed by Ribbands (1953). The bees seal off all cracks or crevices opening into the hive with a mixture of beeswax and plant resins. When the temperature falls in winter, the bees in the hive cluster together. A quiet outer shell of bees acts as an insulator while the bees more centrally located in the cluster move about. By moving or vibrating their wing muscles, the bees generate metabolic heat. A temperature difference as great as 59°C (106°F) has been recorded between the outside environment (−28°C) and the center of a cluster (31°C) of bees in winter. The enclosed and insulated hive and the remarkable social thermoregulation of the common honeybee have enabled this species, which belongs to an essentially tropical group, to invade and live in cold climates.

Honeybees also have an impressive ability to cool down the hive during very hot weather (Lindauer, 1961). Some bees bring droplets of water into the hive while others fan the moist air out the entrance hole. When they are raising brood, the temperature at the center of the brood nest was found by several investigators to have a mean daily range of less than 1°C. Optimal brood development occurs between 34.5 and 35°C (see Ribbands, 1953).

Lüscher (1961) has found that species of termites with larger and better-insulated nests maintain a higher, more constant temperature in the nest than do species having small nests with thin walls. Even more remarkable is the air-conditioning and ventilation system he describes for the large mounds of a fungus-growing termite (*Macrotermes*). The metabolism of the mass of termites and of the fungus gardens in the interior part of the mound keeps this part several degrees warmer than the outer parts of the mound. A slow circulation of air was found within the mound with warm air, containing a relatively high percentage of CO_2, rising by convection from the brood chamber in the center of the nest to an air space ("attic") above, whence the air moves outward to a series of narrow air channels just within the relatively thin-walled vertical, external ridges of the mound. Here the air is cooled by radiation through the walls and sinks. At the same time, at least in dry weather, the air loses part of its CO_2 by diffusion outward through the walls. The cooled and refreshed air then moves centrally to an air chamber ("cellar") just beneath the brood chamber and fungus gardens in the center of the nest, and the whole circulation is repeated.

Mammals and birds, like insects, exercise some control over the temperature within their nests. Kinder (1927) kept rats at different temperatures and found that the amount of nest building was inversely proportional to room temperature. At 40 to 60°F the rats used five to six times as many strips of paper for nest building as they did at 90°F.

Oliver Pearson's article (Paper 12) on the harvest mouse is notable for its imaginative placing of ecology and physiology on a common basis by the use of metabolic measures. By measuring the oxygen-consumption rate under experimental conditions, he was able to compare the relative energy-conserving value to the mouse of its fur and of its nest at different temperatures. This is a method of wide general significance, and Pearson also uses it to gain some idea of the energetic value to the mouse of huddling and of nocturnal habits. Pearson has been director of the Museum of Vertebrate Zoology at the University of California in Berkeley. He and his students have been doing a good deal of work on the ecological and behavioral aspects of bioenergetics in various terrestrial vertebrates. He was one of the first persons to advance the idea of metabolic budgets of animals in relation to their natural life.

H. J. Frith's article (Paper 13) is on the mallee fowl, an Australian bird, which instead of sitting on its eggs incubates them in a mound that it builds of soil and decaying vegetation. This bird and its mound nest illustrate a remarkable degree of behavioral thermoregulation. The temperature of its mound varies systematically with the seasons in such a way that the temperature of the nest close to the eggs is maintained within a narrow range between 32 and 35°C. To obtain this information required many hours and months of observation, often under trying conditions. Nevertheless, throughout the work the author maintained high and uncompromising standards of accurate description and measurement. He then followed up his descriptive work and confirmed his interpretations with an experimental study of temperature control in the mound (1957).

We first met Frith and heard of his interesting work with the mallee fowl in the summer of 1953 at the Delta Waterfowl Research Station at the south end of Lake Manitoba, Canada. We were impressed by the fact that the Commonwealth Scientific and Industrial Research Organization of Australia was sending this young scientist around the world to visit various research stations and institutions to widen his perspective and for some first-hand acquaintance with what was going on in the world of research outside his country. Some twenty years later we met him again in Australia, where Frith was now chief of the Division of Wildlife Research of the C.S.I.R.O., and as secretary-general had just organized the first international congress of ornithology ever held in Australia, which we attended. In the meantime, he had written a book on the mallee fowl (1962) and its peculiar nesting habits, and he has also greatly extended his investigations into the nesting habits of other species of megapodes.

The article by White and Kinney (Paper 14) develops in a sophisticated, quantitative fashion the importance of multiple-factor interaction among physiology, behavior, and the environment for avian incubation, and the key importance of adequate nest insulation. These authors show how a complex system of interactions among behavior, environment, nest, and eggs results in effective regulation of egg temperature. The village weaverbird (*Ploceus cucullatus*) with its finely woven roofed nest is the main experimental example. In this species the male weaves the outer shell of the nest, and the female lines the lower half of the nest and does all the incubating. The investigators found that females with relatively well insulated nests were able to spend a smaller proportion of their time incubating and so had more time available for feeding and other activities than did birds with more poorly insulated nests. This factor of nest insulation may be critical for small passerine birds, which must fast overnight within the nest when external temperatures reach a minimum.

White is professor of physiology at the University of California, Los Angeles, and Kinney is a postdoctoral fellow in the same department (physiology). White, who once studied with S. C. Kendeigh at the University of Illinois, is much interested in thermal aspects of environmental physiology of vertebrate animals. He is also the author of the article "Respiration and Respiratory Mechanisms" in the *Encyclopedia Brittanica*.

The last article in this section (Paper 15), by White, Bartholomew, and Howell, illustrates social homeostasis, with special reference to temperature control during the winter in the large communal nest of the sociable weaver, *Philetairus socius*, of the Kalahari Desert in southwestern Africa. This nest, the most spectacular built by any species of bird, resembles a gigantic apartment house, is built of straws and fine twigs, and has the entrances to the many nest chambers all opening on the underside of the massive structure. During winter nights temperatures in the Kalahari often fall to freezing, but the air temperatures in chambers occupied by the birds, which sleep in their nest, were found to be as much as 18 to 23°C above external air temperature. The temperatures in the chambers depend on both the heat released by the birds and on the insulation supplied by the nest. The more the birds and the bigger the nest, the greater the independence from variations in outside temperatures. The resulting selection pressure has helped stimulate evolution of the high degree of aggregation characteristic of this sociable species as well as the enormous size of the communal nest. The large nest mass has often been likened in appearance to a haystack in a tree. The relatively high temperatures in the chambers at night diminish the metabolic cost of thermoregulation and, the authors calculate, allow an energy expenditure of 40 percent less than would be necessary if the birds roosted in the open. This energy saving in turn enables the birds to survive the cold winter of the Kalahari. One should keep in mind that other ecological factors besides temperature control, for example predation pressure, were probably involved in the evolution of the compound nest.

The authors also returned to study the same nest in the summertime (Bartholomew, White, and Howell, in press). Although air temperatures outside the nest ranged from 16° to 33.5°C, such was the insulating value of the thick nest mass against the desert heat that temperatures in occupied chambers varied over a range of only 7 or 8°C. In winter the observers had found up to four or five roosting adult birds per chamber, with some chambers left empty. In the same nest in summer they found no more than two adults per chamber, and virtually all chambers were occupied.

We have already introduced White. The other two authors are both professors in the biology department of the University of California at Los Angeles. Bartholomew has spent many years investigating the adaptations of different animals to desert environments in various parts of the world, and he has received the Brewster Medal, highest award of the American Ornithologists' Union, for his research on the physiological ecology, behavior, and energetics of desert birds. Howell has also traveled widely in the course of his research and has made many notable contributions to our understanding of bird biology. He has served as president of the Cooper Ornithological Society, chief ornithological society of western North America.

REFERENCES

Allee, W. C. 1931. *Animal Aggregations: A Study in General Sociology.* University of Chicago Press, Chicago.

Bartholomew, G. A., F. N. White, and T. R. Howell. 1975. The thermal significance of the nest of the sociable weaver, *Philetairus socius*—summer observations. *Ibis,* 117: in press.

Collias, N. E. 1939. The California woodpecker and Ritter. *Ecology,* 20(3):423–424. (A book review.)

Collias, N. E., and E. C. Collias. 1967. A quantitative analysis of breeding behavior in the African village weaverbird. *Auk,* 84:396–411.

Collias, N. E., and E. C. Collias. 1971. Some observations on behavioral energetics in the village weaverbird. I. Comparison of colonies from two subspecies in nature. *Auk,* 88:124–133.

Darwin, C. 1859. *On the Origin of Species by Means of Natural Selection, or the Preservation of Favored Races in the Struggle for Life.* John Murray, London. 502 pp.

Frith, H. J. 1957. Experiments on the control of temperature in the mound of the mallee fowl, *Leipoa ocellata* Gould (Megapodudae). *C.S.I.R.O. Wildl. Res.,* 2:101–110.

Frith, H. J. 1962. *The Mallee Fowl.* Angus & Robertson Ltd., London. 136 pp.

Kinder, E. F. 1927. A study of the nest-building activity of the albino rat. *J. Exp. Zool.,* 47:117–161.

Lindauer, M. 1961. *Communication Among Social Bees.* Harvard University Press, Cambridge, Mass. 143 pp.

Lüscher, M. 1961. Air-conditioned termite nests. *Sci. Am.,* 205(1):138–145.

Paynter, R. A. (ed.). 1974. Avian energetics. *Publ. Nuttall Ornithol. Club 15.* 334 pp.

Ribbands, C. R. 1953. *The Behaviour and Social Life of Honeybees.* Bee Research Association, London. 352 pp.

Schmidt-Nielsen, K., and B. Schmidt-Nielsen. 1953. The desert rat. *Sci. Am.,* 189(1):73–78.

Wellington, W. G. 1957. Individual differences as a factor in population dynamics: the development of a problem. *Canadian J. Zool.,* 35(3):292–323.

Wellington, W. G. 1960. Qualitative changes in natural populations during changes in abundance. *Canadian J. Zool.,* 38:289–314.

Wellington, W. G. 1964. Qualitative changes in populations in unstable environments. *Canadian Entomol.,* 96:346–451.

Wilson, E. O. 1963. The social biology of ants. *Ann. Rev. Entomol.,* 8:345–368.

11

Reprinted from *Nat. Hist.* **83**(1), 65–72 (1974)

Tents and Tactics of Caterpillars

A colony of western tent caterpillars has leaders and followers, a snug home, and a bag of survival tricks

Text and photographs by William G. Wellington

The sudden change in behavior of the hundred caterpillars was as unexpected as it was dramatic. Within seconds, the placidly browsing colony of western tent caterpillars stopped eating the leaves of a red alder tree, turned almost in unison, and rushed back to their tent at the base of the branch. In less than a minute, the members of the colony had disappeared inside the silken shelter. We were left standing beside the tree, wondering what had caused the miniature stampede.

And truly it had been a stampede, compared with the sedate, nose-to-tail processions that had become so familiar during the weeks we had been recording the behavioral patterns of this colony. Tent caterpillars that have finished eating normally return to their tent at a leisurely pace. Once there, they mill about on its surface for a few moments before settling in a closely packed cluster to bask and digest their meal. So the suddenly interrupted feeding, the headlong rush,

and the dramatic disappearance caught our attention.

We had been watching this colony since the eggs hatched in early April because we wanted to know if weather changes might affect caterpillar behavior and development—perhaps even survival. We had decked the leaves, the twigs, and the tent itself with a tangle of sensors, which continuously reported temperatures and moisture levels where the insects lived. The colony was in an exposed location, so our first reaction to the unusual behavior was to look at the microclimatic recorder.

The sensors, much more responsive than ordinary thermometers, register minute changes almost immediately. Their recent traces, however, showed no unusual feature nor did they deviate as we watched them for a few moments longer. We were ready to conclude that, despite the insects' uncommon behavior, nothing extraordinary had been happening to the microclimate in which

they lived. Some of us were even beginning to suspect that we had read too much into that behavior, when my assistant pointed to the recorder. All the traces for surface temperature—leaves, bark, and silk—had suddenly begun to drop sharply! A moment later, the air temperature also began to fall, and in less than half an hour there was no doubt that a cold front was overhead.

While rain and wind lashed the area, the caterpillars remained snugly in their tent, impervious to raindrops and wind. They did not reappear until intermittent sunshine again warmed their tent. But we continued to wonder what early warning system had induced the group response so far in advance of the changes our meteorological sensors measured.

Temperature measurements had already shown that the tent often became a sort of a greenhouse, collecting heat in sunny periods and slowing its loss during cloudy intervals. It provided a warmer and

Shortly after hatching, a whole colony of quarter-inch-long western tent caterpillars feeds on a single leaf of a red alder tree. Their silk shelter has begun to form. At first, the larvae feed mainly on the epidermal layer, but two days later, right, they have begun to turn the leaves into skeletons. The long shape of the tent indicates this is an active colony, which probably will build several tents in its lifetime.

Active colonies may abandon their tents several times, making as many as seven before they reach the fourth instar. Each of these tents, larger than the last, is elongate, often club-shaped, owing to the greater activity of the larvae (above). Tent caterpillars of an active colony (upper right) bask on their tent while digesting their meal. An inactive colony (lower right) has a compact tent. Because these larvae feed less frequently, the colony develops more slowly than an active one.

more even environment than the uncertain spring weather outside its walls. So there was no doubt as to why the caterpillars had sought their shelter. The question we could not answer was: how had they detected the impending change?

·The silken shelter the colony sought is a remarkably versatile support system if.one must live on a tree branch in coastal British Columbia in April. Although spring weather everywhere is variable, there is ample support for the belief of Canada's West Coast residents that their springs are, in the best Orwellian tradition, more variable

than others! So the western tent caterpillar needs every adaptational aid it can cull from its evolutionary storehouse to survive in such a hostile environment.

At first glance, the very young larvae seem poorly suited to their harsh surroundings. To move about to seek food, they must have an ambient temperature above 60° F. But the spring air is often below 60° F. Also, the larvae soon perish in very moist or very dry air, and according to local residents, there are only two kinds of Aprils on the West Coast: too dry or too wet.

The conflict between the tent cat-

erpillar's delicate constitution and the harsh weather ordinarily would guarantee extinction, but the silk tent is a remarkably effective shelter against climatic extremes. When the outside air is too dry, the silk walls act as a vapor barrier, slowing the passage of moisture outward and thus reducing the risk of desiccation. When there is rain, the impervious walls shed water and prevent flooding within the shelter. But the tent's prime advantage is the "greenhouse"· function. Because the shelter collects and retains heat, the tiny larvae can continue to feed on enclosed leaves

when passing clouds interrupt the sun's warmth. In years when sunny weather predominates, the young larval develop unhindered, larvae survival is high, and the number of caterpillars is likely to increase. In contrast, cloudy springs with prolonged wet periods are harmful and many colonies are decimated.

As the larvae grow, they spend more time traveling out from the tent to eat. They usually have several feeding periods each day, interspersed with rest periods during which the caterpillars return to cluster on the tent's surface.

Once again, silk sheets have an advantage. Although cold-blooded animals can make brief forays into cool, leafy shade to obtain food, they have difficulty digesting their food in cool places. Because metabolic processes accelerate at higher temperatures, digesting a meal while basking on the warm tent surface is a quick method of transforming lunch into body growth.

Curiously, the tent, which contributes so much to growth and survival, is in one sense a mere byproduct of that basking habit. The young caterpillars continuously trail silk from their mouths as they move about the tree. On the branches, the silk threads form highways that lead to different feeding places. At the basking site, a pad of silk is produced during the restless milling that precedes and follows basking. That pad forms the foundation of the tent. The milling accompanying each subsequent rest period adds successive layers, so the tent grows in size and complexity day by day.

As the tent extends over nearby twigs, its layers begin to form a series of interconnected compartments. Within those compartments, a change in behavior takes place after the caterpillars complete their third molt, usually

Migrating tent caterpillars tend to climb up and travel along train rails, impeding the progress of engines on upgrade slopes.

in the fourth week after they hatch.

Every insect, at certain stages in its life, displays innate responses to unfavorable circumstances, which help it survive the stresses of potentially lethal extremes. To humans an insect's responses may seem a bit roundabout, but from the insect's point of view they generally work very well. For example, defoliators that feed on sunlit leaves may be exposed to dangerous solar heating. As it becomes hot, the insect may not be able to locate a temperature gradient that it could follow to a cooler place on the leaf. While a route of lower temperatures may not be apparent to the insect, there is nearly always a gradient in light intensity that can serve equally well. So an overheated insect can reach a cooler place by moving quickly to the darkest spot it perceives. It returns to a brighter area when it gets cooler.

That is how the young larvae of the western tent caterpillar behave. When they are too hot, they move to the shade; when they are cool they move into the light. When they are small, such changes may involve nothing more than crawling from the top to the bottom of a leaf. As they grow larger, they may have to move along a branch.

Until they pass their third molt, the caterpillars respond negatively to light when overheated. After that molt, an overheated caterpillar responds positively to light and moves to the brightest spot. When it reaches the brightest place, it lifts the forepart of its body until its head is pointed directly at the sun, and maintains that posture until ambient temperatures fall below the level that produced the response.

This reversal in behavior coincides with a seasonal increase in solar heating that can raise the temperature drastically inside the tent. During the hotter days in May, the older larvae customarily spend the midday period inside the tent, moving from compartment to compartment as each chamber overheats. While the sun is highest and most intense, they are not exposed to the dry heat outside.

The last compartment often does not become too hot until after midday. By then, although the air tem-

perature may still be rising, the sun's heat has already begun to decrease. Consequently, when the caterpillars finally respond to overheating by emerging from the tent into the brighter light, the conditions they encounter outside have already improved. They further reduce the heat load on their bodies by their stance, which exposes the least amount of body surface to the sun's rays.

The changed response to light thus enables the western tent caterpillar to take advantage of the tent, despite the disadvantage arising from the more intense solar heating that occurs later in the season. Interestingly, another, unrelated tent dweller, the fall webworm, also displays this change in response among its older larvae, whereas the closely related forest tent caterpillar, which does not construct a tent, does not change with age. Both young and old larvae of the forest tent caterpillar respond to overheating by moving into the shade.

Another kind of response to light becomes important when the maturing western tent caterpillars leave the colony and scatter to feed as isolated individuals. About two days before they are ready to spin cocoons, these isolated larvae drastically change their behavior. For the first time since hatching, they stop eating and become extremely restless, crawling long distances over the ground in what is called "prepupational travel." (That is the state in which they may be found crossing roads or climbing the walls of houses.)

Like the honeybees studied first by Karl von Frisch, the traveling caterpillars navigate by referring to the pattern of plane-polarized light from the sky overhead. Their dependence on polarized light can be demonstrated easily. If you cover them with a polarizer and rotate the polarizer, the larvae will correct their direction of travel accordingly.

Every species of tent caterpillar exhibits this response. Its rigid dependence on the sky's light pattern has led to the forest tent caterpillar's colorful reputation as a train stopper. Whenever large infestations occur in aspen forests, mil-

lions of caterpillars may descend from the trees in a few days and begin their prepupational travel. If the caterpillars reach a railroad track, they climb onto the steel rails, but rarely climb down again. Thus the growing throng atop the rail is channeled along the track. A train encountering the horde on even the slightest grade cannot maintain traction on the slippery rails.

Train crews have mounted brushes and sanders ahead of the engine with no success. Railroads might better prevent the caterpillars from reaching the rails by exploiting their known behavior. On slopes, a dummy rail on either side of the tracks would collect and retain all the inept climbers, thus preventing them from reaching the functional rails.

With a range of marvelous strategies involving behavioral changes in response to heat and light, families of western tent caterpillars survive the fluctuating, often harsh weather of West Coast Canada. But an even more fascinating type of individual behavior is vital for the survival of the species. Each caterpillar, when it is born, tends to fall into a behavioral spectrum consisting mainly of two classes of larvae: dependent and independent. The differences between these two classes become apparent from the first day that the eggs hatch into a squirming, greenish brown mass of tiny larvae.

Many of the tiny caterpillars cannot leave their hatching place without a silk trail to guide them. But some do not need an existing trail. They can create one, and in doing

An active caterpillar colony emerges from its heated tent. At this stage, the larvae cool themselves by pointing their heads at the sun.

pillar existence. The proportions of the different kinds of individuals within a family crucially affect the development of the colony that arises from that family. The wrong proportions—too few active and too many extremely sluggish individuals—mean certain death for all members, active and inactive alike, even in the best environment. Group activity in such colonies is simply too low to foster individual survival.

A small increase in the proportion of active members will produce a somewhat inactive colony that may survive a favorable spring, but not an inclement one. In good years, therefore, such colonies increase in some localities; in bad years, they decrease everywhere.

On the other hand, if 20 percent or more of the hatchlings are independent individuals, a colony will be sufficiently active to withstand all but the harshest springtime. In favorable weather it thrives. In bad years it may be the only kind that survives.

The proportions of the different types of larvae influence colony development mainly through their effect on the length of the clustering periods between meals. A cluster forms because a few caterpillars stop milling and stay perfectly still. The tufts of bristles protruding from their sides catch those of their neighbors, thus halting them. So the restless throng soon becomes a motionless cluster.

The cluster disbands because some individuals in the outermost ring of larvae become sensitive to the drying of their exposed flanks as they finish digesting and become hungry again. Their squirming is transmitted like a wave through the whole cluster via the interlocked bristles, and in a few seconds, the group disbands.

so they lead the rest of the family away from the egg cluster.

The proportions of dependent and independent larvae vary among families. Some families have only a very few active, independent members and a much larger number of comparatively inactive, dependent ones. In the latter, several of the inactive larvae may be so sluggish that they seem almost comatose. In contrast, other families have as many as 40 independent individuals out of some 200 members. Such groups have few exceptionally sluggish larvae; often they have none.

These individual differences persist throughout the life cycle. Highly active larvae ultimately become active moths that can fly for miles before they tire. Less active caterpillars transform into moths with reduced flight capacities. And the most sluggish individuals, if they survive to the adult stage, become moths that can scarcely fly.

Because the differences in activity and behavior are persistent, they influence every facet of tent cater-

The activity rating of the larvae that start the clustering process is unimportant. Any motionless individual can be an anchor point. But the activity level of those on the perimeter crucially affects the duration of clustering. If those larvae are very active types, they quickly become restless as their hunger increases and their exposed sides dry, so clusters to which they belong break up according to their faster digestive rate, not the slower rates of less active members. In contrast, if the perimeter includes only less sensitive, sluggish individuals, the duration of clustering is greatly prolonged. Only near starvation is likely to rouse the worst sluggards and then, only briefly.

In a colony with many more sluggish than active members, the odds are against active individuals often being on the edge of the cluster. The sluggish colony therefore spends much more time clustering than feeding, so the growth of its members soon begins to lag behind the growth of members of an active colony. On the perimeter of a predominantly active colony there are always some agile larvae that can quickly break up the cluster; therefore, the members of an active colony eat often and develop rapidly.

In both types of colonies, active individuals trapped in the midst of a cluster cannot rouse themselves or their fellows. Their interlocked bristles and the proximity of other comatose forms seem to immobilize them, regardless of their hunger. Possibly the higher level of respired carbon dioxide within the cluster is also a factor; the gas quickly stupefies most insects.

The different types of colonies produce different tents. The larvae of active colonies move restlessly up and down a tree branch before they cluster, so their tents become long and narrow. Less active colonies mill about in a more confined space, so their tents are compact; often, they are nearly square or pyramidal.

Very active colonies build and abandon several tents in rapid succession, while inactive colonies enlarge the original one. Infectious disease is slight among active colonies because contaminated tents and diseased cadavers are abandoned quickly. Inactive colonies are more likely to become diseased because the living continually encounter the dead.

Since an active colony moves its tent site from branch to branch, its members constantly have access to fresh, undamaged leaves. The inactive colony is soon eating already damaged foliage, so its food is less nutritious. The larvae of the most sluggish colonies never venture more than a few inches from their tent, even after destroying every leaf in the vicinity, and may starve only inches away from foliage just beyond their limited reach.

The level of colony activity influences the fate of each individual caterpillar. Each individual's level of activity depends, in turn, initially on the order in which it developed within its mother's ovaries. The first eggs obtain more of the maternal food reserves than the last, so the first few embryos are the most vigorous. When an egg mass hatches, the most active larvae emerge from the first-laid eggs; less active individuals emerge from eggs deposited later; and the most sluggish, the moribund, and those too weak to pierce their own eggshells are found among the last eggs deposited.

The gradient in vigor and activity that the female imposes on her progeny is not affected by her own activity rating. Active and inactive females alike produce their best eggs first and the worst last. But if the range of vigor they impose is constant, the proportions of the different kinds of progeny they produce are not. An active female, which has fed well as a larva, enters her pupal stage with much greater food reserves than her inactive sisters. Thus she can produce a greater number of active progeny before she begins to produce less active ones. And differences in the proportions of the various types of progeny are, as we have seen, the basis for the different kinds of colonies.

As female vigor is also expressed in flight, the moths that produce the most active colonies also are the ones that fly farthest to deliver them. Less vigorous moths lay their eggs closer to their own birthplace.

Thus the differences in activity govern the placement of each new generation, as well as its growth and survival.

Spreading the risk by keeping some progeny near the parents' birthplace while scattering others farther afield is a very ancient form of insurance against sudden extinction. Nowadays we call it "differential dispersal," but the version practiced by the western tent caterpillar is only one of many developed by plants and animals evolving in harsh environments.

When any one of these biological strategies is examined closely, some of its strengths may also prove to be weaknesses. For the western tent caterpillar, the production of short-range fliers can get out of hand in an isolated population. If too many active moths of each generation leave, the activity of the remaining residents drops sharply in a few generations. Ultimately, colonies become too sluggish to survive, and the local population collapses.

Despite such local difficulties, the method has kept the species extant in the rugged terrain of coastal British Columbia. And that is no mean feat in a fantastic mosaic of hill and valley climates, many of them unfavorable, and all of them unpredictable. We humans should not be too quick to impose our standards on the caterpillars' solution to their environmental problem, for what may seem like a high price to us may, in fact, be a small one to them, especially in relation to the fringe benefits they derive from differential activity.

When a colony is properly balanced, the active members lead the group to new forage, and they also increase the frequency of the feeding periods by keeping the rest periods brief. The less active individuals not only improve the quality of the tent by weaving a closer mesh than their more restless fellows, but their greater numbers absorb some of the onslaughts by predators or parasites, thus saving the more vigorous few. The system therefore is basically sound for the population at large. And it has provided us with a fascinating example of the impact of individual behavior on group survival. □

12

Reprinted from *Physiol. Zool.*, **33**(2), 152–160 (1960)

THE OXYGEN CONSUMPTION AND BIOENER-GETICS OF HARVEST MICE

OLIVER P. PEARSON

Museum of Vertebrate Zoölogy, University of California, Berkeley

RATES of metabolism or of oxygen consumption have been reported for many species of small mammals, but little effort has been made to relate such measurements to the energy economy of small mammals in the wild. Such effort has been avoided because the rate of metabolism varies so much with changes of the ambient temperature and with activity of the animal. I believe, however, that these variables can be handled with sufficient accuracy so that one can make meaningful estimates of the 24-hour metabolic budget of free-living mice in the wild. In this study I have measured the oxygen consumption of captive harvest mice under different conditions, and from these measurements I have estimated the daily metabolic exchange of wild harvest mice living in Orinda, Contra Costa County, California.

The harvest mice used in the study (*Reithrodontomys megalotis*) are nocturnal, seed-eating rodents living in grassy, weedy, and brushy places in the western half of the United States and in Mexico. In Orinda they encounter cool wet winters (nighttime temperatures frequently slightly below 0° C.) and warm dry summers (daytime temperatures sometimes above 35° C., but nights always cool). They do not hibernate.

MATERIAL AND METHODS

Five adult harvest mice were caught on January 29 and 30, 1959, and were kept in two cages in an unheated room with open windows so that the air temperature would remain close to that outside the building. They were fed a mixture of seeds known as "wild bird seed." Metabolic rates were tested between January 29 and April 1 in a closed-circuit oxygen consumption apparatus similar to the one described by Morrison (1947) but without the automatic recording and refilling features. All tests except the 24-hour runs were made during the daytime and without food. Since harvest mice are strongly nocturnal, several hours had usually elapsed between their last meal and the measuring of their oxygen consumption. When placed in the apparatus, the mice usually explored the metabolic chamber and groomed their fur for about half an hour and then went to sleep on the wire mesh floor of the chamber. One hour or more was allowed for the animals to become quiet and for the system to come to temperature equilibrium. The animals usually were left in the chamber until from five to ten determinations of oxygen consumption had been made, during which they had remained asleep or at least had made no gross movements. Each determination lasted between 9 and 24 minutes. The mice were weighed when they were removed from the apparatus. Oxygen consumptions are reported as volume of dry gas at 0° C. per gram of mouse.

RESULTS

SIZE × RATE OF METABOLISM

Adult harvest mice weigh between 7 and 14 grams. Larger individuals consume oxygen at a lower rate per gram of body weight (Fig. 1). For example, at 12° C. a 12-gram mouse would use only 1.17 times as much oxygen per hour as an 8-gram mouse, although it is 1.5 times as heavy. The various points in the regression of body weight against rate of oxygen consumption can be fitted adequately with a straight line, and from the slopes of such lines illustrating the regression at different ambient temperatures it may be seen (Fig. 1) that at cold temperatures a variation of 1 gram in body weight causes a greater change in metabolic rate than at 30° C. At 1°, 12°, and 24° a change of 1 gram in weight is associated with a change in oxygen consumption of 0.98, 0.48, and 0.35 cc/g/hr, respectively.

At warm and moderate temperatures there was little variation in the measurements of each mouse during any one run (Fig. 1), but at 1° C. the variation was sometimes enormous. Since each measurement was made over a period while the mouse was inactive, the variation must stem from a real difference in the resting metabolism of each mouse at different times. I believe that lability of body temperature is the cause. Harvest mice exposed to cold and hunger in box traps sometimes are found to be torpid and with a cold body temperature. If they are tagged and released, they can be recaptured in good health at subsequent trappings, demonstrating that harvest mice have a labile body temperature and can recover from profound hypothermia. During the metabolic tests at 1° C., especially those with the mouse in a nest, there was a tendency for most of the measurements to lie at one level; but there would be a few very low readings and a few intermediate readings, presumably as the animal entered and emerged from the low-metabolic condition (best shown by the 11½-gram mouse in Fig. 1). In response to cold coupled

with restful surroundings, as in a nest, the animals probably relaxed their temperature control temporarily. This explanation seems plausible in view of the known lability of the body temperature of some rodents such as *Peromyscus* (Morrison and Ryser, 1959), *Dipodomys* (Dawson, 1955), and *Perognathus* (Bartholomew and Cade, 1957) under similar circumstances. Birds permit their body temperature to drop about 2° C. when they sleep at night, and this is accompanied by a drop of as much as 27 per cent in rate of metabolism (De Bont, 1945). The 40 per cent drop shown by some of the mice may have been accompanied by a drop in body temperature of several degrees.

RESTING METABOLISM AT DIFFERENT TEMPERATURES

Since the weights of adult harvest mice vary so much, it is desirable to eliminate the size variable by adjusting all rates of metabolism to a single average size (9 grams). This has been done by using the series of regression lines in Figure 1. Where each of these lines crosses the 9-gram ordinate, that value is taken as the appropriate rate for a "standard" 9-gram harvest mouse and is used in Figure 2.

The middle curve in Figure 2 shows that the minimum rate of oxygen consumption of harvest mice (2.5 cc/g/hr for a 9-gram mouse) is reached at the relatively high ambient temperature of 33° or 34° C. and that there is almost no zone of thermal neutrality. Rate of metabolism almost certainly begins to increase before 36° C. is reached so that the zone of minimum metabolism could not include more than 3°. The critical temperature (33–34°) is remarkably close to the upper lethal temperature. The single animal tested at 37° died after two hours at this temperature but provided several good measurements before entering the

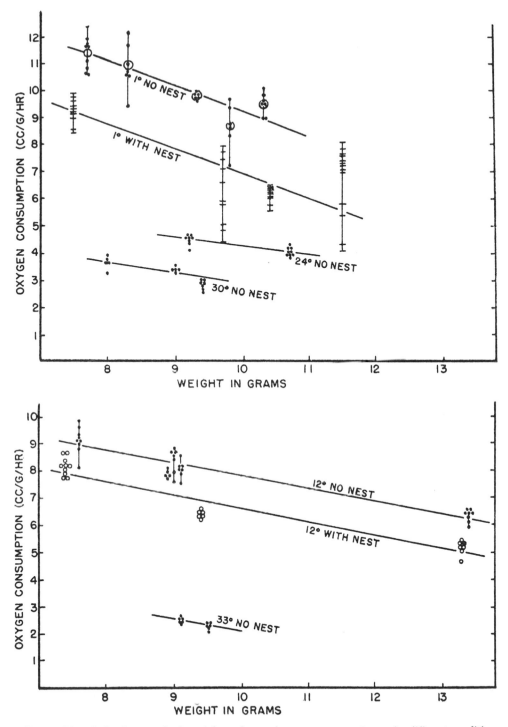

FIG. 1.—The relation between body weight and rate of oxygen consumption under different conditions, showing also the variation in individual measurements. Each cluster or vertical array of points represents a series of values obtained from a single individual.

final coma. Because of the large exposed surface of calcium chloride and soda lime in the metabolic chamber, relative humidity was probably low; heat death would probably occur at an even lower temperature under humid conditions in which cooling by evaporation would be limited.

ered body temperature. Inclusion of these low values causes the apparent decrease of the slope of the two curves between 12° and 1°. No body temperatures, however, dropped to the torpid level. *Reithrodontomys megalotis* is able to maintain its temperature well above the torpid level even when sleeping in cold sur-

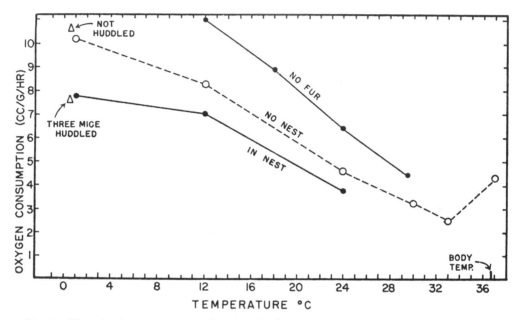

Fig. 2.—The rate of oxygen consumption of resting harvest mice at different temperatures in a nest, without nest, and without fur. All three curves have been adjusted, on the basis of the regression lines shown in Fig. 1, to represent a 9-gram mouse. Triangles indicate rate of oxygen consumption of three mice huddled together without a nest compared with the expected rate for the same three mice singly (average weight 8.5 grams). I am grateful to Martin Murie for supplying the value for deep body temperature, which was the average of many determinations made during the day and night at ambient temperatures between 14° and 27° C.

The increase in rate of metabolism at cool temperatures is almost linear between 33° and 12°; each drop of 1° C. causes an increase in the rate of oxygen consumption of 0.27 cc/g/hr. This rate of change, possibly because of the small size of harvest mice, is greater than that of any of the rodents listed by Morrison and Ryser (1951) and by Dawson (1955). The averages used for the two points at 1° C. include several low values obtained while the animals probably had a slightly low-

roundings. In this respect it differs from the pocket mouse (*Perognathus longimembris*), a mouse with which it should be compared because of its similarly small size. When pocket mice are caged at cold temperatures with adequate food, they either drop into torpor or are continually awake and active. They may even be *unable* to maintain a high body temperature during a prolonged period of sleep at cool temperatures (Bartholomew and Cade, 1957).

The only other report on the rate of oxygen consumption of harvest mice lists a rate of 3.8 cc/g/hr at 24° C. for mice with an average weight of 9.6 grams (Pearson, 1948*a*). This rate is almost 10 per cent lower than the comparable rate obtained from Figure 1 and is below the range of variation obtained at this temperature. The difference may be accounted for by the fact that the mice used in the earlier study were acclimated to a warmer temperature (for discussion of the effect of acclimatization on metabolism see Hart, 1957).

INSULATING EFFECTIVENESS OF FUR

Figure 2 shows also the metabolic effect of removing all the fur (277 mg. in

cent at intermediate temperatures and 24 per cent at 1° C. (lowest curve in Fig. 2). To obtain these measurements, individual mice placed in the metabolic chamber were provided with a harvest mouse nest collected from the wild (shredded grass and down from Compositae), and this the mouse quickly rebuilt into an almost-complete hollow sphere about three inches in diameter. Metabolic rates were counted only when a mouse was resting quietly deep in the nest.

THERMAL ECONOMY OF HUDDLING

The metabolic economy of huddling was measured on one occasion with three mice at an environmental temperature of 1° C. without nesting material. The rate

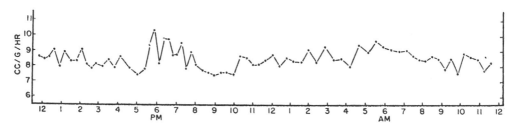

FIG. 3.—Rate of oxygen consumption of a 9-gram harvest mouse for 24 hours at 12° C.

the single 8.8-gram specimen used) with an electric clipper. When calculating the points for the curve in Figure 2, 0.28 grams was added to the naked weight and then this rate was adjusted to that for a 9-gram animal on the basis of the regression lines shown in Figure 1. The rate of metabolism of the naked mouse was about 35 per cent higher at each of the temperatures used, and the rate increased 0.38 cc/g/hr for each 1° C. drop in air temperature.

INSULATING EFFECTIVENESS OF NESTS

When normal, fully furred mice were given an opportunity to increase their insulation by constructing nests, their metabolic rates were lowered about 17 per

of metabolism per gram of huddled mice was 28 per cent less than it would have been for a single one of the mice (Fig. 2). The metabolic saving would probably be greater when more mice were huddled together and less when only two mice were huddled, as is true for feral *Mus* (Pearson, 1947) and laboratory mice (Prychodko, 1958).

24-HOUR OXYGEN CONSUMPTION IN CAPTIVITY

Figure 3 illustrates the rate of oxygen consumption of a mouse kept in the apparatus at 12° C. without nesting material but with food and water for 24 hours. The mouse consumed 1,831 cc. of oxygen to give an average rate of 8.48 cc/g/hr. This is equal to a heat production of about 8.8 Calories per day. In agreement with the fact that activity of harvest mice in the wild is greatest shortly after dusk (Pearson, 1960), the oxygen consumption was greatest at that time. The prolonged low period lasting from about 8:30 to 10:00 P.M. was unexpected in this nocturnal animal.

Pearson (1947) used as an indicator of the nocturnality of different species the ratio of the total amount of oxygen consumed at night (6:00 P.M. to 6:00 A.M.) to that consumed in the daytime. For the harvest mouse described in Figure 3, the ratio is low—1.02—but it should be pointed out that the record was made at 12° C., which is colder than the temperature used for the species in the earlier report. Temperature affects the night/day ratio of oxygen consumption because the difference in amount of oxygen consumed during rest and during activity is proportionately great at warm temperatures and small at cold temperatures.

EFFECT OF ACTIVITY ON METABOLISM

An athlete is able, for short periods, to raise his rate of metabolism to a level 15 to 20 times his basal rate, but small mammals do not match this effort. The peak metabolic effort of mice running in a wheel is only 6 to 8 times their basal rate (Hart, 1950). At 0° C. lemmings running in a wheel at a speed of 15 cm/sec increase their oxygen consumption less than 35 per cent above the level of resting lemmings (Hart and Heroux, 1955). At cool ambient temperatures, such as this, small mammals expend so much energy at rest that a considerable amount of activity causes only a proportionately small increase in oxygen consumption;

TABLE 1

THE 24-HOUR OXYGEN CONSUMPTION (IN CC.) OF A 9-GRAM HARVEST
MOUSE DURING DECEMBER AND JUNE AT ORINDA, CALIFORNIA

| | | DECEMBER | | | JUNE | |
		Without Nest	With Underground Nest		Without Nest	With Underground Nest
Nocturnal habit	4 hr. above ground at 1° C.*	367	367	4 hr. above ground at 12° C.†	297	297
	20 hr. under ground at 10° C.‡	1,548	1,296	20 hr. under ground at 18° C.§	1,152	954
	Activity correction‖	+119	+119	Activity correction‖	+119	+119
		2,034	1,782 cc. (8.55 Cal.)#		1,568	1,370 cc. (6.58 Cal.)#
Diurnal habit	20 hr. under ground at 10° C.‡	1,548	1,296	20 hr. under ground at 18° C.§	1,152	954
	4 hr. above ground at 6° C.**	333	333	4 hr. above ground at 25° C.††	155	155
	Activity correction‖	+119	+119	Activity correction‖	+119	+119
		2,000	1,748 cc. (8.39 Cal.)#		1,426	1,228 cc. (5.89 Cal.)#

* Mean temperature in runways at time of passage of harvest mice in December.
† Mean temperature in runways at time of passage of harvest mice in June.
‡ Underground temperature in December.
§ Underground temperature in June.
‖ Add 40 per cent of the oxygen consumption on the surface at a temperature of 12° C.
Assumed 4.8 Cal. per liter of oxygen.
** Mean half-hourly temperature in runways between 6 A.M. and 6 P.M. in December.
†† Mean half-hourly temperature in runways between 6 A.M. and 6 P.M. in June.

and at cold temperatures the metabolic cost of keeping warm may be so high as to leave little or no capacity for exercise (Hart, 1953). During measurement of the resting metabolism of harvest mice, numerous measuring periods had to be discarded because the mouse was moving around in the metabolism chamber. Such activity rarely raised the oxygen consumption more than 40 per cent above the level of a resting animal at the same temperature. During the 24-hour run at 12° C., the highest metabolic rate occurred during an 11-minute period when the average oxygen consumption was 10.36 cc/g/hr. This is only 40 per cent greater than the lowest rate recorded for that mouse during any one measuring period. The maximum metabolic effort recorded for any harvest mouse was that of an 8.6-gram mouse at 1° C. This animal persisted in gnawing, exploring, and trying to escape from the chamber for more than two hours. During one 10-minute period its oxygen consumption averaged 15.8 cc/g/hr, which is 50 per cent higher than the rate of a resting mouse at the same air temperature and six times the minimum value for the species at thermal neutrality. This is probably not far from the peak metabolic effort of the species.

On several occasions I have watched undisturbed harvest mice carrying on their normal activities in the wild, and I have been impressed by their leisurely approach to life. Hard physical labor and strenuous exercise must occur quite infrequently. Most normal activities of harvest mice are probably accomplished without a rise in metabolic rate more than 50 per cent above what it would be in a resting animal at the same air temperature.

24-HOUR METABOLISM IN THE FIELD

The preceding observations indicate that ambient temperature is a much more important variable than activity in the 24-hour energy budget of harvest mice in the wild. By use of automatic devices that record the temperature in mouse runways whenever a mouse passes by, the temperature encountered by harvest mice during their nightly periods of activity are known (Pearson, 1960). I have also recorded throughout the year the temperature five inches below the surface of the ground. This gives an approximation of the temperature encountered by the mice while they are in their retreats during the daytime. Some of these surface and underground temperature measurements have been used in the calculations summarized in Table 1.

To complete the calculations in Table 1, it has been necessary to estimate how many hours of each 24 the mouse spends on the surface of the ground and how many below the surface. No good data exist, so I have made an estimate based on the behavior of captive animals and on automatic recordings made at the exit of an underground nest box being used by wild harvest mice. Admittedly this estimate (4 hours on the surface each night) could be wrong by 50 per cent or more, but it should be noted that an error of two hours in this estimate would only alter the answer (the total 24-hour metabolism) by about 25 per cent. Assuming that the rate of oxygen consumption during above-ground activity is 40 per cent higher than the rate of a mouse resting at 12° C. (see above), the activity correction used in Table 1 can be calculated.

In 24 hours in December a harvest mouse uses 8.55 Calories, and in June, 6.58 Calories (Table 1), assuming that the mouse has the benefit of a nest. A nest reduces his daily energy budget by about 12 per cent. These estimates of daily metabolic demands seem reasonable when compared with the values actually obtained by measuring the 24-hour oxygen consumption of captive animals, as reported above. The average metabolic impact, or daily degradation of energy, by a single harvest mouse should be somewhere between that in December and that in June, perhaps 7.6 Calories. This is about the same as that of a hummingbird in the wild (Pearson, 1954)—less than half that of a much heavier English sparrow (Davis, 1955).

BIOENERGETICS

In seasons when harvest mice are abundant, there may be twelve of them per acre (Brant, 1953). At that population density, the species would be dissipating at the rate of 91 Calories per acre per day the solar energy captured by photosynthesis, or something like $\frac{1}{2}$ of 1 per cent of the energy stored each day by the plants in good harvest-mouse habitat in the Orinda area. This percentage was calculated using a net productivity of 20,000 Cal/acre/day, which was estimated by assuming 4 Calories per gram of dry vegetation (based on data in Brody, 1945, pp. 35, 788) and an annual crop of 1,800 kg. of dry vegetation per acre (based on Bentley and Talbot, 1951). The harvest mice on this hypothetical acre are causing about the same caloric drain on the environment as all the small mammals in the acre of forest described by Pearson (1948b).

By using caloric units, direct comparison can be made of the metabolic impact of different species, as in the example above. Similarly, the metabolic cost of different activities and different habits can be compared (Pearson, 1954). For example, harvest mice are strongly nocturnal (Pearson, 1960), in spite of the fact that air temperatures are much colder at night and force mice to consume more oxygen and more food than if they were diurnal. Since evolution has permitted nocturnality to persist, it seems logical to assume that the value of nocturnality to harvest mice is greater than the metabolic cost. I estimate that during a 24-hour period in December a 9-gram harvest mouse uses 0.16 more Calories by being nocturnal than it would if it were diurnal (Table 1). In summer, the difference is even greater, 0.69 Calories. The average is 0.42 Calories, or about $3\frac{1}{2}$ grains of wheat. This is a rough estimate of the price each harvest mouse pays for nocturnality. Some environmental pressure makes harvest mice remain nocturnal, and this pressure must be more than 0.42 Calories per mouse per day. If harvest-mouse nocturnality evolved for one reason only—to avoid predation by hawks—then we would have discovered a minimum estimate of the predation pressure of hawks on harvest mice. Surely the situation is not this simple; nevertheless, it is interesting to measure the pressure that makes harvest mice nocturnal even if the cause of the pressure is not known.

SUMMARY

Oxygen consumption of harvest mice reaches a minimum of 2.5 cc/g/hr at an ambient temperature of 33° C., and the zone of thermal neutrality is not more than 3°. Each drop of 1° in ambient temperature causes an increase in the rate of metabolism of 0.27 cc/g/hr. Removing the fur raises the rate of metabolism about 35 per cent, and use of a nest lowers it 17 to 24 per cent. Huddling by three mice at 1° reduces the rate 28 per cent.

Exercise at cool temperatures causes a relatively small increase in the rate of metabolism, whereas change of ambient temperature has a great effect. Making use of the temperatures that harvest mice are known to encounter in the wild, the 24-hour oxygen consumption of a wild harvest mouse was calculated to be 1,782 cc. in December and 1,370 cc. in June. The average (1,576 cc.) is equivalent to about 7.6 Calories per day. A dense population of harvest mice would dissipate about 91 Calories per day per acre, which is about $\frac{1}{2}$ of 1 per cent of the energy stored by the plants each day.

By being nocturnal, harvest mice encounter cooler temperatures, and this habit increases the daily energy budget of each mouse by 0.42 Calories, or about $3\frac{1}{2}$ grains of wheat.

LITERATURE CITED

BARTHOLOMEW, G. A., and CADE, T. J. 1957. Temperature regulation, hibernation, and aestivation in the little pocket mouse, *Perognathus longimembris*. Jour. Mammal., 38:60–72.

BENTLEY, J. R., and TALBOT, M. W. 1951. Efficient use of annual plants on cattle ranges in the California foothills. U.S. Dept. Agriculture, Circular No. 870, 52 pp.

BRANT, D. H. 1953. Small mammal populations near Berkeley, California: *Reithrodontomys, Peromyscus, Microtus*. Doctoral thesis, University of California, Berkeley.

BRODY, S. B. 1945. Bioenergetics and growth. New York: Reinhold Publishing Corp.

DAVIS, E. A., JR. 1955. Seasonal changes in the energy balance of the English Sparrow. Auk, 72:385–411.

DAWSON, W. R. 1955. The relation of oxygen consumption to temperature in desert rodents. Jour. Mammal., 36:543–53.

DE BONT, A.-F. 1945. Métabolisme de repos de quelques espèces d'oiseaux. Ann. Soc. Roy. Zoöl. Belgique, 75 (1944):75–80.

HART, J. S. 1950. Interrelations of daily metabolic cycle, activity, and environmental temperature of mice. Canadian Jour. Research, D, 28:293–307.

———. 1953. The influence of thermal acclimation on limitation of running activity by cold in deer mice. Canadian Jour. Zoöl., 31:117–20.

———. 1957. Climatic and temperature induced changes in the energetics of homeotherms. Revue Canadienne de biol., 16:133–74.

HART, J. S., and HEROUX, O. 1955. Exercise and temperature regulation in lemmings and rabbits. Canadian Jour. Biochem. & Physiol., 33:428–35.

MORRISON, P. R. 1947. An automatic apparatus for the determination of oxygen consumption. Jour. Biol. Chem., 169:667–79.

MORRISON, P. R., and RYSER, F. A. 1951. Temperature and metabolism in some Wisconsin mammals. Federation Proc., 10:93–94.

———. 1959. Body temperature in the white-footed mouse, *Peromyscus leucopus noveboracensis*. Physiol. Zoöl., 32:90–103.

PEARSON, O. P. 1947. The rate of metabolism of some small mammals. Ecology, 28:127–45.

———. 1948a. Metabolism of small mammals, with remarks on the lower limit of mammalian size. Science, 108:44.

———. 1948b. Metabolism and bioenergetics. Scientific Monthly, 66:131–34.

———. 1954. The daily energy requirements of a wild Anna Hummingbird. Condor, 56:317–22.

———. 1960. Habits of harvest mice revealed by automatic photographic recorders. Jour. Mammal. (in press).

PRYCHODKO, W. 1958. Effect of aggregation of laboratory mice (*Mus musculus*) on food intake at different temperatures. Ecology, 39:500–503.

13

Reprinted from *C.S.I.R.O. Wildl. Res.*, 1(2), 79–95 (1956)

TEMPERATURE REGULATION IN THE NESTING MOUNDS OF THE MALLEE-FOWL, *LEIPOA OCELLATA* GOULD

By H. J. FRITH*

[*Manuscript received June* 18, 1956]

Summary

The temperature conditions in the mounds of the mallee-fowl, *Leipoa ocellata* Gould, have been investigated by means of recording instruments. It is shown that the incubation of the eggs is largely achieved by heat generated by the fermentation of organic matter. By manipulation of the covering soil the birds keep the temperature within the range at which egg development takes place. The principal activity is designed to keep the mound temperature down. Towards the end of the season the heat from the Sun becomes important, and at the end of the season active steps are necessary to increase the amount of heat reaching the eggs from this source.

I. INTRODUCTION

The members of the family Megapodiidae are remarkable in that they do not build a nest and brood their eggs, but lay them in a hole in the ground or in a mound of soil or leaves, where they are hatched by the Sun's heat or the heat generated by fermentation. Of the members of the family that inhabit the mainland of Australia, two are found in the moist tropical jungles, and one in the arid interior. The jungle fowl, *Megapodius freycinet* (Dumont), is found only in the jungles of northern Australia; the scrub turkey, *Alectura lathami* Gray, is also a jungle dweller, its range extending, in the rain-forest areas, down the eastern coast. It has also been recorded, occasionally, in the drier areas further inland. The nests of both species consist of large heaps of rotting vegetation from the forest floor, which, under the moist tropical conditions, generate much heat. Although it seems necessary for these species to delay egg laying until the temperature has declined somewhat from the very high temperatures reached early in the fermentation process, it is thought the maintenance of the temperature around the eggs is then a relatively simple matter.

The mallee-fowl or lowan, *Leipoa ocellata* Gould, would appear to require different and more complicated means of incubation in developing and maintaining the correct temperature throughout the season. The species is confined to the dry inland parts of southern Australia,† and is closely associated with mallee scrub, an open vegetation of dwarf eucalypts. Under the conditions of low rainfall, high temperature, and relatively sparse vegetation that exist in this region, the collection of sufficient leaf material and its subsequent fermentation presents a problem. Briefly, the nesting habit of the mallee-fowl consists of excavating a hole in the

---

* Wildlife Survey Section, C.S.I.R.O., Canberra.

† The extension of the range of *L. ocellata* into the coastal "corridor" in the extreme south-western corner of the continent (Serventy and Whittell 1951) provides the only occurrence of the species outside the inland zone.

ground, filling it with vegetable material, and covering the whole with soil. The mound is visited almost daily, and is frequently dug out and the thickness of the soil cover altered. This process is obviously adapted to exercise some sort of heat control on the egg chamber.

The mound-building habit of the mallee-fowl has excited great interest since its discovery, and much has been conjectured and written on the temperature regulation. Few of the statements are based on any experimental data, and none satisfactorily explains the process. Mattingley (1908) considered that the birds must exercise a considerable control on the amount of heat reaching the eggs, and that this control was achieved by opening the top of the mound during the day to allow the Sun's heat to warm the eggs and closing it at night to conserve the heat. Ross (1919) secured a series of temperature readings in the mound, and they seemed to remain fairly constant despite a fluctuating air temperature. However, he failed to report even the time of day when these temperatures were measured, nor did he form any conclusions. Lewis (1939, 1940) noted that the temperature within the mound seemed to remain constant, despite the opening and closing by the birds; he concluded: "if the Lowans' practice of opening and closing the mound practically daily is not for the purpose of utilizing solar heat what other good purpose would be served by such an action?" These observations suggest that the temperature in the mound may be regulated by the working of the birds.

In addition to the methods of temperature regulation, the value of the leaf material buried in the mound has been much discussed, but no work has been done to measure the heat produced by it. Ashby (1922) concluded: "the spare stiff cardboard-like leaves of mallee preclude any possibility of fermentation". Lewis (1939, 1940) came to the conclusion that the main function of the material was to raise the temperature of the mound prior to egg laying. He did not visualize it being an important source of heat in the actual incubation of the eggs.

The present work was undertaken to provide some factual data on the effects on mound temperature of the opening and closing and the true function of the organic matter. The observations and experiments reported in this paper were carried out in an area of typical mallee scrub 27 miles from Griffith, N.S.W. The soils of the area have not been described, but are typical sandy mallee soils underlain by clay, similar to the mallee soils described by Taylor and Hooper (1938). The climate is hot and dry.

II. Mound Construction

The process of construction as observed in a series of mounds during 1953 is briefly described below and illustrated in Plate 1. The nesting mounds are used year after year, being renovated between laying seasons, and only rarely is an entirely new mound found; one was discovered under construction in February 1953. A small depression 2 ft wide and 6 in. deep was scratched by a bird which was subsequently identified as the male. By the middle of May the excavation was 7 ft in diameter and 2 ft deep. At this time the mound was inspected daily for 7 days, and it was found that work was done daily, in the early morning. By mid June the hole was 10 ft in diameter and 3 ft deep, the spoil being neatly heaped around the edge, to a height of 18 in. Towards the end of June the collection of leaf material began. In this case the nearest material was under a clump of mallee 30 yd from the mound, which was located in a small clearing. This material was scraped up and moved to the mound, and before sufficient had been collected it was necessary for the bird to gather it 50 yd from the mound.

When a previously used mound was renovated it was dug out completely and the excavation deepened at the end of hatching in March. It was then normally not visited for 2–3 weeks before the collection of the organic matter started. Lewis (1940) considered that the material was left beside the mound until it was saturated by rain, and then put into the hole. However, in all the mounds observed by the author, when the first few heaps of material reached the side of the excavation they were put into the hole, which was progressively filled.

During July and August collection of the material proceeded until it was mounded up in the excavation to a height of 2 ft above ground level. The exact height varied from mound to mound but was always of this order. During the last week in August the egg chambers were dug and a small amount of soil from the heaps of spoil was scattered over the leaf material. As the season proceeded, the amount of soil covering was gradually increased until mid December. The completed mounds varied from 6 to 18 ft in diameter.

The amount of organic matter relative to soil varied greatly in different mounds. One extreme case was mound 9. This mound was constructed in an area of *Melaleuca uncinata* and was built almost entirely of this material, the fine needle-like leaves being mixed with a very small amount of sand on the surface. Although some sand was scratched on early in the season serious covering did not begin until the end of December at this mound. Similarly, mounds 31 and 18 were of pure mallee leaves with hardly any soil covering until the end of December. In contrast to these mounds, mound 4 contained no organic matter at all. This mound was opened up in the normal manner in March, and seemed to have been abandoned until November 13, when it was discovered to have been recently mounded up and to contain three eggs; the clutch was subsequently completed and is mentioned again in Section V(c).

In between these extremes was mound 6, a very large mound 18 ft in diameter with only a few cubic feet of organic matter, and mound 16 which was similar to it. Mound 5, the mound selected for experiment (see Section IV(a)), was typical of the majority in regard to the amount of organic matter. Measurements of the relative volumes of sand and leaf material were not made.

III. Incubation: Factors Affecting Mound Temperature

In the late winter and spring the temperature of the interior of the mounds rapidly increased owing to the fermentation of the organic matter and the general rise in temperature at this season; but before proceeding to a detailed discussion of the temperature changes in the mound it is necessary to consider two factors: soil temperature and the birds' activity.

(a) Soil Temperature

Alternate solar heating by day and radiational cooling by night cause the surface temperature of the soil to undergo a large diurnal variation which, by conduction of heat, is propagated downward into the soil. This diurnal temperature wave becomes displaced in time according to the depth and also becomes progressively weaker, so that at a depth of 2–3 ft it is quite small. In a mallee-fowl mound the conditions are complicated by the presence of an additional and fairly constant source of heat—the heap of decaying organic matter buried below the surface. The effect of this cannot be accurately forecast without experimental data for the specific case.

When the soil is not being disturbed conduction is the only important method of heat transfer in it, and the two important factors affecting conduction are the heat conductivity and heat diffusivity of that soil. Low values for these factors are associated with soils of low moisture content, and sandy soils have lower values than clay soils. Thus it would be expected that the loose dry sands of mallee-fowl mounds would be very efficient insulating materials.

(b) Birds' Activity

The mounds kept under observation at Griffith were visited and dug into by one or both birds almost daily throughout the incubation season. The frequency of digging varied seasonally and inversely with the average air temperature prevailing, being far more frequent early in the season than later. It was seen that, whereas the mounds were opened almost daily at the beginning of the season, in January the digging was done once a week, and towards the end of the season about every 3 or 4 days. In very hot weather the mounds were opened less frequently than in cooler weather, and in cold weather they were not opened at all.

In addition to this variation in frequency associated with air temperature, the time of day the work was done and the time during which the mound was allowed to remain open varied. Typically, early in the season opening of the mound started soon after dawn and was done rapidly. At this time of the year the mounds were small, and the leaf material covered by only a few inches of soil. Under these conditions the birds worked rapidly and continuously until the vicinity of the eggs was reached. On reaching this objective work was immediately reversed, and the excavation filled in again. On these occasions the excavation was small and conical, being the width of the egg chamber at the bottom and 4–5 ft wide at the top. The work usually required about 1 hr and was always completed before sunrise.

Later in the season, but before the full heat of summer had started, the mounds were opened in a different manner. On these occasions the mound was opened more thoroughly but in a more leisurely fashion. Work would start before dawn, and the birds would work methodically until the eggs were reached. Having then proceeded to cover them with 6–12 in. of soil they would depart, returning at intervals during the day to replace a few more inches of soil in the excavation. The mound was always fully restored by 4 p.m.

In the very hot weather of midsummer the mounds were opened soon after dawn every 6 or 7 days. On these occasions the digging was very thorough and the mounds were opened to within an inch or so of the eggs. On reaching the eggs the work was reversed and the mound completely restored and heaped up high. The process resulted in a complete removal and mixing of the whole of the covering of the eggs, and required 2–3 hours' constant work.

Towards the end of the season, when the average air temperature was declining, the time the mounds were allowed to remain open was still further increased. On these occasions work was not started until the sun was shining directly onto the mound. Usually the birds' operations began at 8–9 a.m. and a large saucer-shaped excavation was made, up to 15 ft in diameter and to within a couple of inches of the eggs. These excavations were allowed to remain open all day. The birds normally returned at about 3 p.m. and proceeded to restore the mound completely; on some occasions they also returned periodically throughout the day and returned small amounts of soil.

The labour involved in these openings of the mounds, particularly in hot weather, caused the birds some distress. They were panting long before completion of the work and were frequently required to rest in the little shade available. It will be shown that the work was of prime importance in regulating the temperature of the egg chamber. (Digging out is illustrated in Plate 2.)

IV. Experimental Method

In the 1952–53 season regular visits were made to three mounds and the activities of the birds were watched from hides. During the later half of the breeding season the bulbs of soil thermographs were buried in the egg chambers of two of the mounds. The other bulbs were buried in the soil beside the mounds at various depths.

In 1953–54, in the light of the previous season's findings, experiments were set up as described under (a) and (b) below.

(a) Artificial Mounds

Two artificial mounds were constructed close to a typical natural mound (mound 5, see Section II). One of these mounds was as nearly an exact duplicate of the natural mound as it was possible to make it. This mound will be referred to hereafter as the "artificial mound." The other mound included no organic matter; whilst it was similar in size and shape to the other mounds, it was built entirely of soil. This mound will be referred to hereafter as the "control".

These three mounds provided the following combination of heat sources capable of affecting the temperature in the mounds:

Natural mound: solar heat, fermentation heat, birds' activity.
Artificial mound: solar heat, fermentation heat.
Control mound: solar heat.

256

The bulbs of soil thermographs were buried in each mound. In the natural mound the bulb was placed among the eggs on a plane through the mid point of the first egg laid. In the other mounds the bulbs were placed at corresponding depths. If the level of any mound altered, this was determined by the routine measurement of the mound's height, and correction made if it was desired.

Two soil heat flux plates were available for the work. One was buried in the natural mound 1 in. below the thermograph bulb, and the other placed in the control in a comparable position. No recording device was available for these instruments. Normally they were read periodically at intervals of 1 hr throughout periods of 36 hr; but on especially interesting occasions this period was extended up to 72 hr. Air temperature was recorded by a thermograph 3 ft from the ground and protected from radiation.

An instrument was devised to record automatically the occasions on which the mound was dug out. This apparatus was quite simply made but was effective. It consisted of a cord fixed to a peg driven into the ground on one side of the egg chamber; the cord was stretched across the chamber and attached by means of a suitable system of weighted levers to the pen arm of a clock-work drum. The cord was 6 in. above the eggs, and while digging in the mound the birds invariably tugged vigorously at it.

(b) Natural Mounds.

Fifteen mounds were selected in the area and visited regularly throughout the season. At each visit the temperature of the egg chamber was taken with an alcohol-in-glass thermometer. The mound was partly opened, and the thermometer pushed into the sand beside the first egg. These temperature readings were designed to secure "spot" information to compensate, in some measure, for the lack of replication in experiment (a). Graduated pegs were driven into the soil beside each mound and the height of the mound and the thickness of the soil cover were determined at each visit.

Hides were erected at five of the mounds. In one of the hides the instruments could be read and the birds watched at the same time. The birds were watched from the hides from dawn to dusk on many occasions, and frequently from before dawn until most activity had ceased at midday.

V. RESULTS

(a) Function of Organic Matter

Figure 1 shows the mean daily temperature in the three experimental mounds throughout the season 1953–54. Daily fluctuations in the temperatures were rather great and tended to obscure some of the features of the graphs, consequently the values have been smoothed by weighting the daily values, the ordinate for each day being the 7-day running mean.

It can be seen in Figure 1 that the temperature in the two mounds containing organic matter differed very greatly from the control. The temperature in these two mounds started to increase in early July at a much greater rate than that—due entirely to solar heat—in the control. In the artificial mound the temperature increased rapidly to 106°F and was maintained at this temperature until mid September. It then began to fall and by mid December was approaching the temperature of the control. (At this date the thermograph was removed from this mound and used for another purpose.) In the natural mound the temperature rose to 92°F, and was maintained at this until mid February, when it began to decrease slowly. After the end of February this decrease was accelerated; and by the end of March the temperature in this mound approximated to that of the control.

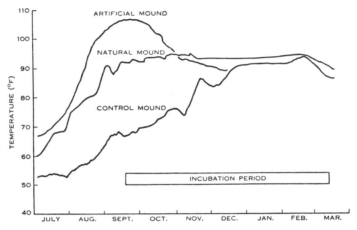

Fig. 1.—Mean daily temperature in the three experimental mounds in the season 1953–54.

From Figure 1 it appears that the organic matter incorporated in the mounds was capable of supplying a considerable amount of heat, and that the differences in temperature between these mounds and the control were due to this. Also in the artificial mound this heat resulted in a much higher temperature than in the natural mound. It may therefore be inferred that the modification of this temperature seen in the natural mound was due to the activity of the birds opening the mound; that is, this activity had the effect of cooling the egg chamber. The rapid decrease in temperature in the artificial mound after mid November to the level of the control indicates that the amount of heat supplied by the organic matter and the Sun was no longer sufficient to maintain the temperature at the former level. As the temperature in the natural mound was still maintained far above the level of the control, it may be inferred that the other variable, i.e. the activity of the birds, was responsible for this. As the figures are available for only one site it will be necessary to support these inferences with other data. This is done in succeeding sections, where it has been found convenient to divide the season into three periods. Although these periods will be discussed separately, it should be remembered that the processes are really associated with air temperature rather than season, and that they grade imperceptibly into one another.

Figure 2 shows the heat flow measurements in the natural mound and the control mound during the period from 4 a.m. October 12 to 10 a.m. October 14,

1953. During this period the mound was under continuous observation. On October 13 the mound was opened by the birds at dawn. The significance of the effect of the opening on the heat flux by causing a great flow of heat to the surface will be discussed in Section V(*b*). For the present it is sufficient to consider only that portion of each graph showing undisturbed conditions, i.e. up to 6 a.m. October 13. It is seen that in the control the normal alternation of positive and negative heat flow occurred, following the diurnal wave of solar heat reaching the surface of the soil. In the natural mound, however, the net heat flow was at all times negative; and this is conclusive evidence of a source of heat below the egg level. There can be little doubt that this heat originated in the organic matter below the eggs.

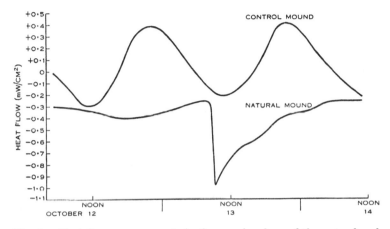

Fig. 2.—Heat flow measurements in the egg chambers of the natural and the control mounds during the period from October 12 to October 14, 1953.

Figure 1 shows that, at this time, the temperature remained about 20°F warmer in the natural mound than at the same depth in the control mound. That the heat fluxes measured in the mounds are consistent with this temperature difference may be shown as follows.

The upward flux of 0·35 mW/cm² (9 × 10⁻⁵ cal cm⁻² sec⁻¹) in the natural mound would require for its transmission a higher temperature at egg level than at the surface by some 10–20°F, values calculated on the basis of a thermal conductivity of between 0·0005 and 0·001 for the 2-ft layer of fine loose dry soil above the flux plates. In the control mound the temperature at the same depth, at this season when the ground is being warmed, would be a few degrees lower than the mean surface temperature* and this agrees with the average small downward heat flux measured in this mound. The heat flux measurements therefore accord with a temperature difference between the interiors of the two mounds of 10–20°F and

---

* The surface of the natural mound would be slightly warmer, on the average, than that of the control, but the differences can be shown to be small (about 2°F). This also operates to make the above 10-20°F an underestimate of the temperature difference between the two mounds. The author is indebted to Mr. E. L. Deacon of the Division of Meteorological Physics, C.S.I.R.O., for assistance with these calculations.

they therefore provide further strong evidence of the importance of the organic matter as a heat source.

The difference in amplitude and phase of the diurnal variations of heat flux in the two mounds (see Fig. 2) can be accounted for in terms of a greater thermal diffusivity of the soil in the control mound than that in the natural mound which had been dug and turned over by generations of mallee-fowl and was dry and in a very fine state of subdivision. The few diggings prior to the construction of the artificial mounds were insufficient to destroy the soil structure as thoroughly; so

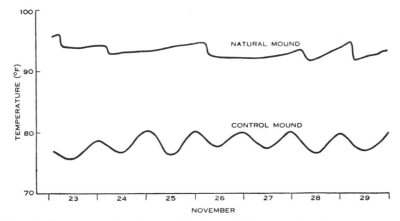

Fig. 3.—Temperature in the egg chambers of the natural and the control mounds during a period of mild spring weather from November 23 to November 29, 1953.

throughout the season the soil of these mounds remained visibly coarser and moister than in the natural mound. The very small diurnal variation of heat flux in the natural mound is evidence of the good thermal insulation provided by the soil blanket.

(b) Maintenance of Temperature

(i) *Spring and Early Summer.*—During the spring and early summer—that is, up to the end of December—the climate was relatively mild and the days were usually bright and clear with daily maxima increasing from about 70°F in September to 95°F in December.

Figure 3 shows the temperature in the natural and control mounds during a typical period of this type of weather, November 23 to November 29, 1953. The natural mound was under continuous observation during this period, and the birds were very active. A chronology of the activity is given below:

November 23

    4.50 a.m.—Male arrived at mound, apparently in no hurry; he slowly started scratching at surface.

    5.12 a.m.—Female arrived and started digging; male increased his rate of work.

    5.30 a.m.—Female departed and was not seen again.

    5.35 a.m.—Male reached eggs and immediately started filling in excavation.

5.43 a.m.—Eggs were covered by about 12 in. of soil and male went away.

10.20 a.m.—Male returned and filled a little more soil into the hole. He then went away; the eggs were covered by 15 in. of soil.

12.45 p.m.—Male appeared but did no work.

3.06 p.m.—Male returned and worked until 5.00 p.m. when the mound was completely restored and mounded over.

November 24

5.30 a.m.—Male arrived whilst still dark and started to dig without any loss of time.

6.00 a.m.—Female arrived and started to dig immediately.

6.50 a.m.—Excavation complete; it was 8 ft in diameter and to the full depth of the eggs. Female departed and the male immediately started to fill the hole in.

7.10 a.m.—Male moved out of sight. The excavation was now half filled-in. 14 in. of soil covering the eggs.

2.00 p.m.—Male returned and completed the filling until the mound was fully restored.

November 25

5.20 a.m.—Male and female arrived but went away without digging.

November 26

4.10 a.m.—Male arrived whilst still dark. He immediately started to dig.

4.35 a.m.—Excavation was apparently complete; it was 5 ft in diameter and to within 2–3 in. of the eggs. This was a small excavation.

4.40 a.m.—Male started to fill in the hole.

5.06 a.m.—Mound three-quarters full. There was 20 in. of soil covering the eggs. Male went away.

9.20 a.m.—Male returned and did a few minutes work and then left.

3.00 p.m.—Male returned and completely restored the mound.

November 27

5.40 a.m.—Male arrived and displayed. Female answered but did not appear.

5.50 a.m.—Male having dug a hole 1 ft wide and 6 in. deep went away again.

November 28

4.20 a.m.—Male arrived and immediately started digging.

4.36 a.m.—Female arrived and also started to dig.

4.52 a.m.—Female went away.

5.20 a.m.—Male had completed the excavation; it was now 8 ft in diameter and to within 2–3 in. of the eggs. He immediately started to fill in.

5.36 a.m.—When the eggs were covered by 12 in. of soil the male went away; he remained, some hundreds of yards off, calling.

6.20 a.m.—Male returned and worked slowly for 10 min and then went away again.

10.03 a.m.—Male returned and worked very slowly for 15 min.

2.20 p.m.—Male arrived again and did a little more.

4.22 p.m.—Male returned and completely restored the mound.

November 29

4.27 a.m.—Male arrived, having been heard calling in the dark for at least 20 min.

5.30 a.m.—Female arrived and started digging.

5.45 a.m.—Female went away.

6.30 a.m.—Male had continued digging and now the excavation was complete, and filling started immediately.

7.00 a.m.—Eggs were now covered by 12 in. of soil, and male went away.

3.20 p.m.—Male returned and, working continuously, completely restored the mound by 4.35 p.m.

It can be seen in Figure 3 that on each occasion that the mound was opened a fall in temperature occurred in the egg chamber. In between openings, the temperature in the egg chamber continually tended to increase, but was controlled by the birds opening up the mound. It had previously been assumed that this type of opening, i.e. digging out in the early morning and finally closing again in the late afternoon, was adapted to *warm* the egg chamber.

Throughout this and similar periods of warm and hot weather, at this mound and the others under observation, the male birds were found to be very attentive to the partially closed mound. On almost every visit to each mound the male was found in the vicinity. When disturbed the birds invariably moved very slowly and carefully away, never more than 100 yd, and then slowly circled the area whilst the eggs were being examined. Immediately the observer left the area they would return and repair the damage to the soil covering. Extreme instances of concern were met with in mounds 15, 18, and 31. At these mounds the males were particularly attentive and rarely moved more than a few yards from the observer. At mound 31 it was the author's habit to remove any new eggs to be measured to the shade of a bush 9 ft from the centre of the mound; and whilst sitting there it was quite usual for the bird to return to the open mound and commence to return the soil. On two occasions whilst the author was digging into the side of this mound the bird was hurriedly filling in the soil again from the other side 5 ft away. This willingness of a normally extremely shy and wary bird to approach a human being in order to maintain a thick covering of soil between the eggs and the Sun's heat indicates that a real danger of overheating the egg exists at this time of the year.

During the period November 5–10, 1953 cold weather prevailed and some rain fell. The temperature in the control mound fell by 6°F owing to the increased conduction of heat to the surface. In the natural mound the birds increased the thickness of the soil covering to 35 in.; and although they visited and walked over the mound daily they made no attempt to open it. Despite the cold weather and the fall in temperature in the control, the temperature in the natural mound rose by 1·6°F. It is considered that, whereas during the active fermentation period it was usually necessary to protect the eggs from the Sun's heat, in very cold weather the greatly increased temperature gradient to the surface was sufficient to conduct heat away from the egg chamber at such a rate there was some danger of too great a heat loss from the egg chamber. This danger was overcome by the birds increasing the thickness of the soil cover and thus more thoroughly insulating the eggs.

If a cold spell were to continue for a sufficiently long time at a period when the organic matter was actively fermenting, it seems probable that the egg chamber would increase in temperature to such an extent that an opening would become necessary. This, perhaps, explains why, during this cold spell, one of the mounds under observation showed very definite signs of having been opened, although no new egg was found. On the other hand, a prolonged cold spell at a time when the fermentation was not very active could lead to a significant reduction in the egg chamber temperature, with some danger to the eggs. Such prolonged conditions have not been encountered by the author; and in the dry inland of Australia cold inclement weather during the breeding season of the mallee-fowl is very unusual, and is always of short duration. It is probably not a serious problem to the birds.

(ii) *Midsummer.*—During January and February the weather was very hot. The daily shade maxima were never much less than 100°F, and on some occasions as high as 112°F.

It can be seen from Figure 1 that at this time of the year the temperature in the natural mound was not very far above that of the control, only a small amount of heat being then produced by the organic material. It is probable that the slight difference in the two recorded temperatures was due partly to this latter source of heat and partly to experimental error.

During this type of weather, activity at the mounds was very slight. The depth of soil over the eggs was increased to 3 ft; and although the birds visited daily, they did not open the mound very often. Digging was limited to once every 6 or 7 days, and was done very early in the morning, at first light. On these occasions both birds participated and worked rapidly, opening the mound completely down to the eggs. The whole surface was removed. Immediately the mound was opened refilling commenced without delay, and the birds seldom paused until the restoration was complete or nearly so. The work was always finished before 9.30 a.m., well before the Sun's heat was very intense. These openings had no immediate effect on the egg chamber temperature.

Temperature readings were taken with alcohol-in-glass thermometers attached to rods and pushed into the mound at various depths immediately before and soon after a typical opening in the hot weather. It was seen that, before the opening, the soil immediately above the eggs was at a higher temperature than that near the surface; and the opening led to a reduction in the average temperature of the soil over the eggs. Such an effect would be expected to result in a reduction in the amount of heat from the Sun subsequently reaching the eggs, and could be useful in delaying a long-term increase in the temperature of the eggs.

On some occasions, when extended periods of very hot weather occurred, the activities of the birds were ineffective in preventing a temperature rise of the eggs. In every mound examined, following a very hot spell in mid December 1953, several of the eggs in the mound at that time subsequently failed to hatch, presumably having been killed by the excess heat. In every mound studied in each year during which observations have been made, chicks which hatch in very hot weather have been suffocated whilst digging to the surface. In some cases they failed to leave the egg satisfactorily, and in other cases failed to reach the surface of the mound.

(iii) *Autumn.*—After midsummer, at the end of February, the temperature in the control mound began to decrease, and a parallel decrease occurred in the natural mound which now had no organic matter to supply heat. The declining temperature was dealt with by the birds by altering the thickness of the soil covering and modifying their activities, in an attempt to maintain the egg chamber temperature at about 93°F.

The temperature in the mounds during the period from noon March 10 to noon March 17, 1954, is shown in Figure 4. The mound was visited by the birds and opened on March 7, 11, 14, and 17. On these occasions the opening was of a different type from those seen previously, the mound being left open during the heat of the day. Work was usually started between 7 a.m. and 8 a.m., and the mound rapidly opened to within a few inches of the eggs.

In the very cold weather, in the spring, the thickness of the soil cover over the eggs was increased by the birds in order to retain the fermentation heat within the egg chamber. During the very hot weather of midsummer also the thickness of the soil cover over the eggs was increased, to 36 in., in order to provide additional

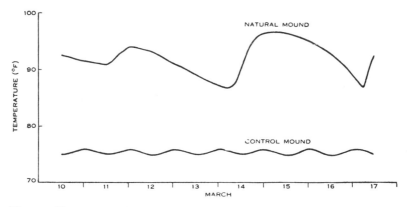

Fig. 4.—Temperature in the egg chambers of the natural and the control mounds during a period of autumn weather with declining temperature from March 10 to March 17, 1954.

insulation to protect them from the Sun's heat. At the end of the hot weather, however, the thickness of the soil cover was reduced by the birds to increase the amount of solar heat reaching the eggs. The thickness was reduced from 36 to 20 in. Although this undoubtedly assisted in warming the egg chamber, more direct methods were also necessary. After digging out, the mound was then allowed to remain open until 4–5 p.m. when the birds returned and rapidly restored it. As might be expected, this type of opening led to an increase in temperature within the egg chamber. Although on each occasion that the mound was opened the temperature was increased by up to 12°F, the effect was very temporary and within 48 hr the temperature had declined almost to its original level. During the whole period covered by Figure 4 a tendency for the average egg chamber temperature to decline can be seen.

Figure 5 shows the heat flow measurements in the two mounds for the period from noon March 10 to noon March 12, 1954. The birds' activity during the period is given below.

March 10

    12 noon.—The mound was flat on top and had not been disturbed since it was refilled in the afternoon on March 9.

    6.20 p.m.—There had been no sign of the birds all day until 6.20 p.m. when the female arrived, stood on the mound for a few moments and then moved away, feeding.

March 11

>   4.30 a.m.—Birds were heard calling as usual in the morning.
>   6.25 a.m.—Male arrived and displayed on the mound; he was answered by female. The male then left.
>   7.10 a.m.—Both birds arrived and started work; they worked swiftly and constantly and by 8.30 a.m. the mound was opened down to within 6 in. of the eggs.
>   8.30 a.m.—Both birds left.
>   9.00 a.m.—Both birds returned and deepened the excavation to within 2 in. of eggs and then left.
>   9.30 a.m.—Female returned for a few minutes but did no work.
>   4.45 p.m.—Male returned and began filling in slowly until half full at 5.10 p.m. when he left.
>   5.30 p.m.—Male returned and began filling in again.
>   6.35 p.m.—The mound was completely restored and the male moved away, feeding.

March 12

>   4.40 a.m.—Birds began calling.
>   6.05 a.m.—Male walked onto mound and then went away, feeding.
> 12.30 p.m.—Observations ceased. The birds had not returned since the dawn visit.

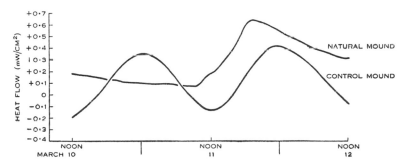

Fig. 5.—Heat flow measurements in the egg chambers of the natural and the control mounds during the period March 10 to March 12, 1954.

It can be seen in Figure 5 that the opening of the mound between 7 a.m. and 8 a.m. on March 11 led to a greatly increased flow of heat into the egg chamber from the Sun, causing the increase in temperature that has already been described. In addition to this direct effect, the filling of the excavation with sand that had been exposed to the Sun during the day had a long-term effect. Following the opening on March 9 there was no noticeable flow of heat from the egg chamber to the surface by the morning of March 11 when the mound was again opened to the Sun. The result was to prevent the establishment of a natural diurnal wave of temperature and to retain the heat efficiently within the mound.

This process of opening the mound to the Sun has long been assumed to be the chief means of incubation of the eggs (Ashby 1922; Lewis 1939, 1940). However, during the 2 years of this study it has been seen to be of very minor importance. In 1953 the first occasion on which it occurred and the egg chamber showed any increase in temperature was February 6; and in 1954 the first occasion was February 12. That is to say the deliberate increasing of the amount of solar heat reaching the eggs was important for about 15 per cent. of the total incubation period; during the rest of the year keeping the eggs cool was more important.

Towards the end of March the mound temperature was regularly falling to 85–86°F in each 48 hr, and on opening was raised to 96°F only to fall to 92°F in the first 24 hr. The last egg hatched on March 16, and the birds continued opening the mound for another week. The last opening, on March 26, increased the temperature from 84 to 93°F, and it fell again to 86°F in the following 24 hr. The mound was then abandoned by the birds.

(c) Mound without Organic Matter

One of the series of mounds kept under observation, mound 4, was constructed without any organic matter, the eggs being laid in direct contact with the soil. Throughout the season this mound was maintained and shaped by the birds in a different way from the others. The eggs were only covered by 20 in. of soil, and the mound was kept broad and flat throughout the early part of the season. During the very hot weather it was mounded up to a height never exceeding 2 ft, compared to a height of 3 ft in the other mounds.

A thermograph bulb was buried among the eggs at a depth of 20 in. on November 23, 1953, and allowed to remain until December 11. The temperature of the mound during this period was found to be maintained at 92°F. There was a diurnal variation of $\pm 0 \cdot 2$°F only. During the same period, in the control mound (which, of course, was not disturbed by birds) the temperature at 3 ft was maintained at 82°F and that at 6 in. at a mean of 86°F. By interpolation it can be assumed the temperature in this mound at 20 in. was about 85°F. It can be seen that the natural mound was maintained at a temperature about 7°F higher than the control mound, and this can only have been due to the activities of the birds. Because of the difference in heat diffusivity of the two mounds (see Section V(a)) this 7°F is also an underestimate of the difference.

At this time of the year, the other mounds in the area were being visited very early in the morning, at intervals of a few days, and the activities of the birds were shown to result in a loss of heat. At mound 4 the activity was different. The mound was visited almost daily by the birds, at about 8 a.m., and opened to within 3 in. of the eggs; there was then a pause of 15–20 min whilst the birds fed quietly in the vicinity, and they would then return and refill the excavation. The refilling process was gradual by comparison with the other mounds, and was never finished before midday. The sand was scratched in to a depth of only an inch or so at a time, and there were pauses of 10–15 min between each burst of work.

The thermograph records showed that, following each such opening, the egg chamber temperatures rose slightly, but never more than $0 \cdot 4$°F, and then slowly declined again. Clearly some heat was provided to the eggs by the opening. On these occasions the surface of the soil in the direct sunlight was very hot at the time of day the mound was opened; and it may well be imagined that the action of scratching in only the heated surface layer would be an effective method of adding heat to the interior of the mound. In this connexion it is of interest to note that Leake (Ashby 1922) was able to hatch mallee-fowl eggs artificially by pouring hot sand into the mound. During the rest of the season the activities at this mound and their results were the same as at the other normal mounds.

The other birds in the area laid their first eggs in the first week of September, but this individual laid its first egg in November (see Section II) and so there was no necessity to use organic matter as a source of heat early in the season. The incubation process was simplified and condensed into the period when sufficient solar heat was available. The implications of the occurrence of this different breeding season will be discussed in a separate paper.

VI. Conclusions

The results of the experiments reported in this paper make it quite clear that the incubation of the eggs of the mallee-fowl is achieved mainly by heat generated by the fermentation of the organic material buried in the mound. Early in the season practically all the heat reaching the eggs is from this source, and fermentation heat remains important until midsummer.

Solar heat, which has been considered by earlier observers to be of over-riding importance, is shown to play only a supplementary role in the incubation process. Early in the season when fermentation within the mound is at its height, active measures are taken by the birds to prevent heat from the Sun reaching the egg chamber. By midsummer, however, the Sun is the chief source of the heat reaching the eggs, and it is still necessary to reduce the amount of heat reaching them. Towards the end of the season, when fermentation has ceased, not enough solar heat penetrates the mound, and it is necessary for the birds to remove the soil to allow more heat to reach the eggs. The period of complete dependence on solar heat therefore represents a very minor phase confined to the last few weeks of the season.

There is little doubt that the ancestors of the mallee-fowl were jungle dwellers, incubating their eggs in mounds built of leaf litter gathered from the rain-forest floor, and largely protected by the canopy from excessive insolation. The onset of aridity in the inland, with its strongly contrasting conditions of low rainfall, high summer temperatures, and lack of shade, necessitated the development in the birds of a pattern of instinctive behaviour adapted to deal with the local factors affecting the construction of the mounds and the temperature changes within them over the long incubation period. It is clear from the observations reported here that the mallee-fowl has developed a complex behaviour pattern that is highly effective for ensuring the successful incubation of its eggs, and is moreover sufficiently adaptable to deal with reasonable variations in seasonal conditions and a wide range of variation in the relative amounts of organic matter and soil incorporated in the mound. There is a strong suggestion that the testing of the egg chamber temperature with the beak is a key factor in determining the nature of the birds' operations on the mound; and experiments are being carried out to see whether "unseasonable" temperatures artificially induced in the region of the egg chamber will stimulate appropriate compensatory activity on the part of the birds.

VII. Acknowledgments

A number of people assisted in the work. The Division of Meteorological Physics, C.S.I.R.O., supplied the heat flux plates and advised in their use and in the interpretation of the results.

TEMPERATURE REGULATION OF MOUND OF MALLEE-FOWL

The construction of the nesting mound of *Leipoa ocellata*. (Photographs by H. J. Frith.)

Fig. 1.—The hole excavated during May 1953.

Fig. 2.—The organic material collected and deposited in the hole during July.

Fig. 3.—The completed mound with the male bird working to regulate the temperature.

TEMPERATURE REGULATION OF MOUND OF MALLEE-FOWL

Temperature regulation in the nesting mounds of *Leipoa ocellata* showing a typical mid-summer opening. (Photographs by R. Carrick.)

Fig. 1.—Male, at dawn, commencing to remove soil cover.
Fig. 2.—Male (left) and female in almost completed excavation.

The work was stimulated by interest shown by Dr. R. Carrick of the Wildlife Survey Section, C.S.I.R.O.

Mr. R. A. Tilt participated in much of the field work and Mr. F. N. Ratcliffe, Officer-in-Charge of the Section, read the manuscript and made many helpful suggestions.

VIII. References

Ashby, E. (1922).—Notes on the mound-building birds of Australia with particulars of features peculiar to the Mallee-Fowl, *Leipoa ocellata* Gould, and a suggestion as to their origin. *Ibis* **4**: 702–9.

Lewis, F. (1939).—Breeding habits of the Lowan in Victoria. *Emu* **39**: 56–62.

Lewis, F. (1940).—Notes on the breeding habits of the Mallee-Fowl. *Emu* **40**: 97–110.

Mattingley, A. H. E. (1908).—Thermometer-Bird or Mallee-Fowl (*Lipoa ocellata*). *Emu* **8**: 53–61, 114–21.

Ross, J. A. (1919).—Six months' record of a pair of Mallee-Fowls. *Emu* **18**: 285–88.

Serventy, D. L., and Whittell, H. M. (1951).—"A Handbook of the Birds of Western Australia (with the exception of the Kimberley Division)." 2nd Ed. p. 67. (Paterson Brokensha Pty. Ltd.: Perth, W.A.).

Taylor, J. K., and Hooper, P. D. (1938).—A soil survey of the horticultural soils in the Murrumbidgee Irrigation Areas, N.S.W. Coun. Sci. Industr. Res. Aust. Bull. No. 118.

Reprinted from *Science,* **186**(4159), 107–115 (1974)

Avian Incubation

Interactions among behavior, environment, nest,
and eggs result in regulation of egg temperature.

Fred N. White and James L. Kinney

"The stable, warm temperature for early avian development is common to all climates, and is maintained through an equally common prescription for temperature regulation by parental behavior" (*1*). We will consider herein the nature of some major ingredients which are formulated into this "common prescription" for incubation of the avian egg.

The importance of regulating the thermal environment of the developing embryo has been emphasized by studies on the domestic fowl by Lundy (*2*). He found that the optimal temperature for embryonic development in temperature-controlled cabinets was between 37° and 38°C, and that no embryos survived continuous incubation above 40.5°C or below 35°C. Similar results

for domestic fowl, pheasant, duck, and quail were reported by Romanoff and Romanoff (*3*). It is generally recognized that 25° to 27°C is a temperature range below which no development occurs, this being the so-called physiological zero temperature. Incubation for any protracted period between physiological zero and the optimum range results in various developmental anomalies. However, maintenance in suspended development below physiological zero for moderate periods is compatible with later development at appropriate temperatures. This correlates with the observed delay in incubation which may occur between the laying of the first egg and the completion of the full clutch. These results emphasize the scope of

the task faced by incubating parents in providing a thermal environment that will ensure successful embryonic development and hatching.

With rare exceptions (*4, 5*), the avian embryo is incubated by an attending adult transferring its body heat to the egg. The mean egg temperature is an approximation of the proper developmental temperature. Huggins (*6*) reported a mean egg temperature of 34.0°C ± 2.38° standard deviation (S.D.) for 37 species representing 11 orders. An extensive study of the house wren, *Troglodytes aedon,* revealed mean egg temperatures well within this range (*7*) while other reports have indicated a range of 34° to 36°C for various species (*8, 9*). The narrow range of mean egg temperatures achieved by diverse species suggests that the thermal requirements for successful development are similar for most if not all birds. This requirement must be met through incubation strategies that compensate for fluctuations in environmental temperature and that allow the attending parents to acquire sufficient food to support metabolism.

The relative responsibilities of parent birds varies from mutualistic sharing of the task to total involvement by one parent for the entire incubation

Dr. White is professor of physiology, University of California, Los Angeles 90024, and Dr. Kinney is a postdoctoral fellow in the Department of Physiology at the University of California, Los Angeles.

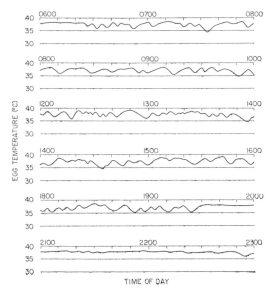

Fig. 1. Portions of a 24-hour record of the egg temperature of the village weaverbird, *Ploceus cuculla-tus*. Note the contrast between the relatively stable nighttime temperatures and those of the daytime hours when the female alternates between attentiveness and activities away from the nest. Variations in nighttime egg temperature are caused by changes in the contact relationship between the brood patch and the egg (adjustments in tightness of sit). Mean nighttime egg temperature is higher than that for the intermittent incubation pattern associated with the daylight hours.

period. The former, bisexual incubation (primary mode) is exemplified by most marine birds, by many nonpasserine birds in various orders, and by most of the more primitive groups and some advanced taxa within the order Passeriformes. Generally, the eggs of such birds are attended at virtually all times by one or the other parent. Exceptions to constant attentiveness during mild weather have been reported among toucans, puffbirds, barbets, ovenbirds, and some antbirds (5). Others engage in single sex intermittent incubation (secondary mode), a strategy in which a single parent, most often the female, divides the time between attentiveness and food gathering, somehow adjusting attentiveness against fluctuations in environmental temperature to regulate egg temperature. House wrens and certain weaver finches are examples of this category. Between these two strategies there exist a number of intermediate situations which we shall refer to as the "intermediate mode." For birds of this category the nonincubating member of the pair may feed the attending mate to a varying extent, depending on the species.

A survey by Van Tyne and Berger (*10*) revealed that, for some 20 families, there appears to be no reliable information on the sex of the parent that incubates the eggs. In 54 percent of the remaining families incubation is performed by both sexes; in 25 percent, by the female alone; in 6 percent, by

the male alone; and in 15 percent by the female, the male, or both. Skutch (5) has reviewed extensively the division of labor between the sexes during incubation.

The bisexual incubation pattern appears to be the primary or primitive mode from which intermittency has been derived. The intermittent pattern is common among passerines, but members of this group which are considered to be more primitive (5) tend to con-

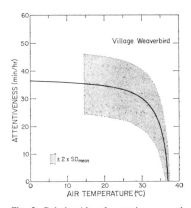

Fig. 2. Relationship of attentiveness and ambient temperature for an aviary-housed colony of village weaverbirds [for derivation of data points, see (*14*)]. There is a large adjustment in attentiveness over the warm range of ambient temperatures but this becomes progressively limited with declining temperature. The attentive curve is described by the equation $Y = (81.8/X - 39) + 38.7$.

form to the bisexual pattern. For intermittent incubators there is usually a nest structure of some complexity, while many bisexual incubators may lay eggs in little more than a slight concavity in the substrate. Adoption of the intermittent strategy appears to be coupled with increasing nest complexity, although exceptions exist. The presence of the bisexual mode among the more primitive members of a recent group, the passerines, while intermittency characterizes more modern representatives which have derived varying degrees of nest complexity, strongly supports the contention that the bisexual mode is the more primitive of the two strategies. It is apparent that the intermittent scheme has been adopted a number of times in avian evolution since it occurs in a variety of groups which are not closely related.

One generalization about birds becomes apparent to those who attempt to divide avian behavior into neat categories: there are almost always exceptions. Such are the unusual strategies of the Megapodidae in which nonmetabolic heat sources are utilized for incubation; the emperor penguin, a species in which the male assumes a sole and constant physical association with the egg for the whole incubation period, relying on stored energy while supporting the egg between the foot webbing and the brood patch; and the peculiar hole nesting of the hornbills in which the female is imprisoned behind a mud wall with her eggs where she is periodically fed by the male through a small opening in her cell. Except for the Megapodidae, the relatively few species of birds that show exceptional incubating behavior probably share a fundamental similarity with other birds regarding the basic mechanisms used in the control of egg temperature.

For single sex intermittent incubation, the task of adjusting incubation time against fluctuations in environmental temperature appears to be met by lengthening the periods of attentiveness (time spent on the eggs per unit time) as ambient temperature declines; however, the precise relationship between attentiveness and environmental temperature has remained unclear. Kendeigh (*11*) reported that at an air temperature of 15°C attentive periods of the female house wren averaged 14 minutes, while at 30°C these periods were reduced to 7.5 minutes. Inattentive periods were approximately the same at all but high temperatures, at which they were lengthened. Qualitatively similar observations have been made for the Holarctic wren, *Troglodytes troglodytes* (*12*). Several reports suggest that at air temperatures near the mean egg temperatures reported by Huggins (*6*), attending birds cease to incubate (*9, 13*). For some birds, attentiveness at high temperatures may take on the form of postural shading of the eggs to prevent overheating (*9*).

In this article we consider experimental evidence which elucidates the relationship between attentiveness and ambient temperature in a single sex intermittent incubator, the village weaverbird (*Ploceus cucullatus*). During incubation the female is neither assisted nor fed by the male. Our study, done in an aviary in Los Angeles, afforded us the opportunity of providing the birds with easily accessible food at all times, so that we could control, to some degree, variations in flight distance and food abundance which might affect sortie time, the amount of time spent away from the nest. We also consider the role of nest insulation as a determinant of attentiveness. Implications concerning sensory perception of egg and ambient temperatures are derived from experiments in which sensory receptor mechanisms have been modified by the application of a local anesthetic. By comparing our data with the available data for other species we obtain a generalized scheme for avian incubation from which we can derive information concerning the zoogeographic distribution of breeding birds.

Attentiveness and Environmental Temperature

The efficacy of incubation was assessed through constant monitoring of the temperature of one of the eggs in a clutch (Fig. 1). Such data also yielded information on the division of time between incubation and activities outside the nest (*14*). Typically, the female enters the nest near sunset and remains with her eggs throughout the night. Nocturnal variations in egg temperature reflect adjustments in her "tightness of sit," specifically the physical relationship between the vascular hot spot on her belly, the brood patch (*15*), and the egg; the mean temperature for the eggs is higher and the variation less at night than during the day (*16*). The intermittent pattern of

incubation commences with emergence in the morning. By comparing the attentiveness, expressed as minutes per hour, with the shaded air temperature, we observed that attentiveness was related to environmental temperature as previously suggested (*7, 11, 12*). The form of this relationship is by no means a simple one, and we found that the data conformed most closely to the equation $Y - A = B/(X - C)$, where Y is attentiveness in minutes per hour, A is the horizontal asymptote of the curve, X is shaded air temperature, C is the vertical asymptote, and B relates to the curvature of the relationship. The equation describes a rectangular hyperbola (*17*) (Fig. 2). We analyzed our data to discover whether there was a relationship between frequency of attentive bouts and environmental tem-

perature and found none, except at the warm extreme of the curve where there was a sharp reduction in attentiveness. We further analyzed the relationship of egg temperature to attentiveness and frequency for an environmental temperature range of $-10°$ to $40°C$ (*18*). The degree to which egg temperature varied was highly dependent upon the attentive frequency and the temperature gradient between egg and environment; however, the mean egg temperature depended almost entirely upon attentiveness.

The attentive curve intersects the temperature axis near $37°C$ where attentiveness becomes essentially nil, a relationship in harmony with several observations from the field for various species (*9, 13*). The range of adjustment in attentiveness is comparatively

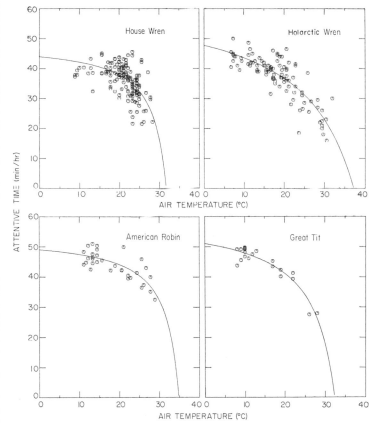

Fig. 3. Attentive behavior of four species of birds in which only the females incubate the eggs (single sex intermittent incubation). Data derived from the house wren (*11*), Holarctic wren (*12*), American robin (*11*), and great tit (*36*). The female great tit may be fed, to a moderate degree, by the male. Values for B, γ, α, A, and C are: house wren, 162, 8.5, 22, 48.6, 35; Holarctic wren, 826, −0.3, 21, 64.2, 50; American robin, 170, 8.3, 25, 535, 38; and great tit, 338, 0.7, 19.7, 60, 38.

large at the warmer end of the curve but becomes increasingly limited at temperatures below that corresponding to the point of maximum curvature (30.1°C) which we have designated α. Thus, as temperature declines, there is a progressive decrease in the time available during a given sortie for activities other than food gathering.

Data on attentiveness as it relates to environmental temperature have been extracted from the literature for four species in which only the females of each pair incubate the eggs intermittently. These data were analyzed by the same method as described for the village weaverbird. Figure 3 illustrates the basic similarity in the attentive curves for these species, the intercepts on the temperature axis suggesting that "recognition" of a critical temperature for cessation of attentiveness is a common phenomenon. However, these data must be viewed against the different methods used by the investigators. Numerous data were collected for the house wren (11), for example, but the ambient temperatures were derived from the mean daily temperatures at a weather station somewhat removed from the study site. For the Holarctic wren (12), the temperature data were more satisfactory, representing hourly means at the site, but only a single individual was studied. Thus, high reliance on the precise form of these curves and their intersects is not warranted.

The more extensive data on the village weaverbird permit a useful comparison of the intercept temperature for zero attentiveness with egg temperatures both during the day and the night. The intercept of 37.0°C (95 percent confidence interval, 1.2°C) for our group coincides closely with the mean maximum daytime egg temperature of 36.7°C ± 1.5° (S.D.) and the mean maximum nighttime egg temperature of 36.7°C ± 1.4° (S.D.). The standard deviations reflect more the variations in the maximum egg temperatures among the animals (19) than great variation for given individuals. There is close coincidence of the peak egg temperatures and the environmental temperature at which attentiveness ceases. This suggests that the sensory mechanisms utilized in reducing attentiveness as the egg reaches a "release" temperature may also be employed in the assessment of the external thermal environment.

Nest Insulation and Attentiveness

The cooling rates of the eggs within different nests provide us with a useful index of nest insulation (20). Among our weaver finches, we found a variety of cooling rates and we have normalized the attentiveness for these birds at 25°C, the mean daytime temperature for our study (21). By plotting cooling rate against the attentiveness at 25°C we obtained a relationship (Fig. 4A) which points up a role for insulation as a determinant of attentiveness. Birds with nests that were relatively well insulated exhibited less attentiveness, and had greater time available for both food acquisition and activities unrelated to food gathering than did birds with less insulated nests. Insulation may well be a manipulatable factor through which breeding range can be extended altitudinally or into colder climates, other critical factors being equal. The variability in the insulation of our weaverbird nests suggests that a range of nest construction potentialities are available for natural selection.

That birds utilizing well-insulated nests spend less time incubating their eggs than do those whose nests are not so well insulated indicates that the mean attentive curve which we have derived (Fig. 2) is in reality a composite of several attentive curves having similar intercepts but different

curvatures. The asymptotes of the attentive curve serve to locate the curve, while the B value dictates the curvature. The insulative component is reflected by B, large B values (22) being associated with high indices of insulation (Fig. 4B).

The massive amounts of data on the regulated mean egg temperatures for various species (6–9), and the observations that some species cease to incubate at or near this temperature (9, 13), make us confident that the intercept on the temperature axis is not a significant major variable. Food abundance and availability have been controlled in our aviary study; however, one can appreciate that variation in the flight distance to food, or in food density, would strongly affect the parent's ability to provide an attentive relationship appropriate to successful incubation. Long flight distance to a poor food source and relatively poor nest insulation would rapidly become incompatible as ambient temperature declined. A comparison of nest materials utilized by *Ploceus cucullatus cucullatus* and its relative, *P. c. graueri*, that lives at a higher altitude, revealed that the nests of *P. c. graueri* contained a much thicker and better insulated lining (23). Similarly, the high-altitude hummingbirds, *Oreotrochilus chimborazo* and *O. estella*, appear to build nests of greater bulk than do their low-altitude relatives (24). The difference between the attentive curves for the wrens, *Troglodytes aedon* and *T. troglodytes* (Fig. 3), may be attributable to nest insulation rather than inherent differences in attentive behavior. The extraordinarily high B value for the Holarctic wren should be indicative of a well-insulated nest. In their description of the nest of the individual utilized in their study, Whitehouse and Armstrong (12) stated that the nest was built "on the disused nest of a robin in a cycle saddle-bag hanging on a nail in a summer-house," a description compatible with the high B value characterizing the attentive curve.

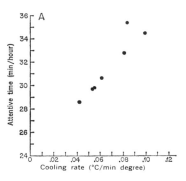

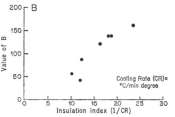

Fig. 4. (A) Relationship of attentiveness to the cooling rates of eggs in nests with different insulative properties among village weaverbirds. Data were normalized to the mean daytime temperature of the study, that is, 25°C. (B) Relation of the B value in the equation describing attentiveness $[Y - A = B/(X - C)]$ to the index of insulation for the various nests.

Mean Egg Temperature and Embryonic Heat Production

Empirically determined rates for heating and cooling of individual nest-egg complexes, and records of egg and environmental temperatures, were utilized to obtain the maximum, minimum, and mean egg temperatures characteristic of the birds in our study (*25*). The mean egg temperature was determined from predicted values for egg temperature based on a few critical values for each cycle. The close correlation between our predicted values and the actual record is shown in Fig. 5. The 24-hour mean egg temperature for the village weaverbird was 35.9°C ± 1.5° (S.D.). During the nighttime, mean egg temperature was 36.5°C ± 1.5° (S.D.) while the daytime mean was 35.3°C ± 1.6° (S.D.). The higher egg temperatures at night are associated with the constant physical relationship of the parent with the eggs during this period. Our data suggest that the mean egg temperature of 34°C ± 2.38° (S.D.) reported by Huggins (*6*) is an underestimate of the 24-hour mean egg temperature for the series of birds in his study. It seems likely that two complications contributed to this underestimate; first, the measurements were made during daylight hours, and, second, some cooling of the eggs may have occurred between the departure of the attending parent and measurement of the egg temperature. Our daytime mean egg temperature of 35.3°C corresponds rather closely to the figure of 35°C which Kendeigh has suggested as characteristic of passerine species (*26*). Dual sex incubators presumably maintain the eggs at a higher temperature by virtue of their essentially constant attentiveness. A temperature of 37.5°C appears to characterize the herring gull egg (*27*).

Upon examination of our data on mean egg temperature it became apparent that, at the higher range of ambient temperatures (30°C and above), the sample size was too small for us to evaluate the exact nature of the curve. To better appreciate this relationship, we determined the mean heating and cooling rates of eggs and calculated the effect of the observed mean attentiveness on egg temperature over a wide range of environmental temperatures (*18*). This curve revealed that mean egg temperature should be relatively stable at 37°C be-

tween point α and the intercept of attentiveness and ambient temperature (31° and 37°C, respectively). We have designated this zone as the zone of stable regulation (Fig. 6). Above an ambient temperature of 37°C, am-

bient and egg temperatures are coincident when incubation ceases. The predicted mean egg temperature for the zone of stable regulation and the mean egg temperature observed were 37°C and 36.9°C, respectively. At ambient

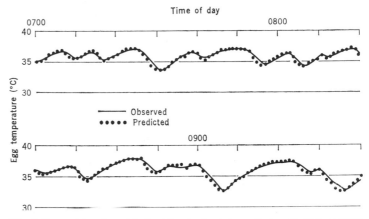

Fig. 5. Comparison of empirically determined egg temperature with that predicted from the mean heating and cooling rates of the egg, ambient temperature, and attentiveness (*25*).

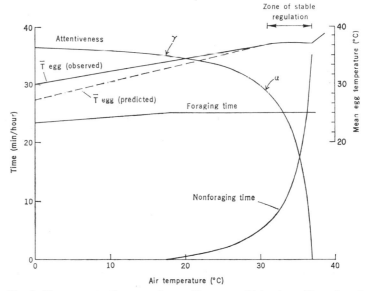

Fig. 6. The mean attentive curve for a village weaverbird colony. The point of greatest curvature, α, corresponds to the lower ambient temperature for stable regulation of egg temperature; γ is the point on the attentive curve at which the mean egg temperature is 34°C. Maintenance below this temperature for a protracted period of time is likely to be incompatible with normal embryonic development. Derivation of the division of nonattentive time as a function of ambient temperature reveals that nonforaging time is adjusted, as a reserve, in favor of stable foraging time and increasing attentiveness at ambient temperatures above γ. Below γ a progressive recruitment from foraging time is mandatory if the attentive relationship is to be maintained.

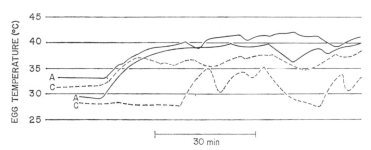

Fig. 7. Effect of injecting the brood patch with a local anesthetic (*A*) on attentiveness and egg temperature. Injection of saline, as a control (*C*), was followed by resumption of a pattern of incubation more characteristic of undisturbed birds. The experiment suggests that a low level of discharge from receptors located in the skin covering the belly intensifies attentiveness and elevates egg temperature.

temperatures below 31°C the predicted mean egg temperature was slightly less than that observed.

Drent (*27*) demonstrated for the herring gull egg that heat production is positively correlated with the weight of the developing embryo. At an incubation temperature of 38°C, heat production exceeded heat loss after day 13, and by day 27 a gradient of + 2.69°C was evident. Qualitatively similar results for the duck egg were obtained by Khaskin (*28*). Thus, embryonic heat production facilitates heating and retards cooling. As a result, the release temperature (the temperature at which the bird can leave the nest) should be reached progressively faster as embryonic development proceeds. This effect probably causes a progressive decrease in attentiveness over the incubation period. Embryonic heat production ought to have an effect similar to an increase in insulation: hence, the *B* value for the attentive curve should increase. The magnitude of this effect on attentiveness is unknown; however, in late embryonic development, it may be sufficient to provide a significant increment in foraging time.

Adjustments in Attentiveness with Environmental Temperature

Nonattentive behavior may be divided into two categories: foraging (behavior related to food acquisition) and nonforaging (social interaction, preening, for example). It is apparent that the time available for nonforaging behavior is maximized at temperatures requiring a minimum of attentiveness. As temperature declines, the ratio of foraging to nonforaging must increase

if metabolism is to be maintained, while attentiveness increases in support of the regulation of egg temperature. This effect of low ambient temperature is inferred from observations made on the Alaska wren when it was incubating at low ambient temperatures and high latitudes (*29*). A high degree of precision in the timing of feeding sorties of 2 to 5 minutes' duration was reported while attentive periods exceeded 20 minutes. Flight was said to be direct. Predictable food supply and good nest insulation may conjoin to allow energetic balance when foraging time is limited.

Birds do not appear to increase their heat production to support incubation except at temperatures which induce shivering thermogenesis (*30*). Rather, it is the percentage of time spent on the eggs that is adjusted. By recruiting time for incubation from nonforaging activities, a bird may adjust attentiveness with no decrement in foraging activities as long as a reserve of nonforaging time remains available. As ambient temperature falls, this reserve must, of necessity, decrease.

Because the greatest fluctuations in egg temperature occur during the daytime in single sex intermittent incubators, it is reasonable to assume that the daytime mean egg temperatures, observed under natural conditions, should represent temperatures compatible with successful development. From available data on daytime mean egg temperatures of passerine birds, 34°C appears to represent a temperature approaching the lower limit for successful hatching. Data on mean egg temperature and attentiveness are available only for the village weaverbird and the house wren. We have located a

point, γ, on the attentive curves for these birds, which corresponds to a mean egg temperature of 34°C (see Fig. 6). At ambient temperatures below this point, the attentive curve is essentially linear. This apparent linearity suggests that the influence of multiple factors as determinants of nonattentiveness has effectively disappeared. The second derivative of the weaverbird attentive curve at γ is − 0.015. For the house wren this value is − 0.009. We think that, in the absence of data on mean egg temperatures, the locus of the second derivative of the attentive curve corresponding to the mean value of − 0.012 should approximate an environmental temperature at which the mean egg temperature is 34°C. This value should become increasingly refined as more data become available for species in which only one sex incubates the eggs intermittently. Therefore, successful incubation should be limited to geographic areas in which the mean daytime temperature exceeds that for γ.

It is reasonable to assume that the nonattentive activities at temperatures below γ are devoted to foraging behavior, because the attentive curve approaches linearity at temperatures below γ and it is unlikely that foraging would be sacrificed in favor of activities not related to the acquiring of energy. For heuristic reasons we have assumed that nonforaging activities have become nil at this temperature. We also assume that foraging time is relatively stable and is sufficient to maintain metabolism over the temperature range in which nonforaging behavior occurs. With this in mind we have calculated the division of time spent in foraging and nonforaging activities at various environmental temperatures (Fig. 6). Over the zone of stable temperature regulation the strategy is to draw heavily on nonessential activities in favor of attentiveness. At temperatures below α, most of the nonforaging time has been utilized and attentiveness is adjusted less and less with declining temperature. This results in a decline in mean egg temperature. Since we have deduced that nonforaging time is nil at γ, any increment in attentiveness at lower temperatures must be gained at the expense of foraging time. The strategy of compromising both attentiveness and foraging time at such temperatures is indicated. These considerations suggest that there is a critical range of temperatures below γ, and that within this temperature range, the energy stores of the attending bird must be utilized if the attentive relationship is to be maintained. A single sex intermittent incubator attending its eggs for a protracted period of time at temperatures below γ would be expected to lose weight, extend the incubation period, or fail to provide the thermal conditions necessary for a successful hatch. For several domestic species, prolonged incubation below 35°C is incompatible with the hatching of viable young. When the zebra finch, *Taeniopygia castonotis*, incubated eggs at an environmental temperature of 14.5°C, all the embryos died. The birds were unable to acquire sufficient energy for the maintenance of the eggs at the temperature appropriate to successful development (*31*).

Sensory Receptors and Attentiveness

The body heat of an attending parent is transferred to its eggs by way of the brood patch. Drent *et al.* (*32*) observed that herring gulls adjusted their contact with artificially heated or chilled eggs, while we observed adjustments in tightness of sit during the nighttime sessions. These observations suggest that the brood patch may contain sensory receptors that provide the parent bird with information concerning egg temperature. We investigated this possibility by anesthetizing the skin of the brood patch, by topical application or subcutaneous injection of 2 percent Xylocaine (Astra). To control for the disturbing influence of capturing the birds and applying the anesthetic, we injected the brood patches of two incubating birds with saline in an identical manner. As shown in Fig. 7, birds with anesthetized brood patches incubated their eggs more intensely than control birds, and their eggs reached the highest temperatures which we have observed. Such birds appeared to spend unusually little time away from the eggs. Although our data are somewhat limited, we conclude that sensory receptors that influence attentiveness reside in the skin of the brood patch, and that a low frequency of discharge in these receptors encourages attentiveness. Whether these same receptors are responsive to the thermal environment when the parent is away from the nest is not known.

Modulation of Attentiveness:
A Summary

We have summarized the probable interactants determining attentiveness for both dual sex (primary mode) and single sex intermittent (secondary mode) incubators in Fig. 8. Cases in which the female incubates the eggs but is periodically fed at the nest by the male, or in which the fat reserves of the attending parent are sufficient to allow protracted incubation, constitute an intermediate mode of incubation.

In the primary mode the principal means of regulating egg temperature is through adjusting tightness of sit. Such adjustments appear to involve detection, through brood patch receptors, of differences between a reference temperature near that of the core of the adult and the surface temperature of the egg. Release from incubation depends upon the intensity of factors competing with the incubation drive. The most important of these competing factors are hunger, the presence and behavior of the relieving parent, or both.

The secondary mode, during nighttime periods, resembles the incubation pattern of primary mode birds in that it is the tightness of sit which is adjusted during this period of protracted attentiveness. The photic environment appears to induce the parent to enter the nest at night. In order to maintain

metabolism, a single sex intermittent incubator must spend periods away from the nest during the day when foraging is possible. Other drives, such as social interaction, preening, and so forth, may also be present. Like Baerends (*33*), we view these drives as factors of varying intensity which compete with attentiveness. Hence, satisfaction of these drives favors the expression of attentiveness. The intermediate mode of incubation, in which attentiveness is relatively protracted, appears to depend upon the amelioration of hunger through utilization of fat stores or feeding of the attending bird by a mate. In this situation the dependency of attentiveness on ambient temperature decreases, and in some cases may show little or no relationship with ambient conditions. Regulation of egg temperature, although provided by one parent, may resemble the pattern of dual sex incubators in which tightness of sit determines egg temperature. Thus, although there is a considerable range of incubating patterns among birds, all species are likely to share common sensory mechanisms for regulating physical contact with eggs.

Because they lay their eggs, birds must behave in such a way that they maintain the eggs at a temperature compatible with embryonic development. When eggs are incubated by both members of the pair bond, the egg temperature is regulated within a rela-

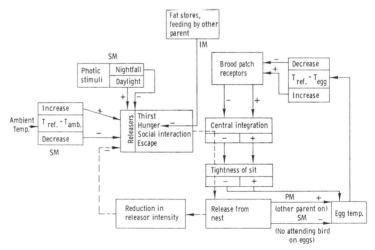

Fig. 8. A generalized scheme for avian incubation. The scheme emphasizes the role of competitive interplay among centrally integrated information related to egg temperature and releasers favoring nonattentiveness. *PM, SM,* and *IM* indicate aspects applying specifically to primary mode, secondary mode, and intermediate mode patterns of incubation.

tively narrow range. Overheating or underheating is prevented by central neural modulation of the tightness of sit. The almost constant attentiveness to the eggs provided by bisexual incubation has certain obvious advantages. Among these is the protection against predation of the eggs by other species or by members of the same species. Further, the constant attentiveness of parents effectively ameliorates the influence of fluctuations in environmental conditions. This is especially important for marine birds which must often forage at great distance from their nesting grounds.

The adoption of the single sex intermittent, or secondary mode of incubation, requires of the attending parent periodic foraging excursions away from the eggs. While unattended, the eggs tend to equilibrate with the environmental temperature, at a rate depending upon the heat-retaining capacity of the egg-nest complex; an attending bird must therefore adjust attentiveness in accordance with the temperatures of its habitat and the degree to which it has insulated its nest. The contribution of nest insulation is best expressed by the value *B* which reflects the curvature of the hyperbolic attentive curve (Fig. 4B). Eggs may be regulated at relatively constant mean temperatures over a rather limited range of warm environmental temperatures by the parent adjusting the time it uses for nonforaging behavior while it is away from the nest. At ambient temperatures below α (see Fig. 6), the mean egg temperature declines because the reserve of nonforaging time has been largely exhausted. At environmental temperatures below γ (Fig. 6), the foraging time must be sacrificed if the attentive relationship is to be maintained. Periodically, environmental temperatures may fall below γ and development will not be affected; however, geographic zones characterized by daily mean temperatures below that indicated by γ are likely to mitigate against successful reproduction. The position of γ is suggested as a means of approximating the lower mean daytime temperature compatible with hatching of the clutch. Furthermore, in the absence of data on the egg temperature, a line projected from a mean egg temperature of 34°C located at γ to 37°C located at α should provide a reasonable estimate of mean egg temperature at any environmental temperature below that corresponding to α.

Other ecological factors being equal, species engaging in single sex intermittent incubation could use several temperature related strategies to extend their breeding ranges into cooler climates. Examples of such strategies are as follows: (i) Nest insulation could be augmented through the use of more or different materials; this would result in a greater *B* value for the attentive curve and a concomitant shift to cooler areas of the curve of both α and γ. (ii) Nesting sites could be selected in close proximity to a predictable supply of food; this would allow closer adherence, by the attending parent, to the attentive relationship necessary for achieving a successful hatch. (iii) The male could become a provider of food for the incubating female or, as in the case of high-altitude hummingbirds in South America, the species could revert to a dual sex mode of incubation (*34*).

Birds engaging in the intermediate modes of incubation may rely upon large fat stores for support of metabolism so that a single parent may indulge in constant attentiveness throughout the entire incubation period. The male emperor penguin is the classic example of this pattern. Pheasants and chickens rely heavily upon their capacity to fast for periods far in excess of that tolerable for small passerine birds. With the exception of infrequent periods away from the clutch for feeding, their pattern of incubation is similar to that seen in bisexual incubators. As with single sex intermittent incubators at night, or bisexual incubators at all periods, adjustments in tightness of sit characterize the way in which egg temperature is regulated. In other species, the nonincubating member of the pair bond may feed the incubating parent. The necessity for foraging is thus reduced to the extent that the energy requirements of the attending bird are met by a helpful mate. The pattern of egg temperatures for such birds should resemble those of bisexual incubators in that the variations around the mean temperature should be slight in comparison to single sex intermittent incubators.

The evolution of nest complexity, although undoubtedly related to the protection of eggs from such factors as wind, rain, and predation, also appears to be related to selective pressures favoring the augmentation of insulation. In addition to attenuating changes in egg temperature while the attending

birds are foraging, the nest structure provides an environment in which the energy reserves of the incubating bird may be conserved at ambient temperatures below the thermoneutral zone (*35*). This factor is critical for small passerine birds which must fast overnight within the nest when the external temperatures reach their minima. In the population of weaverbirds that we studied, the range of cooling rates of eggs in different nests suggests that these birds possess a spectrum of nest building potentialities which influence insulation. This variation is available for natural selection and indeed appears to have been one of the factors allowing for altitudinal extension of the breeding range for *Ploceus cucullatus graueri* and some Peruvian hummingbirds. The insulation of the giant nest of the sociable weaverbird appears to be a critical factor in allowing opportunistic breeding during the cold winter of the Kalahari desert (*35*).

Further studies of the complex interactions among climates, nests, eggs, and parent birds should yield greater insight into the zoogeographic distribution of breeding ranges and provide a more complete appreciation of the "common prescription" for temperature regulation of the avian egg by parental behavior.

References and Notes

1. L. Irving and J. Krog, *Physiol. Zool.* **29**, 195 (1956).
2. H. Lundy, in *The Fertility and Hatchability of the Hen's Egg*, T. C. Carter and B. M. Freeman, Eds. (Oliver & Boyd, Edinburgh, 1969), p. 143.
3. A. L. Romanoff and A. J. Romanoff, *Pathogenesis of the Avian Embryo* (Wiley-Interscience, New York, 1972), pp. 57–90.
4. C. G. Sibley, *Condor* **48**, 92 (1946).
5. A. F. Skutch, *Ibis* **99**, 69 (1957).
6. R. A. Huggins, *Ecology* **22**, 148 (1941).
7. S. Baldwin and S. C. Kendeigh, *Sci. Publ. Cleveland Mus. Nat. Hist.* **3**, 1 (1932).
8. C. W. Kossack, *J. Wildl. Manage.* **11**, 119 (1947); C. R. Eklund, *Bird-Banding* **32**, 187 (1961).
9. T. R. Howell and G. A. Bartholomew, *Condor* **64**, 6 (1962).
10. J. Van Tyne and A. J. Berger, *Fundamentals of Ornithology* (Wiley, New York, 1959), p. 294.
11. S. C. Kendeigh, *Ill. Biol. Monogr.* **22**, 1 (1952).
12. H. L. K. Whitehouse and E. A. Armstrong, *Behavior* **5**, 261 (1953).
13. D. Lack, *Occas. Pap. Calif. Acad. Sci.* **21**, 27 (1945); P. Ward, *Ibis* **107**, 326 (1965).
14. Attentive times were determined from temperature recordings from thermistors or thermocouples positioned to measure temperature in the nest cup or implanted near the center of an egg in the nest cup. Attentive intervals were tested against visual records and were found to be accurate within ± 15 to 20 seconds. Values for attentiveness represent the total time "on" for the number of complete "on-off" cycles required to satisfy a minimum time interval of 60 minutes. The ambient temperature reported with each attentive value is the average shaded air temperature during that interval. Because of the nature of the daily temperature fluctuations, it was

judged that the daily mean temperature was not always representative of the ambient conditions to which the animals were exposed. For intervals greater than about 60 minutes, the variance for ambient temperature increased greatly for some intervals. For intervals shorter than about 40 minutes, variation in cycle length greatly increased the scatter for the attentiveness-temperature relation. The 60-minute interval appears to agree with that employed by Whitehouse and Armstrong (12).

15. R. E. Bailey, *Condor* **54**, 121 (1952).
16. The mean nighttime temperature for all animals ($N = 13$) was 36.5°C and the standard deviation for individual mean egg temperature ranged from 0.3° to 0.8°C. The mean temperature for each egg was derived from 15 to 36 measurements. Comparable figures for daytime were as follows: when $N = 15$, the mean temperature was 35.3°C and the range of standard deviations was 0.8° to 1.7°C; the means were obtained from 19 to 117 measurements.
17. Data were coded for linear regression where $X = 1/(\text{air temperature} - C)$ and constants, B and A, in the equation $Y = BX + A$ were determined by Bartlett's three-group method for model II regression [R. E. Sokal and F. J. Rohlf, *Biometry* (Freeman, San Francisco, 1969), pp. 483–485]. The value of C was determined by a process of reiteration whereby A and B were calculated for each value of C. The constants for the equation of best fit were derived from the values, C, A, and B, which produced the least sum of squares when tested against the uncoded data. For derivation of values for attentiveness (Y) and temperature (X), see (14).
18. From the rates of heating and cooling of eggs and from the attentiveness, the minimum, maximum, and mean egg temperatures can be calculated as a function of cycling rate and ambient temperature if brood patch or body temperature can be assessed or estimated (estimated to be 41.5°C). At steady state, the egg must heat as much as it cools. Therefore $T_{E,0}(i) = T_{E,0}(i + 1)$, where $T_{E,0}$ is the egg temperature at the start of heating; therefore,

$$T_{E,0} = \frac{T_A + (T_B - T_A)e^{B_c(1-Y)/F} - T_B e^{B_h Y/F + B_c(1-Y)/F}}{1 - e^{B_h Y/F + B_c(1-Y)/F}}$$

where $T_{E,0}(i)$ is the egg temperature at the start of heating for some cycle (i); $T_{E,0}(i + 1)$ is the egg temperature at the start of heating for the following cycle ($i + 1$); T_A is ambient temperature; T_B is brood patch or body temperature; B_c is the rate of cooling of the nest-egg complex; Y is attentiveness, determined by (minutes on/60 minutes)/60 minutes; F is the cycling rate (expressed as cycles per minute); and B_h is the rate of heating of the nest-egg complex. The mean egg temperature can be found as a quantity: the integral of egg temperature during heating plus the integral of egg temperature during cooling, divided by the time for one cycle. The following equation yields the steady state value for mean egg temperature as a function of ambient temperature and cycling rate:

$$T_E = 1/F\left[\int_0^{Y/F} T_E(t) + \int_{Y/F}^{1/F} T_E(t - Y/F)\right]$$

$$= T_A + (T_B - T_A)\left[\frac{Y + F\left(1 - e^{B_h Y/F}\right)\left(1 - e^{B_c(1-Y/F)}\right)}{\left(1 - e^{B_h Y/F + B_c(1-Y)/F}\right)\left(\frac{B_c - B_h}{B_c B_h}\right)}\right]$$

where t is time. At steady state, the maximum egg temperature is given by the equation

$$T_{E,m} = T_B - (T_B - T_{E,0})e^{B_h Y/F}$$

19. A single class analysis of variance showed a significant variation for mean egg temperatures among birds ($P < .001$).
20. We use the term "insulation" loosely to mean the heat-retaining properties of the egg-nest complex.
21. A single class analysis of variance showed a significant difference among cooling rates of eggs ($P < .001$). This finding prompted us to take a closer look at the attentiveness-temperature relationship. Data points for attentiveness were corrected to the mean daytime ambient temperature for our study (25°C) according to the equation: $Y_{corrected} = Y_{actual} + B[25 - 1/(T_A - C)]$, where Y_{actual} is the attentive time in minutes per hour and T_A is the corresponding ambient temperature; B is 81.8 and C is 38.7. A one class analysis of variance showed a significant difference among attentive times among animals ($P < .001$).
22. A family of attentiveness-temperature curves was derived for some of the birds in our study for which cooling rates were determined and background disturbances were judged to be minimal. It was not possible to employ the method described in (17) to determine the individual attentiveness-temperature curves because the range of ambient temperatures for a given day was usually too small. However, an estimate of the B value of the curve for each animal was found by fixing two points and a limit. One point represented the mean ambient temperature and the mean attentiveness for a given animal for 1 day. The second point represented the X-intercept for our attentiveness-temperature curve which was derived from the pooled data. We think that this value is of biological significance and preferred to use it rather than a limit, the vertical asymptote. On the other hand, the Y-intercept appears to be a point of mathematical convenience, hence we preferred to set the upper limit of the curve by employing the horizontal asymptote calculated for the pooled data. Thus for the equation $Y - A = B/(X - C)$, we estimated B and C for each animal assuming: $A = 38.7$; that at $Y = 0$, $X = 37$; and that at $X = $ mean ambient temperature, $Y = $ mean attentive time for a given animal on a given day. From the family of attentiveness-temperature curves, the attentive time at $T_A = 25$°C was plotted as a function of cooling rate. The product moment of correlation was 0.942 [$R_{(\alpha=0.01)} = 0.917$]. The individual B values were plotted as a function of 1/cooling rate (insulation index), and the corresponding product moment of correlation was 0.937 [$R_{(\alpha=0.01)} = 0.917$].
23. N. E. Collias and E. C. Collias, *Auk* **88**, 124 (1971).
24. O. P. Pearson, *Condor* **55**, 17 (1953); G. T. Corley Smith, *Ibis* **111**, 17 (1969).
25. Cooling and heating rates were determined for each nest-egg complex on each day. Egg temperatures at the start of heating, at the start of cooling, and at the end of cooling, a value for shaded air temperature, and the time intervals for heating and cooling were recorded for each consecutive attentive and inattentive period. The maximum egg temperature was calculated as the average of all values for egg temperature at

the end of heating. Likewise, the minimum egg temperature was taken as the average of the values for egg temperature at the end of cooling. Minimum and maximum egg temperatures were determined for each animal. The values for heating and cooling rates; the time intervals for heating and cooling; the values for egg temperature at the start of heating, the start of cooling, and the end of cooling; and the ambient air temperature were used to predict the egg temperature at any point in time during the experiment. Maximum temperature was not allowed to exceed the value of egg temperature at the end of heating for each cycle nor was it allowed to fall below the value for egg temperature at the end of cooling. A sampling of the actual temperature record and the corresponding predicted values is given (Fig. 5). The mean egg temperature was calculated for each animal as the integral of the predicted values divided by the total time of the experiment. Equations used in prediction of egg temperature are as follows: for heating, $T_E(t) = T_B - e^{B_h t + A_h}$, where T_E is the predicted egg temperature with respect to time, t; T_B is the estimate of brood patch temperature, 41.5°C; B_h is the heating rate; and $A_h = \ln (T_B - T_{E,0})$. $T_E(t) \leq$ egg temperature at end of heating. For cooling, $T_E(t) = T_A + e^{B_c t + A_c}$, where T_A is the ambient temperature for the cycle, B_c is the cooling rate, and $A_c = \ln$ (egg temperature at start of cooling $- T_A$). During the nighttime, maximum and minimum temperatures were taken at inflection points and the mean temperature was calculated as the mean for hourly temperature readings. Birds were active for almost 12 hours a day, thus the 24-hour average for egg temperature was taken as a simple mean for daytime and nighttime values.

26. S. C. Kendeigh, in *Breeding Biology of Birds*, D. S. Farner, Ed. (National Academy of Sciences, Washington, D.C., 1973), pp. 311–320.
27. R. H. Drent, *Behaviour* **17** (Suppl.), 1 (1970).
28. V. V. Khaskin, *Biofizica* **6**, 57 (1961).
29. A. C. Bent, *Life Histories of North American Nuthatches, Wrens, Thrashers, and Their Allies* (Dover, New York, 1964), p. 166.
30. J. R. King, in *Breeding Biology of Birds*, D. S. Farner, Ed. (National Academy of Sciences, Washington, D.C., 1973), pp. 78–107.
31. A. J. El-Wailly, *Condor* **68**, 582 (1966).
32. R. H. Drent, K. Postuma, T. Joustra, *Behaviour* **17** (Suppl.), 237 (1970).
33. G. P. Baerends, *ibid.*, p. 263.
34. R. T. Moore, *Wilson Bull.* **59**, 21 (1947); E. Schafer, *Biol. Soc. Venezolana Cien. Nat.* **15**, 153 (1954).
35. F. N. White, G. A. Bartholomew, T. R. Howell, *Ibis*, in press.
36. H. N. Kluijver, *Ardea* **38**, 99 (1950).

37. We thank N. E. Collias for the use of his aviary and colony of village weaverbirds. Our colleagues J. M. Diamond, G. A. Bartholomew, T. R. Howell, and N. E. Collias read and criticized the manuscript and made valuable comments. We thank V. Debley, D. Ward, B. Castro, and A. Creese for collecting data and spending many uncomfortable hours in the aviary blinds. The latent interest of F.N.W. in avian incubation was stimulated through lectures delivered by Professor S. C. Kendeigh in 1952–53. A grant from the UCLA computer committee provided support for data analysis. We utilized equipment supplied through grant GB-8523 from the National Science Foundation. J.L.K. is supported by grant 4371G3, Los Angeles County Heart Association.

15

Reprinted from *Ibis*, **117**(2), 171–179 (1975)

THE THERMAL SIGNIFICANCE OF THE NEST OF THE SOCIABLE WEAVER *PHILETAIRUS SOCIUS:* WINTER OBSERVATIONS

Fred N. White, George A. Bartholomew & Thomas R. Howell

Received 15 August 1973

The Sociable Weaver *Philetairus socius*, inhabiting the arid parts of southern Africa, builds the largest and most spectacular of all birds' nests. Unlike most other members of the family Ploceidae, which build large colonies of separate nests, many pairs of Sociable Weavers occupy a single compound nest, each bird contributing not only to the building of individual chambers but also to the maintenance of the structure as a whole. A single nest may be 5–6 m long, 3–4 m wide and 1–2 m thick, weigh more than 1 tonne and be occupied by 100 or more pairs of birds. A large nest is the work of many generations of weavers and may be used at least for decades and sometimes for a century or more (Collias & Collias 1964).

Friedmann (1930) and Collias & Collias (1964) have provided general descriptions of the nest of the Sociable Weaver, but the most detailed account of the nest and of the natural history of the species is that of Maclean (1973). In our account we will refer to the total structure as the 'nest' and to individual units within it as 'nest chambers'; the components of the nest apart from the chambers we will call the 'matrix'. The nest is a non-woven, thatch-like structure composed of grass stems and twigs and is situated in a tree or on a man-made structure such as a utility pole. The nest chambers are more or less spherical and each communicates with the exterior by a short, vertically-directed, tubular passageway. All chambers have individual entrances and are completely separate from each other, although they are usually packed closely together.

All chambers open to the underside and form a single more or less horizontal layer which follows the contours of the nest. The top of the nest is thatched with a coarse layer of larger grass stems and twigs. Adults of both sexes and even juvenile weavers continuously insert more grass stems into the nest and add sticks to the top. Old chambers are filled in, and the nest gradually expands both vertically and horizontally; it may ultimately enlarge to the limits supportable by the tree in which it is located. Sociable Weavers are abundant in the Kalahari Desert of southern Africa, an area characterized by intense summer heat, sparse and erratic rainfall, and cold dry winters during which air temperatures may fall to —10° C at night. Breeding may occur at any season, including the winter months of July and August (Maclean 1970), and is associated with the effects of rainfall rather than photoperiod.

Previous investigators have dealt primarily with the natural history and behaviour of this remarkable species. Our study examines the role of the nest in ameliorating the effects of the physical environment on the birds occupying it. We made our measurements and observations in the vicinity of Twee Rivieren in the Kalahari Gemsbok National Park of South Africa during July and August 1972, when many Sociable Weavers were breeding. We subsequently made additional observations on these birds and their nests during the summer (December) of 1973 (Bartholomew, White & Howell, in press).

METHODS

Air temperatures within the domes of individual nest chambers were measured by long-lead thermistors (Yellow Springs Inc.), the tips of which were guided through the matrix with heavy wire. The wire guides were then secured to the nest and adjacent limbs in order to insure a stable position of the thermistors. Temperatures at various levels through a vertical section of the nest matrix were monitored by means of copper-constantan thermocouples sealed in small, milled-out recesses spaced at 20 or 40 cm intervals in the wall of a hollow fibreglass fishing-pole stock that was thrust through the nest. Air temperature in the shade, measured by a thermistor situated beside the nest, served as the chief environmental comparison temperature. Daily high and low temperatures were recorded with a maximum-minimum mercury thermometer attached to the shady side of the study tree.

A Microbus parked about 15 m from the nest served as a hide and instrument shelter. The presence of the vehicle and the wires running from it to the nest caused no apparent disturbance to the birds.

RESULTS

The dry winter season in the Kalahari is characterized by clear skies and chilling southerly winds. During the period of our study no rain fell. The nest chambers were dry and the interior of the matrix showed no signs of moisture or fermentation. Air temperatures ranged from −2·5°C to 24·5°C. The typical daily cycle spanned a range of 20–25°C.

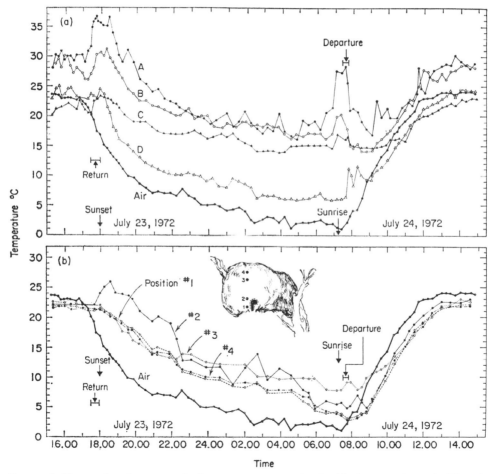

FIGURE 1. Twenty-four-hour record of temperatures associated with the nest shown in Plate 1. (a) Air temperatures in nest chambers. A. B and C occupied by roosting birds. D, unoccupied chamber. (b) Temperatures through the nest matrix. Numbers represent the relative positions of the thermocouples in the structure.

PLATE 16. Sociable Weaver nest utilized for principal study site. The vehicle served as an instrument shelter and hide.

We have summarized the temperature data for a typical large nest (Fig. 1), the conformation of which is shown in Plate 16. This nest contained 70 chambers, two of which were taken over for roosting by a resident pair of Pygmy Falcons *Polihierax semitorquatus*. We estimated the total population of Sociable Weavers resident in this nest at 150 birds.

TEMPERATURES OF UNOCCUPIED CHAMBERS

Measurements from an unoccupied nest chamber (Chamber D, Fig. 1) indicate that the nest buffers the effects of external air temperature. When the external air temperature was falling during the night, the air temperature in the unoccupied chamber paralleled this fall but remained 4–6 °C above it. From about 22.00 hrs until sunrise the difference between the air temperature in this chamber and the external air temperature remained relatively stable. This stability and the persistent maintenance of a temperature difference between the unoccupied chamber and external air may be attributed to the lateral transfer of heat from the occupied nests which were maintained at as much as 18 °C above external air by the heat production of the birds. Note the periodic increments in temperature in the unoccupied nest during the middle hours of the night as heat was laterally transferred from the occupied nests. Note also the marked increases in temperature in this empty chamber following the burst of heat production which accompanied mass departures and returns at sunrise and sunset. During the daylight hours, the temperature in the unoccupied chamber rose with external air temperature but lagged behind until the latter reached a plateau in the early afternoon.

TEMPERATURES IN OCCUPIED CHAMBERS

During daylight hours, birds continuously returned and departed, usually in small groups (4 to 12 birds) and infrequently as individuals. The total number of birds occupying the nest at any one time during daylight hours was small, but active nests containing eggs or young were either continuously occupied or frequently visited by adults. The heat production of the occupants of these active nests is reflected by periodic elevations in chamber temperature (Fig. 1, Chamber A at 09.00–12.00 hrs). Just before sunset, when the birds returned *en masse*, the temperature of the nest chambers rose appreciably despite falling external air temperature. In Chamber A the temperature reached 37 °C as the birds went to roost, an elevation of about 23 °C above ambient. All occupied chambers which were monitored remained substantially warmer than the empty chamber and the external air, and stayed relatively uniform from midnight until sunrise.

We gained the impression, later verified, that the temperature in given nest chambers was directly dependent on the number of birds in each chamber (see below, Fig. 3). During the middle hours of the night periodic increases of temperatures in occupied chambers were repeatedly observed. The oscillations in temperature at 02.00 and 04.00 hrs in chamber A were synchronous with increased levels of vocalization from many parts of the nest. At sunrise, prior to *en masse* departure, a marked increase in activity and vocalization was accompanied by a sharp increase in the temperatures of both occupied and unoccupied chambers.

TEMPERATURES IN THE MATRIX

Temperatures taken through a vertical section of the solid thatch of the nest matrix followed the daily cycle of air temperature. During the night, however, at peak occupancy, the low conductance of the nest materials together with the heat produced by the birds maintained a substantial gradient between the temperature within the matrix and the external air (Fig. 1). The surface of the thatched roof (Fig. 1) cools sharply at night and the low surface temperature minimizes heat loss due to free convection. The shell of still air trapped by the coarse surface layer should minimize heat loss by forced convection and this may be of particular significance during the winter months which are characterized by strong and sustained winds. The loose upper surface also traps heat in the deeper parts of the matrix. The heat retention of the nest as a whole is made more effective by the elongated, downward facing entrance tubes which are offset from the flask-shaped chambers (Fig. 2) and thus minimize thermal radiation to the cold ground.

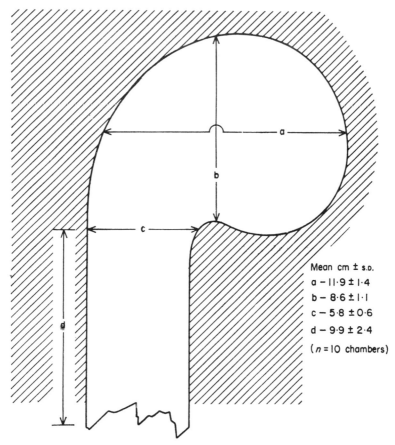

FIGURE 2. Conformation and dimensions of a Sociable Weaver nest chamber. The nest matrix is represented by cross-hatching.

EFFECTS OF NEST SIZE

The thermal importance of nest size, occupancy of individual chambers, and colony size can be appreciated by comparison of temperature data from small and large nests. We instrumented a small four-chambered nest which was being constructed by colonists from a large nest situated about 3 km SSE of Twee Rivieren in Botswana which had reached the maximum size compatible with the structure of its supporting tree (*Boscia albitrunca*). The small nest was also in a *B. albitrunca* and was about 0·5 km from the home nest. It measured 69 cm from top to bottom, 150 cm in diameter on one axis, 72 cm on the other, and was 290 cm above the ground. Sixteen or 18 birds, most of which returned to the main nest at night, were involved in the construction of the small nest. Three of the four chambers were instrumented in the manner previously described. In contrast to the large nest described in Figure 1, the temperatures in the chambers of the small nest during daylight hours were essentially the same as the external air temperatures (Fig. 3).

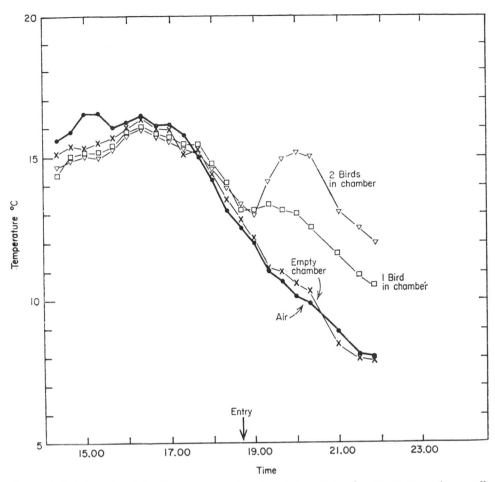

FIGURE 3. Relationship of chamber occupancy to external air and chamber temperatures in a small Sociable Weaver nest (total, 4 chambers) which was under construction.

Only three birds, two of which shared one chamber, roosted in the nest during the night. As the night air cooled, the temperature in an unoccupied chamber remained virtually identical with that of the outside air. The air in the chamber with one occupant remained about 2·5 °C above external air temperature. The chamber occupied by two birds warmed to 5 °C above ambient and remained four or more degrees above ambient temperature during the period of observation. The temperatures recorded in this small nest were much lower than those observed in the large nest (Fig. 1) at similar ambient temperatures.

DISCUSSION

Sociable Weavers are among the very few passerine birds permanently resident in the desert that are both sedentary and highly colonial. They form large and dense local populations, the members of which rarely travel more than 1·5 km from the nest while foraging (Maclean 1973). A large sedentary population obviously requires local food resources that will not be exhausted. This is a situation difficult to achieve in the Kalahari Desert where food production is erratic because of seasonal, scant and unreliable rainfall.

In this ecological situation any strategy that reduces the energy requirements of a species should be favoured by natural selection. Kendeigh (1961) demonstrated that the

House Sparrow *Passer domesticus* conserved energy by using artificial nest-boxes in winter. Our observations strongly suggest that the nest of the Sociable Weaver, through its ameliorating effect on the influence of environmental temperature on metabolism, diminishes the impact of the birds on the locally available food supply. This may have one or several results: (1) increased environmental carrying capacity, (2) increased population density, (3) diminution in time spent foraging, and (4) independence from seasonality of reproduction. Clearly many parameters of the ecology of the Sociable Weaver are affected by the nest. To an extent that is unique among birds, virtually every aspect of its life cycle is linked closely to the nest at all times of the year. Some of the more obvious elements of this complex system are considered below.

NEST SIZE

All individuals in a colony participate in the construction and maintenance of the nest at all times of the year throughout the life cycle. This continuous nest-building should be selectively advantageous because the insulative effectiveness of the nest increases with size. In addition, the metabolic heat input to the nest increases with the number of occupants. Hence, the larger the population, the greater the amount of heat produced; the larger the size of the nest, the greater the amount of this heat that is retained. Consequently, the energy demands for thermoregulation associated with cold winter nights are minimized in large colonies. As a result, large populations occupying very large nests can breed at any season of the year provided that animal food for the young is available. Thus, year-round opportunistic breeding should be a corollary of large colony and nest size.

The reduced metabolic cost of thermoregulation associated with nest insulation and colony size should result in selective pressures favouring the construction and maintenance of very large nests. It is of interest that most of the birds which we observed constructing small, new nests returned each night to the large nests of the parent colony.

The tendency of these birds to build ever larger nests is shown by the fact that many nests in the area of Twee Rivieren have reached the maximum size compatible with the trees supporting them. Under natural conditions the upper size limit of a Sociable Weaver nest appears to be set by the configuration and strength of the structure on which it is built. Nests located on the crossbars of telephone poles often fill all available space. Nests built in tree aloes or kokerbooms, *Aloe dichotoma*, often fill them so completely that they look like gigantic straw toadstools. The massive branches of the camelthorn acacia (*Acacia giraffae*) often break under the weight of huge nests.

Although it appears to be advantageous for Sociable Weavers to have large nests, their continuous nest-building may result in many unoccupied chambers, particularly during the winter. The construction of an apparent excess of nest chambers need not, however, be maladaptive. We found that the birds do not necessarily roost in specific chambers on consecutive nights unless they are breeding. Consequently, they can control the nocturnal thermal conditions to which they are exposed by varying the number of individuals roosting in a given chamber. The more birds, the higher the air temperature in the chamber. During cold weather we found that four or even five birds may crowd into a single chamber. Converse adjustments are made in warm weather (Bartholomew, White & Howell, in press).

METABOLIC CONSIDERATIONS

The potential importance of the interaction of metabolic heat production and nest insulation to the energy budget of the Sociable Weaver can be assessed by analogy with the House Sparrow, a closely related form of similar weight (Barnett 1970). If we assume that the energetics of temperature regulation in the Sociable Weaver parallel those of the House Sparrow, we can estimate the metabolic savings that result from spending the night in a nest chamber rather than roosting in the canopy of a tree. By analogy with the observations of Hudson & Kimzey (1966) on House Sparrows we can reasonably assume that in the Sociable Weaver the lower critical temperature is about 20 °C, that standard energy metabolism in the thermal neutral zone is 3 cm³ O_2 g^{-1} h^{-1}, and that below 20 °C oxygen consumption is inversely related to air temperature and increases 0·2 cm³ O_2 g^{-1} h^{-1} °C.

As shown in Figure 1, the temperatures of occupied chambers remained relatively stable throughout the night at levels at or near the presumptive thermal neutral zone. However, recurrent increments of heat production associated with physical activity periodically introduced additional heat in excess of that produced by resting birds. Thus, the tendency for chamber temperature to drift passively toward that of the external environment was periodically reversed. These bouts of activity, which resulted in an increase in heat production, were relatively short compared to the quiet periods but still had a marked effect on chamber temperature because the insulation of the nest allowed only a slow loss of heat from the nest chamber. We estimate that the period of increased heat input lasted for only about 20% of a given cycle. The actual increase in metabolic rate during the phase of increased heat input appeared to be relatively small in terms of peak metabolism since it raised the air temperature in the chamber only about one-fifth as much as the more intense period of activity that preceded departure from the nest in the morning. The bursts of increased heat input maintained an average level of chamber temperature that afforded significant reduction of energy expenditure during most of the night (Fig. 4).

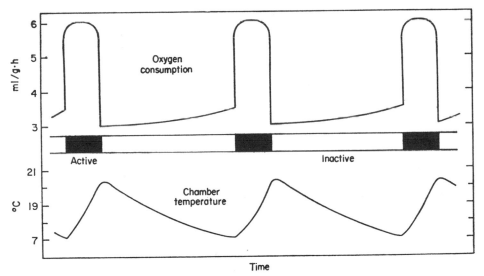

FIGURE 4. Schematic representation of the effect of periodic nocturnal activity on chamber temperature and oxygen consumption.

For heuristic purposes we shall assume that the very high chamber temperatures resulting from the activity preceding morning departure reflect a metabolic rate that is five times standard, while the more modest recurrent elevations in chamber temperature during the night result from a rate that is twice standard. We shall base our calculations on the data from chamber A (Fig. 1). From 18.00 to 01.00 hrs, the birds in this chamber remained within their thermal neutral zone, thus consuming 3 cm³ O_2 g^{-1} h^{-1}. From 01.00 to 06.00 the birds spent 80% of their time resting quietly at a mean air temperature of

18 °C. Under these conditions they would consume $3\cdot4$ cm^3 O$_2$ g^{-1} h^{-1}. During the remaining 20% of the time (the periods of increased heat input), their metabolism was twice standard, or 6 cm^3 O$_2$ g^{-1} h^{-1}. Summing the oxygen consumption from 18.00 to 06.00 hrs yields a total of $40\cdot6$ cm^3 g^{-1}. If the birds had roosted in the open during this same period at the external air temperatures shown in Figure 1, their cumulative oxygen consumption based on the mean temperature for each hour would have been $71\cdot8$ cm^3 g^{-1}. Thus, by huddling together in their chambers the birds diminished oxygen consumption by 43%.

The average weight of Sociable Weavers is $27\cdot5$ g (Maclean 1973) and there were approximately 150 birds in the colony. If the conditions measured in Chamber A were typical for the other chambers in the nest, and we use the usual caloric equivalent for oxygen ($4\cdot8$ kcal per litre), the energy expenditure of the entire colony during the night would have been 804 kcal as compared with 1422 kcal if the birds had roosted in the open air (to convert kcal to kJ, multiply by $4\cdot186$).

These figures can be given more ecological immediacy by converting the kilocalories to the weights of seeds or insects that would be required to sustain the metabolic rates under the conditions observed.

The caloric content of grass seeds averages about $4\cdot8$ kcal per gramme dry weight (Cummins & Wuycheck 1971). Air-dried seeds have a water content of about 10%, and the metabolic coefficient of small birds is about 80% (Kendeigh 1959). Using these figures the energetic cost of roosting in the open for one night as compared with huddling in a nest chamber amounts to the consumption of $(1422-804)[1\cdot1/(4\cdot8\times0\cdot8)] = 177$ g seeds. If the birds ate only insects having a water content of about 70% and caloric content of $5\cdot5$ kcal per gramme dry weight (Cummins & Wuycheck 1971), the daily savings in food resulting from roosting in the nest would be $(1422-804[1\cdot7/(5\cdot5\times0\cdot8)] = 239$ g. If the average weight of the insects were 50 mg, the members of the colony would have to capture about 4500 insects per day (30 per bird) to supply the energy required to roost at night in the open as compared with roosting in the nest. These figures ignore the additional cost of foraging and are thus conservative estimates of energy savings. The conditions on which these calculations are based would reasonably apply to about 50–60 days per year. The strategy of roosting in the nest rather than in the open, in effect, increases the resources of the area in which the birds forage by an amount in excess of $8\cdot5$ kg of seeds during this season. These considerations assist in an understanding of the high density of this species and may partially account for the often small intercolony distances as well as the reproductive success at virtually all seasons of the year.

The selective advantages associated with a given nest are available generation after generation. We have here a situation in which a given generation reduces, through its contribution to the nest structure, the energetic cost of thermoregulation to subsequent generations over many decades.

ACKNOWLEDGMENTS

This study was supported by a grant from the National Geographic Society, Washington, D.C., and carried out with the cooperation of the National Parks Board of South Africa and the Percy FitzPatrick Institute of African Ornithology. We are grateful to Dr G. de Graff, Dr W. Roy Siegfried, Dr A. S. Brink, Mr Christoffer LeRiche, and Mr and Mrs I. J. P. Meyer for advice and assistance. Dr Gordon L. Maclean facilitated our field work by introducing us to the avifauna of the Kalahari and the biology of the Sociable Weaver. Maxine White and Elizabeth Bartholomew aided in the collection of data.

Summary

The Sociable Weaver of southern Africa builds the largest of all birds' nests. Individual nests contain many chambers and may be occupied by a colony of a hundred or more pairs. Most aspects of the biology of this species are linked, either directly or indirectly, to the nest and its construction. One readily demonstrable function served by this enormous nest is the amelioration of the impact of low environmental temperatures.

During winter in the Kalahari Desert, sustained chilling winds are common and night-time temperatures often fall to freezing. Under such conditions, air temperatures in the unoccupied chambers remained several °C warmer than the external environment. Air temperatures in occupied chambers, warmed by the body heat of the birds, were as much as 18–23 °C above external air temperature, and during most of the night approximated the predicted zone of thermal neutrality.

Except when breeding, the birds do not necessarily roost in the same chambers on consecutive nights; on a cold winter night as many as five birds may crowd into a single chamber. Chamber temperatures increased directly with the number of occupants.

The roosting birds periodically showed bursts of activity. The heat released by their increased metabolism during these intervals of activity caused recurrent sharp elevations of the temperatures in the chambers and the adjacent matrix of the nest. Since the temperatures in the chambers depend on both the insulation supplied by the nest and heat released by the birds, the more the birds and the bigger the nest, the greater the independence of the system from outside temperatures.

The giant communal nest of the Sociable Weaver contributes to successful residence in the Kalahari Desert in part by allowing the birds to roost during the winter at temperatures much higher than those of the external environment. The relatively high temperatures in the chambers during the night diminish the metabolic cost of thermoregulation, allowing an energy expenditure 40% less than would be necessary if the birds roosted in the open. This energy saving, in effect, increases the carrying capacity of the local environment and allows an increase in population density. It could diminish the time and energy spent foraging, and it clearly contributes to the year around reproduction of the species.

The reduced metabolic cost of thermoregulation that is associated with communal roosting in large nests has contributed to selective pressures favouring the construction and year around maintenance of giant nests that remain in use for decades. One consequence of this situation is that the energetic savings and other advantages attributable to the large nest accrue to the extant population from the contributions of the many preceding generations that built and maintained the nest.

REFERENCES

BARNETT, L. B. 1970. Seasonal chages in temperature acclimatization of the House Sparrow, *Passer domesticus*. Comp. Biochem, Physiol. 33: 559–578.

BARTHOLOMEW, G. A., WHITE, F. N. & HOWELL, T. R. (in press). The thermal significance of the nest of the Sociable Weaver, *Philetairus socius*—summer observations. Ibis.

COLLIAS, N. E. & COLLIAS, E. C. 1964. Evolution of nest-building in the weaverbirds (*Ploceidae*). Univ. Calif. Publs. Zool. 43: 1–239.

CUMMINS, K. W. & WUYCHECK, J. C. 1971. Caloric equivalents for investigations in ecological energetics. Mitt. int. ver. limnol. 18.

FRIEDMANN, H. 1930. The sociable weaverbird of South Africa. Nat. Hist. 30: 205–212.

HUDSON, J. W. & KIMZEY, S. L. 1966. Temperature regulation and metabolic rhythms in populations of the House Sparrow, *Passer domesticus*. Comp. Biochem. Physiol. 17: 203–217.

KENDEIGH, S. C. 1959. Energy responses of birds to their thermal environments. Wilson Bull. 81: 441–449.

KENDEIGH, S. C. 1961. Energy of birds conserved by roosting in cavities. Wilson Bull. 73: 140–147.

MACLEAN, G. L. 1973. The Sociable Weaver, Parts 1–5. Ostrich 44: 176 261.

MACLEAN, G. L. 1970. The breeding seasons of birds in the southwestern Kalahari. Ostrich supp. 8: 179–192.

Fred N. White, Department of Physiology, George A. Bartholomew & Thomas R. Howell, Department of Biology, University of California, Los Angeles, California 90024, U.S.A.

Part IV

MECHANISMS AND ONTOGENY OF BUILDING BEHAVIOR

Editors' Comments
on Papers 16 Through 21

16   **PETERS**
      Maturing and Coordination of Web-Building Activity

17   **PEAKALL**
      Synthesis of Silk, Mechanism and Location

18   **LEHRMAN**
      Excerpts from *Interaction Between Internal and External Environments in the Regulation of the Reproductive Cycle of the Ring Dove*

19   **COLLIAS and COLLIAS**
      An Experimental Study of the Mechanisms of Nest Building in a Weaverbird

20   **COLLIAS and COLLIAS**
      The Development of Nest-Building Behavior in a Weaverbird

21   **WILSSON**
      Excerpts from *Observations and Experiments on the Ethology of the European Beaver* (Castor fiber *L.*)

In this part we focus on the mechanisms and ontogeny of external construction by selected terrestrial arthropods and vertebrates, comparing the two from the viewpoint of general problems and basic principles. Topics considered are the factors underlying normal sequences in the building of a structure, the relative importance and relationship between maturation and practice, the interaction between internal and external factors in construction, and the role of specific sign stimuli in building behavior.

One general principle that helps to determine building sequences is the idea that the nature of the work done by a builder is determined by the work already done. This idea has been conceptualized by Grassé (1959, 1967) in his term "stigmergy." Grassé is the editor of a multivolume French treatise on zoology, has spent a good part of a lifetime working on termites, and has been president of the French National Academy of Sciences. Stigmergy, as Grassé defines it (1959, p. 79), is "the stimulation of the workers by the very performances they have achieved." He illustrates the concept with experiments done with termites in equatorial Africa, for example on *Cubitermes* sp. When he placed 350 worker termites along with some larvae, a queen, and some soil in a covered petri dish, the building activities of the workers were uncoordinated at first, and they placed pellets of soil here and there at random. But once pellets, apparently by chance, happened to be placed in contact with each other or to form lines in certain locations, these accumulations became highly attractive to the termites and rapidly grew to pillars, as each termite adding to the structure tried to place its pellet as high up as it could reach. The tops of adjacent pillars were then built toward each other and became connected to form arches. Other factors, besides stigmergy, are important in directing building activities, and Grassé found that the vicinity of the queen was a strong stimulus to building, and so led to the construction of the royal cell about the queen. Others have emphasized the importance of gradients of humidity, air currents, and chemicals to building orientation (see Wilson, 1971).

Crowding is probably an important factor in leading to growth of the colony nest. Skaife (1961, pp. 15–16), who spent many years studying the South African black-mound termite (*Amitermes atlanticus* Fuller), described how the mound grows. The workers do not venture out onto the exterior of the mound where they might be quickly attacked and carried off by ants. Instead, the termites make tiny holes close together in the outer wall, holes just big enough for them to put their heads through. They thrust moistened sand particles through these holes and these stick together, forming a moist bulge outside the mound. More termites go into the bulge, make holes in it, and repeat the process, and so the exterior of the mound grows as thousands of termites work together. Stigmergy enters in insofar as the termites close the holes they have made but does not explain why they make the holes in the first place.

Although a number of general principles must enter into the determination of building sequences, the concept of stigmergy is no doubt one of the most important of these. We have discussed an example of it in termites, which are highly social insects. Next we wish to give an example of stigmergy in a spider, which is a solitary animal, although the term "stigmergy" was not used in this early example from the observations and experiments of Hingston (1920) on an orb-web spider (*Araneus nauticus*) in India. Hingston was an officer in the Indian Medical Service and a fellow of the Royal Zoological Society. According to Savory (1952, p. 24) in his book *The Spider's Web,* Hingston's was the first account of observations on web spinning accompanied by experimental interference with the spider's actions. Some highlights may be mentioned.

Before describing his experiments, Hingston gives a clear account of the normal pattern of spinning an orb web in the spider. Savory (1952) also describes with illustrations and in some detail how an orb web is made, and a condensed description is given by Kaston in Paper 6 of this volume. Hingston observed that, after completing the radii of the web, the spider returned to the center of the web and using its forelegs like a pair of dividers checked the distance between radii. Apparently, legs too widely expanded means that a radius or more is absent. Each time that Hingston cut a radius previously laid down, the spider ran out a replacement until twenty-five radii had been so destroyed and replaced. The spider then ceased replacing radii and went on to the next stage of its work, construction of the temporary spiral working from near the center of the web outward. The spider measures the distance along a given radius between successive turns of this spiral by using its own length, placing a foreleg on the previous turn while its spinnerets at the rear of its abdomen touch the radius and fasten the thread. Similarly, it uses the span, from the tip of one foreleg to its spinnerets, to measure the correct distance between the successive turns of the viscid or catching spiral. This spiral is laid down from the outer toward the inner part of the web and replaces the temporary or scaffolding spiral. The latter is eaten by the spider, thus conserving valuable silk. When Hingston cut away one segment of the viscid spiral, parallelism of the next turn was lost in that sector of the web, since the spider measured distance first from the preceding turn of the viscid spiral within the sector, and then from the still intact spiral in the next sector.

The construction of an external structure depends on development of the ability to build by the builder. In turn the ontogeny of building ability depends on physiological maturation, as of the nervous system or of special glands that furnish building materials or that are concerned with the ability to build. In some animals practice in building might conceivably be needed, but evidence for such experiential factors seems to be minimal for arthropods.

The article by Peters (Paper 16) on the maturing and coordination of web-building activity in the spider *Zygiella x-notata* provides an excellent example of building ability essentially dependent on maturational factors and apparently independent of any need for practice in building. The first web constructed by a young *Zygiella* after it has left the cocoon or egg sac under natural conditions is a fully regular orb web. Normally, the spiderlings spend two and one half weeks inside the egg sac after hatching, during which time they continue to mature physically. If they are removed from the egg sac or cocoon before they would normally leave and are then placed in frames, they build very irregular webs at first, but as they become older, they build more and more regular orb webs.

Hans Peters and his students have been studying spiders in Germany for many years. He is the author of a classic analysis (1931) of the stimuli involved in prey capture by the common garden spider or cross spider, *Araneus diadematus*.

Further analysis of the maturational factors in the ability to build requires more knowledge of the physiological and anatomical basis of such ability. The article by Peakall (Paper 17) describes recent progress in understanding the synthesis of silk, which takes place in specialized glands inside the spider's abdomen. Different kinds of silk may come from different glands. Peakall also describes some of the molecular basis for the synthesis of silk. As a readily available single protein, silk has attracted attention from molecular biologists interested in the general problem of protein synthesis. In addition, Peakall has discovered a special receptor organ next to one of the silk glands that may function to inform the spider of the state of accumulation or depletion of the supply of silk in the gland. Silk is well worthy of further investigation in the present context, since it is the substance most generally and widely used among animals (i.e., by land arthropods) for making highly organized external constructions that involve behavioral manipulation, rather than external constructions produced merely by physiological secretion, such as shells or mucous tubes.

Much work has been done on the physiological basis of the development and maturation of building ability and other hive duties in the worker honeybee. Ribbands (1953) and Wilson (1971) review literature on the relationship between age and type of work done by the honeybee. Although age plays some part in determining the task carried out by each worker, there is a wide latitude in the age at which different workers commence the same duty. Pioneering studies by Rösch (1927) showed maximum

295

development of wax glands in his bees when they were sixteen to eighteen days old, provided they were given opportunity for comb building. Bees taken from comb-building clusters were most frequently of this same age, whereas bees engaged in other duties within the hive averaged younger in age. Wax glands degenerate as older bees become field bees and foragers. However, Rösch also found that when there was a shortage of wax-producing bees, the wax glands regenerated in some of the field bees, which then built comb again. There is therefore a good deal of flexibility and adaptability provided by the interaction between the internal and external factors that govern building behavior in honeybees.

Michener (1974, p. 64) has emphasized that the external construct itself may play an important role in helping to determine sex and caste in the honeybee (*Apis mellifera*). It is assumed, with some justification, that cell size and shape influence the queen, so that she ordinarily lays unfertilized eggs in cells of the right size to produce drones and fertilized eggs in the smaller cells that will produce workers. Artificially large worker cells cause larvae in them to be fed more and produce larger workers than normal. Fertilized eggs are laid in the very large, peanut-shaped queen cells, and the larvae in these cells are given a much greater and richer food supply than are other types of larvae.

There are reciprocal effects between nest-building behavior and endocrine physiology in birds. This point is well illustrated by the investigations of Lehrman and his associates on ring doves (*Streptopelia risoria*). The late Daniel S. Lehrman was the director of the Institute of Animal Behavior at Rugers University and was widely known for his contributions to our understanding of hormones and behavior in birds. He was also a member of the National Academy of Sciences. He has summarized some of the most significant work of his laboratory in Paper 18.

An isolated female ring dove does not lay eggs, but she will ovulate in about a week if placed with a male, a nest bowl, and nesting materials. Presence of a male alone is not so effective in inducing ovulation. The nest bowl and nest materials evidently enhance male courtship in stimulating the secretion of gonadotrophin by the pituitary gland of the female. This hormone stimulates the ovaries of the female. It was recently found that treatment with both estrogen and progesterone was necessary to stimulate nest building by ovariectomized ring doves (Cheng and Silver, 1974). Testosterone propionate will stimulate a high level of nest-building behavior in castrated ring doves (Martínez-Vargas, 1974).

Presence of a nest bowl and nest materials also helps initiate incubation. Lehrman suggested that nest building may stimulate progesterone secretion, and he confirmed the observation of Riddle and Lahr (1944) that progesterone treatment will stimulate onset of incubation by paired ring doves. However, estrogen is also needed for incubation by ovariectomized doves (Cheng and Silver, 1975). The physiological importance and sources of progesterone in birds require some further investigation.

Robert Hinde (1965) and his co-workers have found that estrogen injections will stimulate nest building out of season by the female canary. The nest in turn helps stimulate the ovaries, presumably by way of the pituitary, and lack of a nest cup or of nest materials delays ovulation. The female village weaverbird (*Ploceus cucullatus*) will not lay eggs in the absence of a nest, whether males are present or not. Females with nests but without males never lay eggs either (Victoria, 1969).

There has been some study of the endocrine basis of nest building in laboratory mammals. In the female mouse at low estrogen levels, addition of progesterone results in a significant increase in the size of nest built, but similar hormonal treatments are without effect in the adult male. However, if the male is castrated on the day of birth, when he reaches maturity his response to treatment with estrogen and progesterone is similar to that shown by the female in both nest-building behavior and sexual responsiveness (Lisk et al., 1974). It appears that the normal desensitization of the male to the female hormones is brought about by androgen during a critical stage in development, and this androgen is secreted by the testis of the newborn.

An example of stigmergy from a vertebrate is the building of the nest by the village weaverbird. Many birds will nest in the same tree in this gregarious species of Africa. The male weaves the outer shell of the nest and then invites a female to enter by means of a special wing-flapping display at the entrance. If she accepts she mates with the male, and after laying two or three eggs, incubates them and undertakes most of the care of the nestlings. The male in the meantime builds other nests to which he attempts to attract additional mates.

Although nest building in birds is a classical example of instinctive behavior, it was for a long time a relatively neglected field of study. Until the study on nest building in a weaverbird reprinted here (Paper 19) was published, no systematic and comprehensive experimental analysis of the precise stimuli and sequences involved in nest building by birds seemed to have been attempted, with the exception of the investigation by Robert

Hinde (1958) on the building of the relatively simple cup nest of domesticated canaries. Our analysis of the building sequences in the village weaverbird gave a demonstration of the principle of stigmergy in a vertebrate. At the time, Grassé's useful term and his work on termites were unknown to us.

At about the same time as we carried out this study of the mechanisms of nest building in the village weaver, we also attempted to raise as many young birds as we could in order to observe and to experiment on the development of nest building in this species. Although nests built by a young male raised in isolation had the species-specific form, we also found that considerable practice was needed before young village weaverbirds could build a nest at all. Young birds deprived of nest materials for about a year were unable to make nests before going through a prolonged period of practice. Since this finding was contrary to the usual notions of "instinctive behavior" in birds then prevalent, we have reproduced this article here, giving the evidence for our conclusions (Paper 20). Elsie C. Collias, who had primary responsibility for this project, received her Ph.D. in zoology from the University of Wisconsin and is a research associate of the Los Angeles County Museum of Natural History and of the University of California at Los Angeles.

In later years we followed the subsequent history of some of the birds reported on and also analyzed the problem further with other village weavers (Collias and Collias, 1973). We found, for example, that the crude appearance of the nest of first-year males largely results from the fact that they have not yet learned to properly select and prepare suitable nest materials. The males mature in their second year, and if by then they have learned to build proper nests, subsequent prolonged deprivation has little or no effect on their ability to build a good nest.

Of all mammals, the beaver is the most noted, best known, and ecologically the most significant for its construction activities. The two species, the American (*Castor canadensis* Kuhl.) and the European (*C. fiber* L.), are so closely related that they have been considered by some taxonomists to belong to a single species.

L. H. Morgan's book *The American Beaver and His Works* (1868) is the classic study of this animal in nature and was the first comprehensive attempt to get accurate and precise information on the beaver, devoid of the many myths, fables, and legends that had surrounded this animal since earliest times. Morgan observed the beaver and their constructions for some twelve seasons in northern Michigan at a time when this country was first

being opened up in order to build the Marquette and Ontona-
gon Railroad to transport iron ore from the mines on the south
shore of Lake Superior. Morgan, who took up the study of beaver
construction for recreation, was one of the directors of the railroad company. He was also a trout fisherman, and was impressed
in that undisturbed wilderness by the sheer numbers of beaver
dams, often only 2 to 3 feet high, which at short intervals
obstructed the streams and over which he and his companion
were often compelled to draw their boat. An area of 48 square
miles with its streams and beaver works was mapped for Morgan
by the civil engineers of the railway company, and he himself
made precise measurements of the size of many different beaver
dams, lodges, and canals. In the mapped area he found no instance where a dam had been built across a stream having a
greater depth than 2 feet at the site of the dam when the water
was at its lowest level. He concluded that beaver dams are necessarily confined to the sources of the principal rivers and to the
small tributaries of these rivers. The importance of beaver dams
in helping to control water runoff and to reduce the intensity of
floods in forested country is evident. In large streams the beavers
live in burrows along the banks, do not build dams across the
river, and are often known as bank beavers. The largest dam that
Morgan measured ("Grass Lake Dam") was 261 feet long along
the crest and 6 feet 2 inches high. He estimated this dam to
contain upward of 7,000 cubic feet of solid materials. Larger dams
in other parts of the country have since been described, but the
amount of work by beavers that was involved in this particular
dam, which formed a lake 60 acres in extent, is impressive
enough.

Morgan determined that a dam is built or maintained by a
single family of beavers at any given time and that many generations of beaver may be involved over many years in the construction of a particular dam. He also notes (p. 103) that in making
repairs in the dam each beaver generally works independently of
the others. A 1970 reprinting of Morgan's pioneer book is available from Burt Franklin, New York.

E. R. Warren (1927) in his book *The Beaver* has a critical review of much of the older literature, including observations by
various naturalists in different parts of North America. Warren
himself made many observations on beavers and their constructions, particularly in Yellowstone National Park. He develops the
idea of the great abundance and importance of beaver meadows
and traces the origin of several such meadows. Beaver ponds

flood out and kill many trees, these ponds then gradually silt up, are abandoned by the beavers, and within a few years may become beaver meadows. He also has a good review of canal digging by beaver. Such canals may be hundreds of feet long, occur at different levels, and may draw water from different sources for the canal within which the beaver transports logs and branches to its pond.

The American beaver once occupied most of the North American continent from northern Mexico north to the limit of deciduous trees, especially of aspen trees. Over most of the area it was exterminated by the end of the nineteenth century, largely by overexploitation for the fur trade. It has been reintroduced in many localities and is now returning over a good part of its former range (Miller, 1972). The European beaver was similarly exterminated by the fur trade over most of its extensive range in Eurasia but is now returning in large parts of the European Soviet Union, where the Russians have carried out many experiments in breeding beavers and setting them loose in the wild. A small remnant of the original Scandinavian stock persisted in Norway and has provided animals for reintroduction into Finland and Sweden (Wilsson, 1968).

The most precise and detailed account of the behavior of the beaver is the monograph by Lars Wilsson (1971) on the ethology of the European beaver. Most of the section on building of lodges, dams, and winter stores is reproduced here (Paper 21), with special reference to the motor patterns of dam building. Wilsson used careful observation aided by analysis of films; he analyzed the sign stimuli to building behavior; and his study seems to be the first scientific and experimental study of the ontogeny of building behavior in the beaver.

Most of Wilsson's study of the European beaver was done in connection with Stockholm University and was aided by the Swedish National Sciences Research Council and the United States Air Force Office of Scientific Research. In the summer of 1963 he visited the United States, where he observed no essential differences in behavior of the American beaver and its European counterpart. In addition to his monograph (covering the period 1957–1965) and various scientific articles, he has written a charming popular book, which has been translated into English (1968), about his observations on beaver.

His studies were done mainly at the Faxälven river system in Sweden, where the light during midsummer at these far northern latitudes is good enough so that even at midnight the beavers, which are nocturnal animals, were readily visible when out in the open. Most of his more analytical studies, however, were made on captive beaver. These beaver were kept in outdoor enclosures, in each of which a small stream ran and was blocked off by wire netting at the inlet and outlet; he also observed beaver in a stream tank in a fish hatchery, in indoor terraria, and in his home.

Although young beaver show dam-building behavior by six or seven months of age or in their first winter, they do not build a complete dam until their second autumn. Practice does not seem to be very necessary, since young beaver which Wilsson kept in still water until their second autumn built their first dam, which was leakproof, soon after being placed in a stream in an outdoor enclosure or in the stream tank. In the latter tank he could control the flow of water experimentally. When the level of the water was raised, the young beavers built up the dam and also worked much more than usual on their lodge, which was in danger of being flooded over. The most important stimulus was the sound of water pouring over the dam. When the sound of rushing water was played back to the beavers from a tape recorder through a loudspeaker placed on or near the dam, the beavers built the dam at that place. Even a young beaver raised in isolation from running water and kept in a wooden tank of still water reacted promptly with building behavior on first hearing the sound of rushing water. Visual and tactile cues are probably also of some importance to dam-building behavior, since, when Wilsson lowered the level of water in the stream tank below the level of the dam, the two young beavers in the tank reacted with building behavior, filling in the indentations and depressions that were exposed in the dam, although the sound of running water was reduced when the water level was lowered.

Some learning is involved in dam building, since animals in the enclosures ceased to react to the rushing water at the inlet of the brook, where they had to build downstream of their work and could not get a result in the form of a lasting construction and a decrease in noise from running water. The beavers then built their dam just upstream from the outlet of the brook. At this place any materials deposited by the beaver were held against the growing dam by the current, instead of being washed away.

Some general conclusions are suggested from a comparison of the mechanisms and ontogeny of building behavior in the highly specialized examples of arthropod and vertebrate architects that are considered in this section:

1. Weaverbirds and beavers resemble the highly social in-

sects and the orb-web spiders, and may differ from many solitary insects in having less dependence in their building on internally fixed sequences of behavior and more dependence on construction work already laid down, as in the principle of stigmergy. However, more study is needed of the solitary insects from this viewpoint.

2. Birds that build complex nests, such as the true weaver-birds, may need practice in order to be able to build a complete species-typical nest; they appear to differ in this regard from arthropods, even including highly social insects such as termites and bees, where practice is unnecessary or of minor importance. Beaver seem to need little practice in dam building before they can build an effective dam, and this fact accords with the relatively uncomplicated structure of the dam.

3. Birds, honeybees, and orb-web spiders all resemble one another in that their building behavior depends on the interaction between such internal factors as special glandular secretions and external factors such as the developing construction itself and the external environment.

4. Specific sign stimuli from the construction work itself and from the external environment are important in guiding the location and types of work done in architects as diverse as termites and beaver.

5. Termites and beaver also resemble each other in that many generations of individuals are often involved in the building of a single external construction.

REFERENCES

Cheng, M-F., and R. Silver. 1975. Estrogen–progesterone regulation of nest building and incubation behavior in ovariectomized ring doves (*Streptopelia risoria*). *J. Comp. Physiol. Psychol.*, 88:256–263.

Collias, E. C., and N. E. Collias. 1973. Further studies on the development of nest-building behaviour in a weaverbird. *Anim. Behav.* 21:371–382.

Grassé, P.-P. 1959. La reconstruction du nid et les coordinations interindividuelles chez *Bellicositermes natalensis* et *Cubitermas* sp. La théorie de la stigmergie: essai d'interprétation du comportement des termites constructeurs. *Insectes sociaux*, 6:41–83, 7 pls.

Grassé, P.-P. 1967. Nouvelles expériences sur le termite de Müller (*Macrotermes mülleri*) et considerations sur la théorie de la stigmergie. *Insectes sociaux*, 14:73–102.

Hinde, R. A. 1958. The nest-building behaviour of domesticated canaries. *Proc. Zool. Soc. Lond.*, 131:1–48.

Hinde, R. A. 1965. Interaction of internal and external factors in integration of canary reproduction. Pp. 381–412 in *Sex and Behaviour*, F. A. Beach (ed.). John Wiley & Sons, Inc., New York.

Hingston, R. W. G. 1920. Geometrical spiders. Pp. 82–107 in *A Naturalist in Himalaya*. Witherby, London.

Lisk, R. D., J. A. Russell, S. G. Kahler, and J. B. Hanks. 1973. Regulation of hormonally mediated maternal nest structure in the mouse (*Mus musculus*) as a function of neonatal hormone manipulation. *Anim. Behav.*, 21:296–301.

Martínez-Vargas, M. C. 1974. The induction of nest-building in the ring dove (*Streptopelia risoria*): hormonal and social factors. *Behaviour*, 50(1–2):123–151.

Michener, C. D. 1974. *The social behavior of the bees*. Harvard University Press, Cambridge, Mass. 404 pp.

Miller, J. W. 1972. Return of the beaver. *Nat. Hist.* 81:64–73.

Morgan, L. H. 1868. *The American Beaver and His Works*. J. B. Lippincott Co., Philadelphia. 330 pp.

Peters, H. 1931. Die Fanghandlung der Kreuzspinne. *Z. Vergleich. Physiol.*, 15:693–748.

Ribbands, C. R. 1953. *The Behavior and Social Life of Honeybees*. Bee Research Association, London. 352 pp.

Riddle, O., and E. L. Lahr. 1944. On broodiness of ring doves following implants of certain steroid hormones. *Endocrinology*, 35:255–260.

Rösch, G. A. 1927. Über die Bautätigkeit im Bienenvolk und das Alter der Baubienen. *Z. Vergleich. Physiol.*, 6:264 298.

Savory, T. H. 1952. *The Spider's Web*. Frederick Warne & Co. Ltd., London. 154 pp.

Skaife, S. H. 1961. *Dwellers in Darkness*. Doubleday & Co., Inc., New York. 180 pp.

Victoria, J. K. 1969. Environmental and behavioral factors influencing egg-laying and incubation in the village weaverbird *Ploceus cucullatus*. Ph.D. thesis, University of California, Los Angeles. 92 pp.

Warren, E. R. 1927. *The Beaver*. Monogr. Am. Soc. Mammalogists 2. The Williams & Wilkins Co., Baltimore. 177 pp.

Wilson, E. O. 1971. *The Insect Societies*. Harvard University Press, Cambridge, Mass. 548 pp.

Wilsson, L. 1968. *My Beaver Colony*. Translated from the Swedish by Joan Bulman. Doubleday & Co., New York. 154 pp.

16

Reprinted from *Am. Zool.*, **9**(1), 223–227 (1969)

Maturing and Coordination of Web-Building Activity

Hans M. Peters

*Zoophysiologisches Institut, Abteilung für Physiol. Verhaltensforschung,
Universität Tübingen, Germany*

Synopsis. Experiments are reported in which 114 *Zygiella x-notata* Cl. spiderlings were taken out of the cocoon before the time at which they normally left (2 weeks after the second molt); 67 were placed on wooden frames and 47 on wire frames. Both groups built partial, irregular thread-patterns at first, and, as they grew older, constructed more and more regular geometric orb-webs: the transition can be traced through several stages. The shape of the web, particularly the presence of the free-sector, was dependent on the frame; webs built in a circular wire frame lacked the free-sector. It is assumed that the accomplishment of full, regular orb-webs is essentially based on the progressive histological differentiation of the central nervous system in the spiderlings during the developmental phase under observation; the wire frame could influence the shape of the web through making construction of a retreat difficult.

My student, Gertrud Mayer (1953), came to the conclusion that the behavioral pattern of the spider constructing its orb-web is completely innate ("rein angeboren"). This view was based on the following observations on *Araneus diadematus*: Miss Mayer put newly hatched spiderlings into small glass tubes, diameter 2.5 mm. In these tubes the spiderlings could barely turn and were unable to perform any complicated movements. Some of them succeeded in performing the second molt[1] within their narrow tube. When they had reached the age at which young garden spiders usually start building webs, the prisoners were released. They were able to spin a fully normal orb-web, immediately after leaving their prison. Thus, the former but less critical experiments of Eliza Petrusewiczowa (1938) were confirmed.

Years ago I became aware by chance that spiderlings precociously taken out of the cocoon, *i.e.*, during a phase in which they normally do not yet build webs, often construct very atypical webs. I suggested that the mechanisms of coordination, un-

derlying web construction, attain their full maturity during the time the spiderling spends in the cocoon. Thus, they would be able to build a regular orb-web immediately when they normally leave the cocoon.

Together with Victor von Harling, these experiments were extended, using *Zygiella x-notata* which in Germany starts laying eggs in autumn.

Figure 1 shows the typical cocoon, artificially opened from above. We took the spiderlings out of the cocoons and put them individually on small frames, where they could spin their webs. We used two kinds of frames, (1) square frames of wood, the length of the beams being 70 mm (inside), (2) circular frames, diameter 70 mm, of wire 1-1.5 mm in diameter. The sort of frame influenced very much the behavior of the spiderlings.

The results which I will discuss are based upon 11 cocoons and 114 animals, 67 on wooden frames and 47 on wire frames.

During normal development at 20-24°C, the young *Zygiella* hatch and leave the eggs three weeks after the cocoon was deposited. Four days later the second molt takes place. After two more weeks the spiderlings begin leaving the cocoon and soon start constructing a first web. The earliest web of such a spider was observed one day after the spider left its cocoon. To test our hypothesis that the behavioral pattern of

---

Dedicated to the memory of my pupil, Victor von Harling, who so tragically lost his life shortly after finishing his essential share in these investigations.

[1] G. Mayer called it "Kokonhäutung". At that time, we did not realize that this is the second molt. As Fr. Meier (1967) found with *Araneus cornutus*, the first molt is closely connected with hatching.

web-building becomes mature while the spider is hidden in the cocoon, we took samples of spiderlings out of the cocoon day by day, beginning with the day of the second molt and ending with the thirteenth day thereafter, and put them on our frames. It would have been useless to test them earlier, *i.e.*, before the second molt, because they seemed unable to move around and produce silk threads.

FIG. 2.

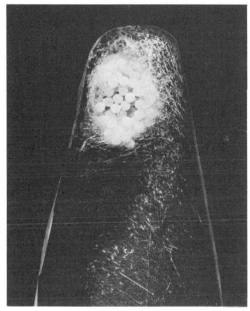

FIG. 1.

I wish to emphasize that the first web constructed by a young *Zygiella,* which has left its cocoon under natural conditions, is fully regular, like that of *Araneus diadematus*. It shows the characteristic feature of the *Zygiella*-web, the so-called "free-sector" (Fig. 2). The "signal-thread" runs through the "free-sector" and connects the retreat with the hub of the web.

When spiderlings are taken out of the cocoon and put on wooden frames, nothing happens during the first two days after the second molt except that they occasionally move around, leaving a "security-thread" or "drag-line" behind. On the third day the young spider starts to construct a retreat. A fixed place, first delineated by some threads, is established on the frame, usually in a corner. These threads become more and more numerous, and finally a silken tube to shelter the spider is constructed (Fig. 3).

This behavior of retreat-building, starting with the third day after the second molt, could be observed in almost all spiders taken out of the cocoon before they would have left it by themselves. As is already shown (Fig. 3) there can be irregular threads leading from the retreat to the wooden frame. These irregular networks can contain short tracks of viscid threads here and there. They correspond to the viscid spiral of true orb-webs in their physical properties, *i.e.*, they are very elas-

FIG. 3.

tic. Microscopic check reveals that they are provided with droplets of glue.

Besides such networks we also observed others which were characterized by a more or less clear centralization. Networks like these are a transition to others in which we can distinguish radii, frame-threads, the viscid spiral, and the auxiliary spiral. But, as far as the placement of these components is concerned, there can be a high degree of irregularity.

FIG. 5.

FIG. 4.

Figure 4 shows a web which was built 12 days after the second molt. It is extremely small and extremely irregular. The viscid spiral is confined to the periphery. We can recognize a "free-sector", but this sector is not truly free, because it is crossed by several threads.

Figure 5 shows a web woven 7 days after the second molt, and the "free-sector" is not very clear. Furthermore, the side parts of the frame and the distances between the radii and the viscid threads are extremely irregular.

An example of a web built 11 days after the second molt (Fig. 6) is peculiar in that the viscid threads are confined to the outer parts of the web. In the inner parts, the auxiliary spiral has not been replaced. Similar webs were observed frequently. We believe that this phenomenon—lack of viscid threads in the central parts of the

web—is related to another property of these premature webs, i.e., deficiency of glue in the viscid threads. This substance abounded in the outer circles of the spiral, and it decreased towards the center of the web. In a few instances there was no glue on the whole catching spiral.

We noted above that spiderlings which had been taken out of the cocoon usually started spinning activity by constructing a retreat. The general pattern of threads which can be recognized as an orb-web appears usually on the fifth day or later.

FIG. 6.

The number of spiders constructing regular webs increases more and more with each day, and, conversely, the number of animals confining themselves to the construction of a retreat, and perhaps some irregular threads in connection with the retreat, decreases.

By the thirteenth day after the second molt, almost all of our subjects had constructed their first geometric orb-web.

What did those spiderlings do which had been placed upon a wire frame instead of a wooden frame after they had been taken from the cocoon?

The broad beams of the square wooden frames, and especially their four corners, certainly greatly encouraged the construction of a retreat. The thin circular wire frames, on the other hand, did not offer such opportunities, and spiders placed on this sort of frame usually failed to build a retreat. We believe that this is why spiders which were given a wire frame usually constructed full orb-webs, i.e., webs without "free sectors". Spiders which had constructed a web of this kind were perched in its center.

Figure 7 shows an example of such a full orb-web, built 15 days after the second molt. The relatively older age of the spider is the reason for the regularity of this

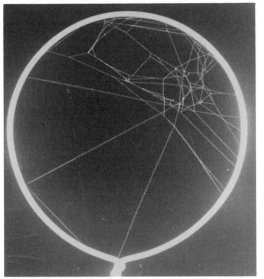

FIG. 8.

web. Needless to say, all that has been said about the maturation of web-building activity from observations on "sector-webs" could be demonstrated as well from observations on spiderlings placed upon wire frames. For example, Figure 8 shows a well centered but very irregular threadwork built five days after the second molt.

These are the observed facts from our studies. I can only offer some suggestions about their interpretation. One may assume that the maturation of web-building activity is a rather complex process, taking place separately in the different parts of the young organism. When the viscid spiral is confined to the outer zones of a premature web, the reason for this could be that the glandulae flagelliformes (Sekiguchi; gl. coronatae Peters) do not yet produce a sufficient amount of silk. The lack of glue droplets on the threads could be interpreted in the same way as a deficiency in the function of the glandulae aggregatae in this phase of development. On the other hand, the reasons for many other anomalies of premature webs should be sought in the field of neural functions. The early webs seem to demonstrate that the central nervous system is not yet completely differentiated when the spiderling is still within its cocoon. At that time the

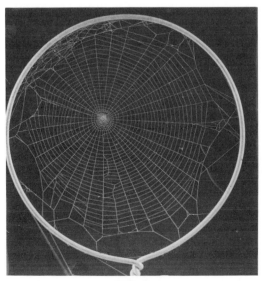

FIG. 7.

neural basis of certain functions would still be lacking or at least be underdeveloped. Furthermore, it seems that the neural system as a whole is not yet able to integrate the various functions to the degree necessary for performing the complex behavioral pattern characteristic for orb-weavers.

It would be worthwhile to describe these functional deficiencies in a more precise and defined fashion. The problem could be approached from two different directions. First, by studying more deeply the behavioral characteristics. Second, and such a study should be combined with the first one, by an investigation of the differentiation of the spider's nervous system in

the phase of development under consideration, as has been started with another viewpoint by Fr. Meier (1967).[2] By comparing the histological stage of the central nervous system with the behavioral state of the spiderlings, one could perhaps localize certain functions and, in this way, gain more insight into the neural basis of web building. One can easily recognize the convergence of this aspect with the laser studies of Witt (as reported in this symposium).

REFERENCES

Mayer, G. 1953. Untersuchungen über Herstellung und Struktur des Radnetzes von *Aranea diadema* und *Zilla x-notata* mit besonderer Berücksichtigung des Unterschiedes von Jugend- und Altersnetzen. Z. Tierpsychol. 9:337-362.

Meier, Fr. 1967. Beiträge zur Kenntnis der postembryonalen Entwicklung der Spinnen - Araneida, Labidognatha - Unter besonderer Berücksichtigung der Histogenese des Zentralnervensystems. Rev. Suisse Zool. 74:1-127. `

Petrusewiczowa, E. 1938. Boebachtungen über den Bau des Netzes der Kreuzspinne (*Aranea diadema* L.). Trav. Soc. Sci. Lettr. Wilno (Trav. Inst. Biol. Wilno No. 9), 13:1-24.

---

[2] In the discussion, P. N. Witt has drawn attention to this paper. Unfortunately, our observations can be related to the beautiful results of Fr. Meier only in a very general sense. When we try to apply the findings of Meier in *Araneus cornutus* to *Zygiella*, we have to take into account that at the time of hatching, the CNS contains only undifferentiated neuroblasts. Histogenesis, *i.e.*, formation of the various types of neurons and their connections, would take place during the time the *Zygiella* spiderlings normally stay within their cocoon.

17

Reprinted from *Am. Zool.*, **9**(1), 71–79 (1969)

Synthesis of Silk, Mechanism and Location

DAVID B. PEAKALL

Department of Pharmacology, Upstate Medical Center, Syracuse, New York

SYNOPSIS. The location and function of the five or six sets of silk glands of *Araneus diadematus* (Cl) are discussed. The structure and function of the three major parts of the ampullate gland indicate a synthesizing, collecting, and possibly structuring section. Two methods of stimulation of the ampullate gland, namely emptying the gland and cholinergic stimulation, are known. In both cases there is an initial secretory stage followed by rapid synthesis of new protein. The sequence of events following stimulation by both methods is described, based on studies of the incorporation of labeled protein and RNA precursors and on autoradiographic studies. Characteristic changes occur in the fine structure during the stimulatory cycle. Several experiments show that the spider has information on the amount of silk available to it for use in web-building. A structure which may act as a biological transducer has been located in the ampullate gland.

Spiders have evolved to make the maximum use of that remarkable protein, silk. To judge from the wide geographic distribution, the number of species as compared to the other orders of the Class Arachnida, and the numbers of individuals per unit area of some species, this exploitation of silk has been successful. Although the spider has successfully made use of silk, it has in so doing placed itself in the position described rather picturesquely by the French zoologists, André and Lamy, "Spiders lead lives bristling with difficulties and are always on the horns of a dreadful dilemma: no food without webs and no webs without food."

To minimize the dangers of running out of silk, the silk gland is prepared for immediate action. Not only is the cytoplasm full of droplets of silk ready to be secreted into the lumen, but the major amino acids of which the silk is comprised are present in high concentrations ready for new synthesis to start. Spiders economize on usage of silk as far as possible by redigesting every scrap of it. The provisional spiral of the web, for example, is eaten as the final sticky spiral is laid down. Examination of serial sections made of the spider just after its web is completed shows that those glands which produce silk for the web are largely empty. The spider's operating margin is small.

The ampullate glands of *Araneus sericatus* comprise 3-5% of the body weight of the spider on a wet-weight basis, and are able to produce protein equivalent to 10% of the gland's weight every web-building cycle. It is this frequent, rapid production of a single structural protein that makes these glands an attractive model for the study of the regulation of protein synthesis. Another useful feature is that the amino acid composition is unusual, the two amino acids, alanine and glycine, comprising some 57% of the protein. Thus, it is possible to follow changes in the amino acid pools that would be difficult if the composition of the protein were more normal.

Although studies of the structure and function of the silk glands of the spider date back to Blackwell's observations in 1839, there is still no final agreement on the number of sets of silk glands. The types reported for *Araneus diadematus* and *Araneus sericatus*, together with their number, location, and function are listed in Table 1. The larger glands are shown dissected out of an adult female *sericatus*

Present address: Division of Ecology and Systematics, Langmuir Laboratory, Cornell University, Ithaca, N. Y. 14850

Supported by research grants PN-5234 from the National Institutes of Health and GB-5736 from the National Science Foundation. Portions of these studies were carried out while the author was an Established Investigator, American Heart Association.

TABLE 1. *Number and position of each type of silk gland and the function of the silk. The number of glands is based largely on the work of Warburton (1890), the modification below the line on the studies of Sekiguchi (1952). The function of the silk is based on Apstein (1889), Warburton (1890), Comstock (1948), Sekiguchi (1952), and Peakall (1964).*

| Type of silk gland | Classification of spinnerets | | | Function of silk |
	Anterior	Median	Posterior	
Ampullate	2	2	—	Drag-line, scaffolding of web, base-line of catching spiral?
Aggregate	—	—	6	Viscid thread of catching spiral
Cylindrical or tubuliform	—	2	4	Egg sac
Piriform	200	—	—	Attachment discs
Aciniform	—	200	200	Swathing bands
Aggregate	—	—	4	Viscid thread of catching spiral
Flagelliform*	—	—	2	Base-line of catching spiral

* Named coronata independently by Peters (1955).

in Figure 1. Personally, I cannot locate the flagelliform glands either histologically or by dissection. The studies on the synthesis of silk have been largely confined to the ampullate glands, and it is these glands that are considered unless otherwise stated. Limited observations suggest that the mode of stimulation of the aggregate gland is similar to that of the ampullate gland, whereas that of the cylindrical gland is quite different (Peakall, 1965).

The ampullate gland consists of three morphologically distinct parts, the tail, the sac, and the duct (Fig. 1). The main, probably the sole, function of the long tail of the gland is the synthesis of silk protein. The fine structure of, and the biochemical changes in, these cells during the protein-synthesizing cycle will be discussed later.

An important function of the sac of the gland is the storage of protein (Fig. 1). The volume of the lumen of the sac is twice that of the tail. The rate of synthesis of protein in the epithelial cells of the sac is only a quarter of that found in the tail. Since the volume of the epithelial cells of the tail is ten times that of the sac, the amount of protein synthesized in the sac is only a few percent of the total protein synthesized by the gland. However, electron microscopy indicates that the function of the epithelial cells of the sac may be complex. Five types of droplets have been found (Bell and Peakall, 1969). One of these appears to be identical with the silk protein droplets found in the tail, and another is similar to the material secreted

in the transitional zone between the sac and the duct. A third type is similar to lipid, and this can be correlated with increased lipid staining observed under the light microscope in the region of the sac as compared to the tail. The function of the other two types of droplets is unknown.

In the transitional zone between the duct and sac there is a region where the epithelium is composed of secretory cells. Electron micrography shows that these

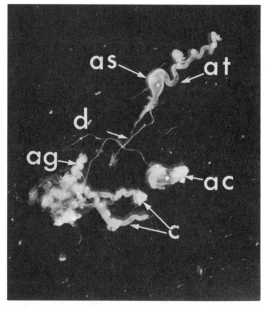

FIG. 1. The aggregate glands (ag), cylindrical glands (c), and two ampullate glands dissected out of an adult female *Araneus sericatus*. One ampullate gland is coiled (ac), the other is straightened out to show the various parts: at, tail; as, sac; d, duct. No staining. × 5

TABLE 2. *Two ways of stimulating synthesis of silk in ampullate glands.*

1a.	Acetylcholine binds with receptor on outer (basal) membrane of gland	Signal across outer epithelial membrane causes release of presynthesized fibroin	Atropine-sensitive Puromycin-insensitive
1b.	Emptying of the gland	Signal across inner (luminal) epithelial membrane causes release of presynthesized fibroin	Atropine-insensitive Puromycin-insensitive
2.	Signal after release of presynthesized fibroin from epithelium	Increase in nuclear size; change in concentration of RNA; increased incorporation of amino acids	Puromycin-sensitive

droplets have an electron density different from silk. One possible explanation of these droplets would be that they are a polymerase. It has been shown for the orbweb builder, *Nephila madagascariensis*, that there is a change of molecular weight from 30,000 in the sac to 200-300,000 in the completed fiber (Braunitzer and Wolff, 1955).

The duct is some five times longer than would be necessary if it had the sole function of connecting the main part of the gland to the spinnerets. The electron microscope reveals that the wall of the duct consists of microvilli arranged radially to the lumen of the duct (Bell and Peakall, 1969). It seems likely, on morphological grounds, that one function of the duct is absorption of water. Preliminary experiments have shown that the water content of the silk in the sac is considerably higher than that in the final fiber.

The long duct may also be of importance in the preliminary stages of orientation of the silk molecules. It is known that the silk in the sac is an α-configuration, whereas in the fiber it is in the form of a β-pleated sheet (Ambrose, *et al.*, 1951; Lenormant, 1956; Warwicker, 1960). However, Wilson, (1962, and this symposium) considers that the main orientation of the fiber occurs between the control valves at the base of the spinneret and the spigot. A further possibility is that it is necessary to have enough material in the duct to make a single radius. It has been noted that the time spent in the center of the web is long compared with that taken to lay down the radii (Witt, *et al.*, 1968). It is possible that the pauses between construction of radii may be necessitated by the need to pump silk into the duct. These various possibilities are not mutually exclusive.

Two ways of stimulating the ampullate gland to produce protein have been found (Peakall, 1966). These are emptying the silk from the gland and administering cholinergic agents. The first mode of stimulation is insensitive to atropine, whereas the second mode is blocked by this anticholinergic agent. In both cases the first stage of the cycle of stimulation is the secretion of preformed protein droplets from the epithelium into the lumen. This is followed by accelerated synthesis of new protein. The two steps, secretion and synthesis, can be separated by pretreatment with a blocker of protein synthesis, such as puromycin or actinomycin D. In this case the stimulation causes the secretion of preformed protein, but this is not followed by the synthesis of new protein. The modes of stimulation and the two-stage cycle of events following stimulation are listed in Table 2.

The sequence of events following stimulation of the ampullate gland will now be considered in more detail. Autoradiographic evidence (Peakall, 1958) shows that acetylcholine binds to the cell membrane but does not enter the cell. The first histological change observed is that the protein stored in the cytoplasm moves towards the luminal end of the cell and is discharged into the lumen. This is accompanied by fusing of droplets of protein; often the shape of the nucleus is distorted by this action (Fig. 2a, b). This change is in full swing within ten minutes of stimulation, and experiments with actinomycin D, which blocks the DNA-dependent synthesis of RNA, show that the signal for increas-

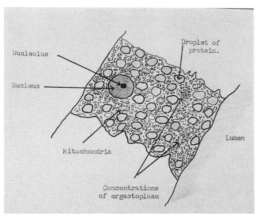

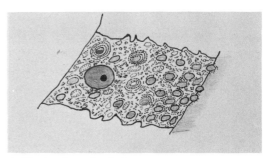

d. 60 min after stimulation.

FIG. 2. Diagrammatic representation of the structural changes occurring in the epithelial cells of the tail of the ampullate gland with stimulation. Based on light and electron microscopic studies.
a. Appearance of cell before stimulation.

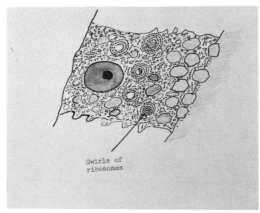

b. 5-10 min after stimulation.

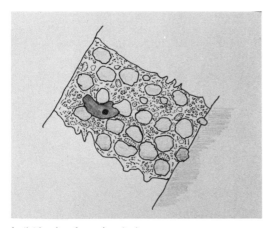

c. 20 min after stimulation.

ing the rate of synthesis occurs within this time. The amount of incorporation of C^{14}-alanine in an hour following stimulation with acetylcholine was measured autoradiographically. The results clearly show that the synthesis of protein is inhibited only if actinomycin D is given within three minutes of stimulation (Table 3). After this time, presumably enough messenger RNA is present to continue with the synthesis of new protein. The half-life of messenger RNA has been found to be approximately four hours. Experiments were carried out in which new synthesis of messenger RNA was blocked 15 minutes after stimulation, and the amount of incorporation of C^{14}-alanine was measured at various time intervals and compared to controls (Table 4).

The movement of ribosomes during the stimulatory cycle has been examined in the light and electron microscopes, and by

TABLE 3. *Autoradiographic counts to show the degree of incorporation of C^{14}-alanine following stimulation with acetylcholine and subsequent administration of actinomycin D. Acetylcholine (1 mg/kg) and C^{14}-alanine (5 μc) were injected into the body fluids at the start of the experiment. Actinomycin D (0.1 μg/kg) was injected at stated intervals after the injection of acetylcholine.*

Interval between Ach and actinomycin D injections (minutes)	Autoradiographic counts per mm² (× 1940). Average of 100 counts with standard deviation; number of spiders used in parentheses.
0	2.0 ± 1.7 (2)
1	1.4 ± 0.9 (2)
2	3.3 ± 1.3 (2)
3	3.9 ± 1.6 (2)
4	26.9 ± 5.7 (2)
5	25.3 ± 6.0 (3)
10	33.7 ± 5.5 (2)
Control (no actinomycin)	31.6 ± 6.3 (4)

TABLE 4. *All spiders stimulated by injecting acetylcholine (1 mg/kg) into the body fluids. Experimental animals given actinomycin D (1 γ/kg) by the same route 15 minutes later. C¹¹-Alanine was given at various intervals after the blocking of RNA synthesis by actinomycin. The amount of incorporation in 1-hr periods was measured at various times and expressed as cpm/γ RNA. Each figure is the average, with standard deviation, of four spiders.*

| | Incorporation cpm/γ RNA | | | |
	1–2 hrs	2–3 hrs	4–5 hrs	6–7 hrs
Experimental	256 ± 35.5	147 ± 23.9	92 ± 17.8	23 ± 4.5
Control	349 ± 41.7	248 ± 31.0	189 ± 25.3	106 ± 16.4
Experimental % of control	73	59	49	22

following a pulse of labeled orotic acid with light autoradiography. Ten minutes after stimulation, the RNA is concentrated in the basal half of the cell (Figs. 3a and 4a). Twenty minutes after stimulation it is more generally distributed but with some concentration in the luminal half (Fig. 3b), which is particularly clear in the autoradiograph (Fig. 4b). After an hour there is little localization of RNA (Figs. 3c and 4c). Electron microscopy shows that new protein droplets are associated with characteristic whorls of ergastoplasm (Fig. 2c). It is of interest that no Golgi apparatus is found in the tail of the gland. It is possible that the by-passing of this packaging step is an adaptation for allowing rapid synthesis.

The course of protein synthesis has been followed by giving a short pulse of labeled alanine at the time of stimulation. It is found that active synthesis and secretion into the lumen continue for 4-8 hr and then droplets of protein are formed which remain in the cytoplasm. The length of time for active synthesis and secretion varies with the mode of stimulation. If the stimulation is caused by emptying of the gland this phase of the cycle lasts for about eight hours; with cholinergic stimulation the duration is only about four hours. The difference is presumably due to the fact that the lumen of the gland remains full of protein in the case of cholinergic stimulation. The importance of the cholinergic mechanism is unknown. It is possible that it acts as a fine control in regulating the production of protein and is the means whereby external stimuli can affect the amount of silk available for web-building.

After droplets of protein are formed in the cytoplasm, the rate of synthesis of protein decreases greatly. However, preliminary work on the composition of the amino acid pools suggests that the accumulation of the most prominent amino acids, alanine and glycine, continues until their levels are high. In glands in which active synthesis of new protein is under way, the level of these two amino acids is low, presumably due to high demand, whereas, in the resting state the levels of alanine and glycine are high. Thus, the gland is poised for action.

There are four experiments which independently suggest that the spider has information on the amount of silk available for web-building (Witt and Reed, 1965; Witt, et al., 1968). Since the value of an incomplete web as a means of catching food is limited, the importance of this information is obvious. The experiments, details of which can be found in the references cited above, are (1) starvation, in which the spider makes a smaller, wider-meshed web; (2) additions of weights to the back of the spider, when a thicker but shorter thread is used in a full-sized web; (3) oral administration of physostigmine, when more protein is used to make a larger web, the increase being especially great in the sticky spiral; and (4) the burning out of radii (Koenig, 1951; Reed, this symposium). In the last case the spider replaces the radii for a while, but finally ignores the oversized angle and goes on to building the spiral. This experiment indicates that the spider may have a continuous feed-back of information on the supply of silk.

The debate about which gland supplies

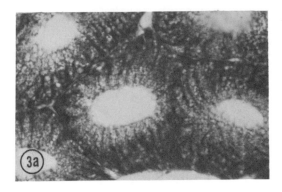

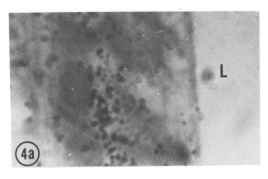

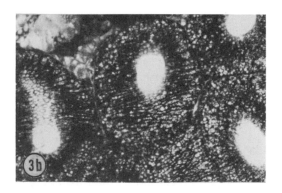

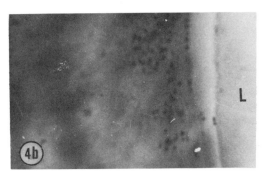

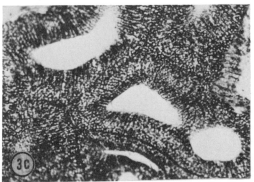

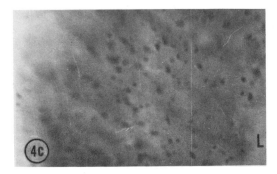

FIG. 3. Cross-sections of the tail of the ampullate gland. Sections treated with DNase (0.1mg/ml) for 3 hr at room temperature. Sections were then stained with Azure B (40°, 3 hr, pH 4.0) and cleared with t-butyl alcohol. × 200
a. 10 min after stimulation.
b. 20 min after stimulation.
c. 60 min after stimulation.
FIG. 4. Autoradiographs of a cross-section of a sin-

gle epithelial wall of the tail of the ampullate gland. Five minute pulse of C^{14}-orotic acid given at the time of stimulation. Sections coated with Ilford I.4 emulsion, lightly stained with Azure B. L, lumen of gland. × 1200
a. 10 min after stimulation.
b. 20 min after stimulation.
c. 60 min after stimulation.

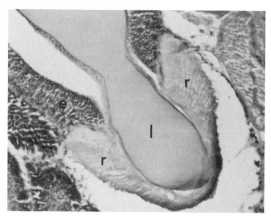

FIG. 5. Cross-sectional views of the ampullate gland showing a structure considered to be a receptor sensitive to the amount of silk stored in the gland. l, lumen of gland; e, epithelium; r, receptor; t, tail of ampullate gland; n, nuclei at edge of receptor; c, connective tissue. Stained with hematoxylin and eosin.

a. Showing the receptor in contact with the lumen. × 200

the silk for the base-line of the sticky spiral does not affect the interpretation of these experiments to any considerable extent. If the material used for the base-line of the sticky spiral comes from the ampullate gland, then the importance of the feed-back of information is clear for all four experiments. Even in the last experiment, that of burning out radii, the value of the feed-back mechanism still remains, assuming that the provisional spiral comes from the ampullate gland. It is of interest that Koenig (1951) finds that the sticky spiral is minimal following the burning out of radii. This suggests that the same gland is involved both in constructing radii and forming sticky spirals, although it is possible that the small spiral is due to disturbance of the spider by the experimentor.

The major question is the mechanism by which the feed-back of this information is mediated. None of the changes that occur in the cell during the cycle of secretion and synthesis that follows the stimulation of the gland appear to be related to the amount of silk present in the lumen. Moreover, whereas most of the silk is synthesized in the tail of the gland, it is stored

in the sac portion. Serial sections of the abdomen through the sac of the ampullate gland were examined. A structure which could be the receptor for the transmission of this information is shown in Figure 5a-c. In Figure 5a the "receptor" is in contact with the silk in the gland in a section of the lumen. The appearance of the "receptor" is very different from the normal epithelium in sections stained with hematoxylin, where the "receptor" is largely stained with eosin. The "receptor" has a striated structure and completely lacks the droplets of protein which are the predominant feature of the normal epithelial cells. Figure 5b is a photograph of another section of the same spider, 100 microns away from the section shown in Figure 5a. At this stage, the receptor is not in contact

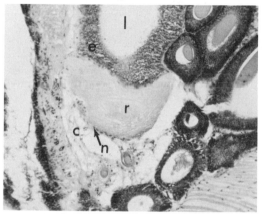

b. A section 100 μ away from (a); receptor no longer in contact with lumen. × 140

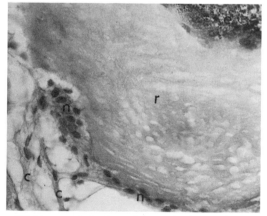

c. Higher-powered view of part of (b). Peripheral nuclei and connective tissue can be seen. × 455

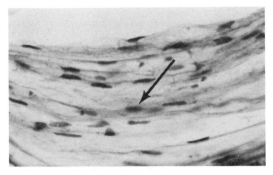

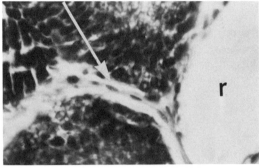

FIG. 6. Comparison of main nerve cord in the pedicel (6a) with the fiber attached to the receptor (6b). Both fibers show elongated nuclei. Stained with Trichrom. ×300

with the lumen of the gland, but the peripheral nuclei and associated fibers can be seen. These features are more clear under higher magnification (Fig. 5c). Cross-sections of the coiled tail of the ampullate gland are also seen in Figure 5b. Staining with silver showed only poorly-stained fine fibers, which may represent the innervation of this structure. Staining with trichrome shows fibrous material with nuclei that were similar to main ganglia passing through the pedicel (Fig. 6a, b).

The structure described appears to have the necessary structural requirements to act as a biological transducer. The striated structure would be capable of transmitting pressure changes which could be changed to nerve impulses. In general, it is similar to a Pacinian corpuscle, a structure in mammalian skin and muscle which converts pressure into nerve impulses (Loewenstein, 1960).

The orb web-building spider presents some unique opportunities for inter-

disciplinary studies, some of which are illustrated in this symposium. The silk glands, as discrete organs producing a single structural protein, are a good model for the biochemist to study protein synthesis. The vital nature of these glands makes their role in the over-all economy of the animal of especial interest. The web produced by this structural protein makes a permanent, measurable record of a complex behavioral pattern. The pattern can be modified by many factors, a few of which affect the synthesis of silk directly. Such a complex operation as the construction of an orb-web must require the utilization of a great deal of the spider's central nervous system. The relationship of changes in pattern in the orb-web to chemical changes in the spider's brain might well be a fascinating and rewarding subject for study.

REFERENCES

Ambrose, E. J., C. H. Bamford, A. Elliott, and W. E. Hanby. 1951. Water-soluble silk: an α-protein. Nature 167:264-265.

Apstein, C. 1889. Bau und Funktion der Spinndrüsen der Araneida. Arch. Naturg. 55:29-74.

Bell, A. L., and D. B. Peakall. 1969. Changes in fine structure during silk protein production in the ampullate gland of the spider (*Araneus sericatus*). J. Cell Biol. (In press)

Blackwell, J. 1839. On the number and structure of the mammulae employed by spiders in process of spinning. Trans. Linnaean Soc. London 18:219-224.

Braunitzer, G., and D. Wolff. 1955. Vergleichende chemische Untersuchungen über die Fibroine von *Bombyx mori* und *Nephila madagascariensis*. Z. Naturforsch. 10b:404-408.

Comstock, J. H. 1948. The spider book. Comstock Publishing Co., Inc.

Koenig, M. 1951. Beitrage zur Kenntnis des Netzbaus orbiteler Spinnen. Z. Tierpsychol. 8:462-493.

Lenormant, H. 1956. Infrared spectra and structure of the proteins of the silk glands. Trans. Faraday Soc. 52:549-553.

Loewenstein, W. R. 1960. Biological transducers. Scientific American 203:98-108.

Peakall, D. B. 1964. Composition, function, and glandular origin of the silk fibroins of the spider, *Araneus diadematus* Cl. J. Exptl. Zool. 156:345-350.

Peakall, D. B. 1965. Differences in regulation in

the silk glands of the spider. Nature 207:102-103.

Peakall, D. B. 1966. Regulation of protein production in the silk glands of spiders. Comp. Biochem. Physiol. 19:253-258.

Peakall, D. B. 1968. Autoradiographic localization of acetylcholine in the ampullate gland of the spider. J. Pharmacol. Exptl. Therap. 160:81-90.

Peters, H. M. 1955. Ueber den Spinnapparat von *Nephila madagascariensis*. Z. Naturforsch. 10b: 395-404.

Sekiguchi, K. 1952. On a new spinning gland found in geometric spiders and its functions. Ann. Zool. Jap. 25:394-399.

Warburton, C. 1890. The spinning apparatus of geometric spiders. Quart. J. Microscop. Sci.

Warwicker, J. O. 1960. Comparative studies of fibroins. 2. The crystal structures of various fibroins. J. Mol. Biol. 2:350-362.

Wilson, R. S. 1962. The control of the dragline spinning in the garden spider. Quart. J. Microscop. Sci. 103:557-571.

Witt, P. N., and C. F. Reed. 1965. Spider-web building. Measurement of web geometry identifies components in a complex invertebrate behavior pattern. Science 149:1190-1197.

Witt, P. N., C. F. Reed, and D. B. Peakall. 1968. A spider's web. Problems in regulatory biology. Springer-Verlag, Heidelberg.

18

Reprinted from *Sex and Behavior*, F. A. Beach, ed., John Wiley & Sons, Inc., New York, 1965, pp. 355–369, 378–380

INTERACTION BETWEEN INTERNAL AND EXTERNAL ENVIRONMENTS IN THE REGULATION OF THE REPRODUCTIVE CYCLE OF THE RING DOVE

Daniel S. Lehrman

MAJOR FEATURES OF SEXUAL AND PARENTAL BEHAVIOR

The ring dove (*Streptopelia risoria*), a small relative of the domestic pigeon, is unusually well suited for psychobiological investigation. It breeds very well in captivity, exhibits behavior patterns which do not appear to have degenerated from the dove wild type in spite of hundreds of generations of domestication, and has been the subject of attention by investigators who have provided a good background of useful information both about its reproductive behavior (Craig, 1909) and about its reproductive physiology (Riddle, 1937).

If a male and female of this species, both of which have had recent breeding experience, but not with each other, are placed together in a breeding cage containing a nest bowl and a supply of nesting material, a regular and predictable sequence of changes in behavior can be observed. The principal activity observed in the first day or so is a characteristic pattern of courtship behavior, in which the male bows and coos at the female (Figure 1). After some hours, the birds indicate their selection of a nest site by crouching in a concave place (e.g., the inside of a nest bowl) and uttering a characteristic coo. Once the site

The research program described in this chapter is supported by research grant MH-02271 and by Research Career Award MH-K6-16,621, both from the National Institute of Mental Health. Grateful acknowledgment is made to the following graduate students who are participating in this work: Mr. Philip N. Brody, Mr. Carl J. Erickson, Miss Miriam Friedman, Mr. Barry R. Komisaruk, Miss Sheila Miller, and Mrs. Rochelle P. Wortis.

[*Editors' Note:* A second printing of *Sex and Behavior*, F. A. Beach, ed. was published by Robert E. Krieger Publishing Co. of Huntington, N.Y. in 1974.]

Figure 1. Male ring dove (*a*) bowing and (*b*) cooing to female, initial courtship behavior

has been selected, both the male and the female actively participate in building a nest there. The male characteristically gathers most of the nesting material and carries it to the female, who spends most of her time standing on the nest, where she takes the nesting material from the male and builds it into the nest. The male may do some of the nest building and the female some of the collecting, but the pattern is predominantly as I have described it.

After several days of nest-building activity, a fairly abrupt change occurs in the behavior of the female. She becomes noticeably more attached to the nest, and more difficult to dislodge from it; it is possible to pick her up while she is standing on the nest and to lift her, in which case she may actually lift the partly built nest clutched in her claws! This kind of behavior usually indicates that the female is about to lay an egg. She lays the first of her two eggs at about 5:00 P.M. and the second egg at about 9:00 A.M. on the second day following. At some time between the laying of the first and second eggs, the female begins to sit on the eggs. Usually on the day when the second egg is laid, the male also begins to sit. The birds take turns in incubating the eggs, the male sitting for about six hours in the middle of the day, the female continuously for the remaining 18 hours. The eggs hatch after about 14 days, and for several days following hatching the parents continue to sit on the young, following approximately the same schedule as that on which they had incubated the eggs. At first the parents feed the young by regurgitating to them a substance produced by the lining epithelium of their crops ("crop milk") (Patel, 1936). The young leave the nest at 10 to 12 days of age, but continue to beg for food from the parents. Starting when the young are 14 to 15 days old, the parents gradually become less and less willing to respond to the begging of the young, while the squabs gradually develop the ability to peck for food from the ground. When the young are about 20 days old, the bowing-coo courtship behavior of the adult male increases in frequency, the parents begin to build a new nest, lay new eggs, and the cycle is repeated. This cycle—of courtship, nest building, egg laying, incubation, and care of the young—lasts 6 to 7 weeks, and may continue throughout the year (at least in our laboratory, where there are no seasonal changes in temperature or length of the day).

Now, these regular, highly predictable changes in behavior are not merely casual or superficial changes in the birds' preoccupations. They represent striking changes in the overall atmosphere of the events going on in the breeding cage. At its appropriate stage each of the behaviors I have mentioned represents the *predominant* activity of the animals at the time. Further, these changes in the behavior are not solely responses to external stimuli. The birds do not build the nest merely because the nesting material is available. In fact nesting material may be in the cage throughout the cycle, but nest-building behavior will still be concentrated at the stage that I have described as the nest-building stage. Similarly, reactions to the eggs and to the young appear to occur only at appropriate stages in the cycle.

These cyclic changes in behavior thus represent, at least in part, changes in the internal condition of the animals, rather than merely changes in their external situation. In fact, the changes in behavior are associated with equally striking and equally pervasive changes in the anatomy and physiological state of the birds. For example, when she is first introduced into the cage, the oviduct of a female dove may weigh about 800 milligrams. Eight or nine days later, when she lays her first egg, the same oviduct may weigh on the order of 4000 milligrams, an increase of some 400 per cent. The crops of both male and female weigh about 900 milligrams when the birds are introduced into the cage. When they start to sit on the eggs ten days later, the crop will weigh about the same. But two weeks later, when the eggs hatch, the crops may weigh about 3000 milligrams, an increase of about 250 per cent. Equally striking changes in the condition of the ovary, the weight of the testes, the length of the gut, the weight of the liver, the histology of the pituitary gland, etc., are correlated with the behavioral cycle (Schooley and Riddle, 1938).

Now, if a male dove or a female dove is placed *alone* in such a cage, no such cycle of behavioral or anatomical changes takes place. A female dove alone in a cage does not lay two eggs every six or seven weeks; she lays no eggs at all. A male dove alone in a cage never shows any interest in nesting material, in eggs, or in young.

The cycle of psychobiological changes which I have described is therefore one which occurs more or less synchronously in each member of a pair of doves living together, but will not occur independently in either of the pair. The subject of the present chapter is our analysis of the origin of this cycle, and of the mechanisms by which it is regulated.

EXTERNAL STIMULI AND THE INDUCTION OF INCUBATION BEHAVIOR

Let us first consider the problem of what makes these birds capable of sitting on eggs. In a preliminary experiment (Lehrman, 1958a), we introduced pairs of ring doves into breeding cages each containing a nest bowl in which was a nest and two eggs. Each member of the pair had been in isolation for from three to five weeks since its last breeding experience. Under these conditions birds do not sit on the eggs until after 5 to 7 days, during which they first court, then exhibit nest-building behavior (Figure 2, Group 1).

We considered the possibility that the latency (5 to 7 days) which preceded the onset of incubation behavior might represent the amount of time required for the birds to become habituated to a strange cage, and to recover from the disturbance of being handled. Pairs of birds were therefore placed in cages containing no nest bowl or nesting material and with an opaque partition separating the male and the female. After the birds had spent seven days in such a cage, the opaque partition was removed, and a nest containing eggs was simultaneously added. Now, if the 5-to-7 day latency period in the original group was due to the trauma of being handled and to the fact that the birds

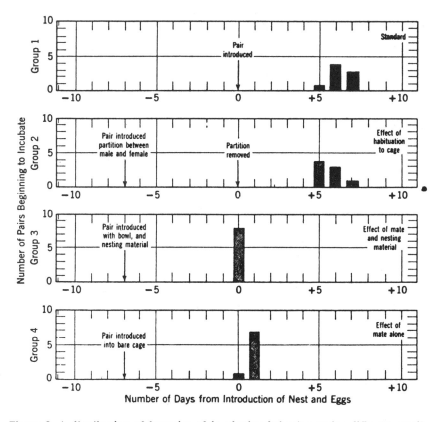

Figure 2. A distribution of latencies of incubation behavior under different conditions of association with mate and availability of nesting material. Nest and eggs introduced at 9 A.M. of Day 0. (From Lehrman, 1958a.)

were in a strange cage, the birds in this group (Figure 2, Group 2) should sit immediately when the partition is removed. However, they did not sit until from 5 to 7 days after removal of the partition.

Another possibility is that stimuli coming from the eggs induce the birds to become interested in sitting on eggs, but that it takes from 5 to 7 days for this effect to reach a threshold value at which the behavior actually changes to incubation. In order to test this possibility, we placed pairs of birds in cages each containing an empty bowl and a supply of nesting material, but no eggs. After seven days the nest bowl was removed with its partially or fully built nest, and replaced with a new nest bowl containing a nest with two eggs. Under this treatment all birds sat on the eggs almost immediately (Figure 2, Group 3). It appears that the stimuli which cause our subjects to change from birds that are not interested in sitting on eggs to birds that are interested in sitting on eggs come, not from the eggs, but from some combination of the presence of the mate and of the nesting situation (nest bowl and nesting material).

In an attempt to separate the effects of the presence of the mate from those of the presence of the nesting situation, we introduced pairs of birds into cages containing no nest bowl or nesting material, and then, after seven days, introduced a nest containing two eggs. These birds did not sit immediately; but neither did they wait 5 to 7 days. Instead the birds sat after approximately one day, during which they engaged in intensive nest-building behavior (Figure 2, Group 4). A fifth group of males and females placed alone in cages containing nest and eggs exhibited no incubation behavior at all.

It appears from these experiments that at least two changes in condition are induced by the treatments we have described. First, stimuli arising from association with a mate cause them to change from birds interested primarily in courtship to birds interested primarily in nest building. Secondly, when the birds are interested in nest building, stimuli arising from the presence of the nest bowl and nesting material appear to facilitate the transition from interest in building a nest to interest in sitting on eggs.

A further demonstration of the effect of external stimuli derived from the mate and/or the nesting material on the development of incubation behavior is found in a second group of experiments (Lehrman, Brody, and Wortis, 1961). We placed pairs of birds in breeding cages for varying numbers of days, in some cases with, and in other cases without, nest bowl and nesting material. Then we introduced a nest and eggs into the cage, and allowed the birds three hours to sit. If neither bird sat within three hours, the test was scored as negative, and both birds were sacrificed for autopsy. If either bird sat within three hours, that bird was removed, and the other bird was given an additional three hours to sit. This experiment therefore provided evidence showing, independently for the male and for the female, the development of readiness to incubate as a function of the number of days with the mate, with or without the opportunity to build a nest. Figures 3 and 4 show the results of this experiment for the females and for the males respectively.

It is apparent from these graphs that association with the mate gradually brings the birds into a condition of readiness to incubate, and that this effect is greatly enhanced by the presence of nesting material. The effect of exposure to the nesting situation is not to stimulate the

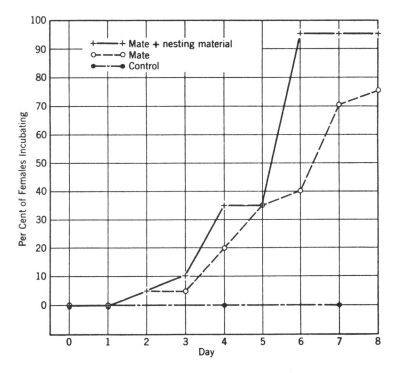

Figure 3. Effect of association with a mate or with a mate plus nesting material on the development of incubation behavior in female ring doves. Each point is derived from tests of 20 birds. No individual bird is represented in more than one point. The abscissa represents the duration of the subject's association with a mate or with a mate plus nesting material, or (for the control group), the time spent alone in the test cage, just before being tested for response to eggs. "Day 0" means that the bird was tested immediately on being placed in the cage. (From Lehrman, Brody, and Wortis, 1961.)

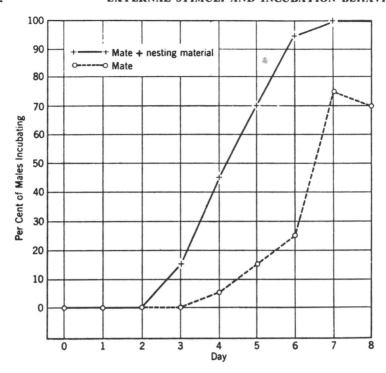

Figure 4. Effect of association with a mate or with a mate plus nesting material on the development of incubation behavior in male ring doves. See the caption for Figure 3.

onset of readiness to incubate in an all-or-none way. Rather, it has an effect upon the development of readiness to incubate which is synergistic with the effect of stimulation provided by the mate (Hinde and Warren, 1959).

It will be noted that by the third day the presence of nesting material makes a difference in the number of males incubating, but a comparable effect is not seen in females until the sixth day. In this connection it is helpful to recall that when the nest is built the male gathers nesting material and carries it to the female. It may be that the condition of the female is affected by the presence or absence of nesting material, starting about the sixth day, because the behavior of the male toward her has been different, depending upon the presence or absence of nesting material, starting at about the third day.

It is clear from this group of experiments that external stimuli normally associated with the breeding situation play an important role in inducing a state of readiness to incubate. We may now inquire into the nature of this state.

HORMONAL INDUCTION OF INCUBATION

We next attempted to induce incubation behavior by injecting hormones into the birds, instead of subjecting them to particular types of external stimulation. For this experiment (Lehrman, 1958b), we treated birds just as we had in the experiments illustrated in Figure 2, except that some of the birds were injected with hormones while they were in the isolation cages, starting one week before they were due to be placed in pairs in the test cages. The results are shown in Figure 5.

Group 1 of Figure 5 is a replication of Group 1 of Figure 2. Here again, birds placed in pairs in a cage containing a nest and eggs sat on the eggs only after an interval of 4 to 7 days, during which they first courted, and then built a nest. When doves were injected with progesterone while in the isolation cages, they sat upon the eggs almost immediately on their introduction into the test cage (Figure 5, Group 2). When the injected substance was an estrogen (diethylstilbestrol), the effect on most birds was to make them sit on the eggs after a latency period of 1 to 3 days during which they engaged in intensive nest-building behavior (Figure 5, Group 3). Testosterone had no effect upon incubation behavior.

Since prolactin has been reported to induce incubation behavior in the domestic chicken (Riddle, Bates, and Lahr, 1935), we have also attempted to induce ring doves to sit on eggs by injecting this hormone (Lehrman and Brody, 1961). We find that prolactin is not nearly as effective as progesterone in inducing incubation behavior, even at dosage levels which induce full development of the crop (the growth of which is under the control of prolactin). For example, when the total dosage of prolactin was 400 I.U., only 40 per cent of the birds sat on eggs, even though their average crop weight was about 3000 milligrams (compared with about 900 milligrams for uninjected control birds). Injection of 10 I.U. of this hormone caused significant increases in crop weight (to 1200 milligrams), but elicited no increase in frequency of incubation behavior, as compared with the performance of control birds. Since, in a normal breeding cycle the crop begins to increase in weight only *after* incubation begins, it seems most unlikely that prolactin plays a role in the initiation of normal incubation behavior in this species.

It appears, therefore, that estrogen may be associated with nest-building behavior, and that progesterone is probably the hormonal initiator of incubation behavior.

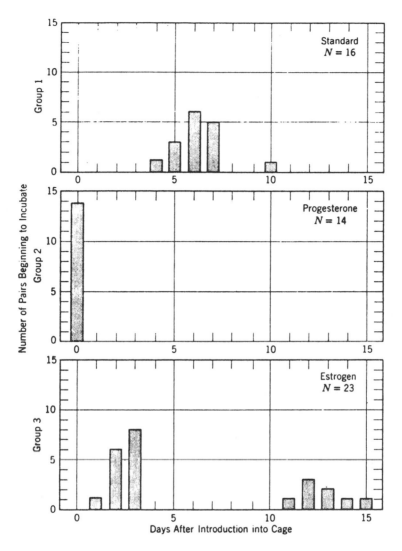

Figure 5. A distribution of latencies of incubation behavior in groups of ring doves subjected to different hormone pretreatments. Birds introduced into experimental cages between 9 A.M. and 9:30 A.M. on Day 0. N = number of pairs. (From Lehrman, 1958*b*.)

EXTERNAL STIMULI AND GONADOTROPIN SECRETION

The foregoing experiments strongly suggest the conclusions that stimuli coming from the mate and from the nesting situation induce readiness to incubate, and that this readiness depends on gonadal hormones. It seems logical to assume that these external stimuli must influence readiness to incubate via stimulation of the secretion of gonad-stimulating hormones by the birds' pituitary glands. Figure 6 shows the percentage of female doves that ovulated after varying periods of association with a male, with and without nesting material. It can be seen that the effect on ovulation is almost exactly coincident with the effect of these stimuli on the onset of incubation behavior in the female, as shown in Figure 3. In individual females there is a very high degree of association between the occurrence or nonoccurrence of ovulation and that of incubation behavior (chi square = 200.4, $p <$ .001).

We carried out these observations because of our interest in the

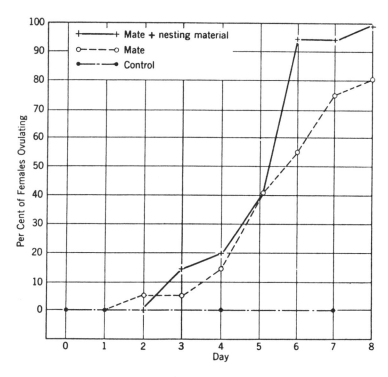

Figure 6. Effect of association with a mate or with a mate plus nesting material on the occurrence of ovulation in female ring doves. See the caption for Figure 3.

coincidence of ovulation and of the onset of readiness to incubate. However, it is not surprising that external stimuli provided by the male can induce ovarian activity in female doves. Harper (1904) observed that ovarian development in the domestic pigeon followed introduction of a male into her cage; and Craig (1911) showed, for the ring dove, that ovulation occurred after association with the male without the necessity of actual copulation. Matthews (1939) showed that a female domestic pigeon can be induced to lay an egg in reaction to visual and auditory stimuli provided by a male from whom she is physically separated.

Mr. Carl Erickson and I have verified this finding in the ring dove, and are presently investigating some of the conditions and properties of this effect. Female ring doves with previous breeding experience are placed in a cage containing similarly experienced male doves, from which they are separated by a glass divider. During the seven days following their introduction into the cages, the oviduct weights of the females rise from a mean of approximately 900 milligrams to approximately 3100 milligrams. If, on the other hand, the "stimulus" males have been castrated two months before the beginning of the experiment, then the oviduct weights rise to only 1800 milligrams. This represents a striking and significant depression of the stimulating effects of the males. Apparently the stimulating effect when the male is present involves not merely the presence of a second bird on the other side of the glass plate, but rests, at least in part, on some (presumably) behavioral characteristics of the male; characteristics which are dependent on the presence of male hormones (Carpenter, 1933). This investigation is continuing.

It is clear from these data that stimuli from the mate (and from the nesting situation) actually stimulate the secretion of gonadotrophic hormones by the female's pituitary gland. It should not be presumed, however, that secretion of each of the pituitary products occurs quite independently as a response to a different external stimulus. The fact is that certain functional relationships within the endocrine system are equally important in some cases. A good example is seen in the process of ovulation. This process is partly set in motion by stimuli from the mate, and further facilitated by stimuli arising from the nesting situation. However, once it has begun the process may continue to completion even if the initiating stimuli are withdrawn soon after ovarian activity has been set in motion, so to speak. Figure 7 shows the result of an experiment (Lehrman, Wortis, and Brody, 1961), in which female ring doves were placed in cages with males, with or without nest bowl and nesting material, and left there for seven days, although

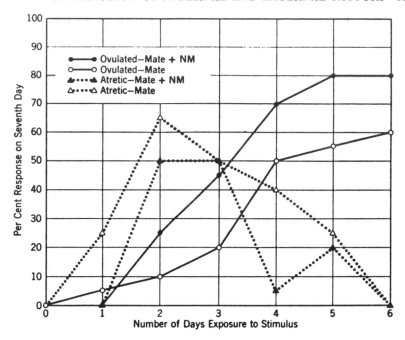

Figure 7. Number of birds ovulating (solid lines) and number of birds developing atretic follicles (dotted lines) by the seventh day, as a function of duration of exposure to a male (light lines) or to a male and nesting material (heavy lines) during the first six days. Atretic follicles are follicles which have begun to develop as a result of stimulation by the male, then degenerated after stimulation was withdrawn. (From Lehrman, Wortis, and Brody, 1961.)

the male (and the nesting material) were removed from the cage at various times after their introduction.

The longer the male was left with the female from the beginning of the 7-day test period, the more likely the female was to go on to ovulate by the end of the 7-day period, after the male had been removed. However, once ovarian development was well under way, the continued presence of the male was not necessary. In this situation the effects of external stimuli clearly are interactive with those of physiological events within the animal; events probably similar to those underlying cyclic and "spontaneous" ovulation.

EXTERNAL STIMULI AND PROLACTIN SECRETION

Since growth of the crop occurs only under the stimulation of prolactin (Riddle and Bates, 1939), increases in weight of the crop may be used as an indicator of the occurrence of prolactin secretion. It will be recalled that the crop increases in weight by about 250 per cent during the 14-day period when the birds are sitting on the eggs. Patel (1936) found that the crops of domestic pigeons would not increase in weight if the birds were removed from the breeding cages at the time the eggs were laid, and thus deprived of the opportunity to sit on eggs. If a male pigeon is removed from the cage while the birds are incubating, the female will continue to sit most of the time. Patel removed male pigeons from the breeding cages just after the eggs were laid, but placed the male in a cage immediately adjacent to the breeding cage, so that he could see his mate sitting on the eggs. Under these conditions, the crops of the male pigeons grew just as if they were sitting on the eggs themselves. Obviously, visual stimuli, and perhaps other external stimuli associated with the incubation situation, induced the birds' pituitary glands to secrete prolactin.

Saeki and Tanabe (1955) found that the pituitary glands of domestic chickens contained large amounts of prolactin as long as the hens were allowed to sit on their eggs. Apparently this phenomenon is fairly general among birds, even including species which do not feed their young by regurgitating a crop secretion.

Working with ring doves in our laboratory, Miss Miriam Friedman has verified Patel's finding that the crop of the male will continue to develop if he is separated from the nest, the eggs, and the incubating female, by a glass plate. The plate is inserted as a cage divider several days after the beginning of incubation. If, on the other hand, the male and female are thus separated from the time of their introduction into the cage, the female will usually build a nest and lay and incubate the (sterile) eggs, but males examined after the female has been incubating for 13 days show *no* crop development. Further studies are now being made of the preconditions for the induction of prolactin secretion by stimulation of this kind.

EFFECTS OF PROLACTIN ON DOVE BEHAVIOR

I have indicated that prolactin is relatively ineffective in inducing incubation behavior in doves, but that progesterone appears to be capable of producing this effect. However, since prolactin appears to be secreted *during* the incubation period, Mr. Philip Brody and I have investigated the question of whether prolactin is capable of *maintaining* incubation behavior which has already been induced by other means.

We tested male and female doves for incubation responses by introducing pairs into a cage containing an artificial nest with two eggs. If they have been in isolation cages for several weeks since their last breeding, birds thus tested never sit on eggs. On the other hand, if they are placed in pairs, allowed to court, build a nest, and lay eggs, and then tested by being moved in (reassorted) pairs to strange cages, all birds sit on the eggs. Thus, under our present methods of testing, ring doves are invariably ready to incubate on the day that the second egg is laid. Similar tests after seven days of incubation and after 12 days of incubation (of the normal 14-day incubation period) yield similar results; i.e., all birds incubate.

Now, if the birds are removed from the breeding cages on the day that the second egg is laid and placed in individual isolation cages, then recombined into pairs after seven days of such isolation and tested for incubation behavior, only 20 per cent of the birds incubate. After 12 days of such isolation, the number of birds ready to incubate is decreased to five per cent. This means that readiness to incubate disappears when the birds are not exposed to the eggs or the incubation situation.

If prolactin (400 I.U., divided evenly) is injected daily for seven days in birds which have been isolated from the day the second egg is laid, 80 per cent of these birds will be found ready to incubate at the end of the treatment period. If this treatment is spread over 12 days, 72.5 per cent of the birds sit. This is to be compared with the figures of only 40 per cent of the birds which sit after seven days of treatment starting when they have been in isolation for several weeks, and only five per cent positive responses when the same amount of prolactin is divided into 12 daily doses, starting when the birds have been in isolation for several weeks.

These data suggest that, even though prolactin is incapable of initiating incubation behavior in birds treated when they are in "sexual" condition, this hormone is capable of *maintaining* incubation behavior in birds which are already (through the influence of progesterone) at the beginning of their incubation period.

One result of the secretion of prolactin during the incubation period is that the crop is well developed at the time the egg hatches and is therefore able to produce crop milk, which the parents regurgitate to the young. I have therefore investigated the effect of prolactin upon the birds' readiness to feed young (Lehrman, 1955).

Twelve adult ring doves with previous breeding experience, kept in isolation cages since their last breeding, were each injected with 450 I.U. prolactin over a 7-day period. Each of these birds, and each of 12 control animals similarly treated except that they were not given any prolactin, were placed, one bird at a time, in a cage with a 7-day-old squab. Ten of the 12 prolactin-injected birds fed the squabs, while none of the controls did so.

[*Editors' Note:* Material has been omitted at this point.]

REFERENCES

Carpenter, C. R. (1933). Psychobiological studies of social behavior in Aves. I. The effect of complete and incomplete gonadectomy on the primary sexual activity of the male pigeon. *J. comp. Psychol.*, **16** (1), 25–94.

Craig, W. (1909). The expression of emotion in the pigeons. I. The blond ring doves. *J. comp. Neurol.*, **19**, 29–80.

——— (1911). Oviposition induced by the male in pigeons. *J. Morph.*, **22**, 299–305.

Farner, D. (1955). The annual stimulus for migration. In A. Wolfson (Ed.), *Recent Studies in Avian Biology.* Urbana: University of Illinois Press, 198–237.

Gemsell, C. A. and Heijkenskjöld, F. (1957). *Acta Endocrinol.*, **24**, 249–254.

Harper, E. H. (1904). The fertilization and early development of the pigeon's egg. *Amer. J. Anat.*, **130**, 421–432.

Harris, G. W. (1955). *Neural Control of the Pituitary Gland.* London: Edward Arnold and Company.

Hinde, R. A. and Warren, R. P. (1959). The effect of nest building on later reproductive behavior in domestic canaries. *Anim. Behav.*, **7**, 35–41.

Lehrman, D. S. (1955). The physiological basis of parental feeding behavior in the ring dove (*Streptopelia risoria*). *Behaviour*, **7**, 241–286.

——— (1958a). Induction of broodiness by participation in courtship and nest-building in the ring dove (*Streptopelia risoria*). *J. comp. physiol. Psychol.*, **51**, 32–36.

——— (1958b). Effect of female sex hormones on incubation behavior in the ring dove (*Streptopelia risoria*). *J. comp. physiol. Psychol.*, **51**, 142–145.

——— (1959). Hormonal responses to external stimuli in birds. *Ibis*, **101**, 478–496.

——— (1961). Hormonal regulation of parental behavior in birds and infra-human mammals. In W. C. Young (Ed.), *Sex and Internal Secretions.* Baltimore: Williams and Wilkins, 1268–1382.

——— (1962). Interaction of hormonal and experiential influences on development of behavior. In E. L. Bliss (Ed.), *Roots of Behavior.* New York: Harper, 142–156.

Lehrman, D. S. and Brody, P. (1961). Does prolactin induce incubation behaviour in the ring dove? *J. Endocrinol.*, **22**, 269–275.

Lehrman, D. S., Brody, P., and Wortis, R. P. (1961). The presence of the mate and of nesting material as stimuli for the development of incubation behavior and for gonadotropin secretion in the ring dove (*Streptopelia risoria*). *Endocrinology*, **68**, 507–516.

Lehrman, D. S. and Wortis, R. P. (1960). Previous breeding experience and hormone-induced incubation behavior in the ring dove. *Sci.*, **132**, 1667–1668.

Lehrman, D. S., Wortis, R. P., and Brody, P. (1961). Gonadotropin secretion in response to external stimuli of varying duration in the ring dove (*Streptopelia risoria*). *Proc. Soc. exp. Biol. and Med.*, **106**, 298–300.

Lofts, B. and Marshall, A. J. (1959). The post-nuptial occurrence of progestins in the seminiferous tubules of birds. *J. Endocrinol.*, **19**, 16–21.

Marshall, A. J. (1959). Internal and environmental control of breeding. *Ibis*, **101**, 456–478.

Matthews, L. H. (1939). Visual stimulation and ovulation in pigeons. *Proc. Roy. Soc.* London, **126B**, 557–560.

Patel, M. D. (1936). The physiology of the formation of "pigeon's milk." *Physiol. Zoöl.*, **9**, 129–152.

Riddle, O. (1937). Physiological responses to prolactin. *Cold Spr. Harb. Symp. quant. Biol.*, **5**, 218–228.

Riddle, O. and Bates, R. W. (1939). The preparation, assay and actions of lactogenic hormone. In E. Allen, C. H. Danforth, and E. A. Doisy (Eds.), *Sex and Internal Secretions* (2nd Ed.). Baltimore: Williams and Wilkins, 1088–1117.

Riddle, O., Bates, R. W., and Lahr, E. L. (1935). Prolactin induces broodiness in fowl. *Am. J. Physiol.*, **111**, 352–360.

Riddle, O. and Lahr, E. L. (1944). On broodiness of ring doves following implants of certain steroid hormones. *Endocrinology*, **35**, 255–260.

Saeki, Y. and Tanabe, Y. (1954). Changes in prolactin potency of the pituitary of the hen during nesting and rearing in her broody period. *Bull. nat. Inst. agric. Sci.*, Tokyo, Ser. G., **8**, 101–109.

—————— (1955). Changes in prolactin content of fowl pituitary during broody periods and some experiments on the induction of broodiness. *Poult. Sci.*, **34**, 909–919.

Schooley, J. P. and Riddle, O. (1938). The morphological basis of pituitary function in pigeons. *Am. J. Anat.*, **62**, 313–350.

19

Reprinted from *Auk,* 79(4), 568–595 (1962)

AN EXPERIMENTAL STUDY OF THE MECHANISMS
OF NEST BUILDING IN A WEAVERBIRD

Nicholas E. Collias and Elsie C. Collias

Nest building is a classic example of what has been called instinctive behavior. But such an explanation of the phenomenon is not sufficient. How does a bird build its nest? This question has intrigued many persons, and a partial answer, based upon a more or less close description of the process of nest building, has been furnished many times. But the basic question as to the nature of the forces that move and guide a bird at each stage in the act of building its nest has remained largely unanswered. This report is an attempt to answer this question for one species of bird, *Textor cucullatus* (Müller), the Common Village Weaverbird of Africa (also variously known as *Ploceus cucullatus* (Müller), the Hooded Weaver, or the Black-headed Weaver), which builds a highly organized and complex nest.

The results of some of our experiments were presented before the Ecological Society of America and before the American Ornithologists' Union in 1960, and an abstract has been published (Collias and Collias, 1959). We are here reporting the results of these and more extended investigations on the same problem in some detail.

Methods and Materials

The methods used included a survey of the literature, detailed study of the breeding behavior and nest building of the Village Weaverbird in its native habitat in Africa, experimental modification of breeding behavior, study of nests and of nest building under controlled aviary conditions in California, attempts to duplicate the nest building mechanisms of weaverbirds by the authors, and study of abnormal nests and the conditions under which such nests are made.

Early studies of the general breeding biology of the Village Weaverbird are those by Benson (1945), von Grzimak (1952), and Chapin (1954). A more detailed study, making use of color-banded individuals and with emphasis on nest building, is one we made in the eastern Kivu District of the Congo on *Textor cucullatus graueri* (Collias and Collias, 1957a, 1957b, 1959). Crook (1960) has also recently described some of the details of nest building in the race *cucullatus* of Senegal, where he finds an essentially similar pattern of behavior to that we have found in the race *graueri*.

The birds that we used for our aviary studies in California belonged to the race *cucullatus* and came from Senegal. Each bird was marked with a distinctive combination of two colored leg bands, the same on each leg, so it could be immediately identified so long as one leg was visible. The birds were named from their colored leg bands; thus, male BW had a black over a white band on each leg. The birds were fed on the parakeet seed mixture readily available from pet shops. Fresh lettuce, cuttle bone, grit, and meal worms were regularly provided for the birds. Crickets were also furnished to birds rearing broods.

The outdoor aviary in which the birds were maintained during the breeding season (May–October) was 5.2 meters high, 5.2 meters wide, and 9.2 meters long, with a wooden shelter at the northwest side. A palm tree (*Phoenix canariensis*) approximately three meters high was planted at one end of the aviary, and an African acacia tree of unknown species, about three meters high, was planted near the other end. The latter tree eventually proved to be the more popular with the birds. Palm fronds and the giant Mexican reed grass (*Arundo donax*) were furnished to the birds as materials for weaving of the nest by the males, while blue grass (*Poa pratensis*) and chicken feathers were provided for the lining of the nest by the females.

The normalcy of the situation was attested to by the birds' fledging a number of broods and raising the young to independence successfully during the second and third years of the study.

In this species the male makes the outer shell of the nest. During the three years of the study, the 11 to 12 males in the aviary built over 250 nests, the great majority of which were normal nests, essentially no different in any detail from nests of the various races of the species built in Africa. During the coldest part of the year the birds were kept in small cages indoors.

When we were in the Congo, Dr. James P. Chapin suggested to us that we might try ourselves to make an artificial weaverbird's nest. Our field studies then left little time for the attempt, but we recently tried it. By carefully observing the basic techniques used by the males in making the outer shell of the nest and imitating the essential features of these techniques, we produced a reasonable facsimile of a normal nest, and this attainment provided a good check to our understanding of the basic mechanisms used by a weaverbird in making his nest. Figure 1 illustrates a nest built by a male weaverbird in our aviary, and Figure 2 illustrates the "weaverbird nest" built by the authors. As far as we know, no person has ever before successfully constructed a bird's nest of any real degree of complexity.

Another technique that we often used in attempting to gain some insight into the mechanisms whereby a weaverbird makes its nest was the study of the conditions under which abnormal nests were produced. Such nests were most common at the start and at the ending of the breeding season and in individuals presumed to have low motivation to build. For example, male BW, who built the bottomless or "canopy" type

Figure 1.   **Nest built by a male weaverbird in an outdoor aviary at Los Angeles.**

nest illustrated in Figure 3, was one of the more suppressed males in the aviary and built fewer nests by far than did most of the other males during the three years of the study. We found that the study of abnormal nests was an invaluable technique in giving new insights into the basic stimulus situations to which the birds were responding at the various stages of nest building. Such nests provided many natural experiments that helped guide the more systematic and deliberate planning of later experiments.

Among the experiments carried out, the details and results of which are described below, were experiments on selection of nest materials of different colors, modification of substrates for analysis of factors determining site of nest attachment, defect experiments relating to the importance of sequences in building and repair, covering of artificial holes in the nest with wire frames of differing mesh size in order to determine the significance of the pre-existing framework to weaving, hooding of parts of incomplete nests as an aid to the elucidation of factors terminating certain

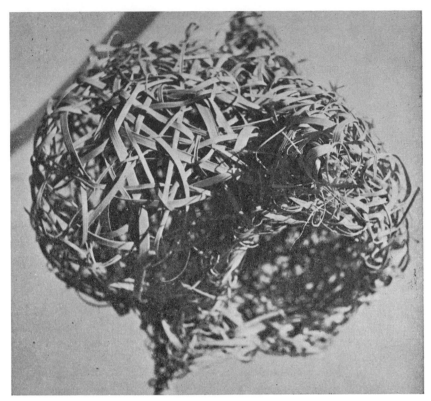

Figure 2. A "weaverbird nest" built by the authors of this article.

stages, and experimental shifts in orientation of nests in an attempt to ascertain factors governing the normal polarity and orientation of the nest.

DESCRIPTION OF NEST BUILDING AND THE
GENERAL MECHANISMS OF WEAVING

The following description and illustrations of the nest and nest building apply equally well to nests we observed built in nature (Collias and Collias, 1957b, 1959), or in our aviaries. All drawings were made by N. E. Collias.

The outer shell of the nest is woven by the male of long strips torn by him from the leaves of giant grasses or palms. The general external appearance of the nest, which is ovoidal or kidney-shaped in form with a bottom entrance, is illustrated in Figure 1. Just within the roof of the external shell the male thatches a special ceiling of short, broad strips of grass leaves (Figure 4). The ceiling is not woven, and in some parts of Africa may be thatched of dicot leaves in addition to the use of strips of grass leaf. However, the birds in our aviary generally restricted themselves to

Figure 3.   An abnormal, canopylike nest built by a male weaverbird while perching on the bare twig below the canopy.

Figure 4.   A normal nest from below, having the bottom half of the nest removed in order to reveal the thatched ceiling of short, broad strips of grass lining the inside of the roof. Antechamber to the right.

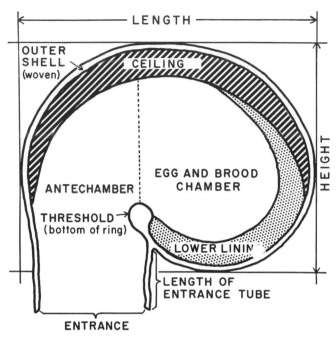

Figure 5.   Longitudinal section of a weaverbird nest to show the inner construction. The lining of the lower half of the nest is put in by the female.

grass leaf strips for their ceiling, rarely adding an acacia leaf or a *Poa* grass head. Figure 5 represents a longitudinal section through a nest and shows the normal extent of the ceiling, as well as that of the lining of the floor of the nest that is put in by the female. This bottom lining is thatched of short strips of palm or grass leaf, grass heads, and feathers. A nest of the Village Weaver is usually about 14–17 cm long by 11–13 cm high. About the entrance in brood nests the male often weaves a short tube, 4–8 cm long.

There are five stages to the building of the outer shell of the nest by the male (Figures 6 and 7): (1) initial attachment, (2) roof and egg or brood chamber, (3) antechamber, (4) entrance, and (5) entrance tube. The ceiling is often started long before the egg chamber or the antechamber has been completed. Figure 7 illustrates the principal stitches used by a Village Weaver in the attachment and outer shell of the nest. In actually weaving his nest the male uses certain basic general mechanisms. He tends: (1) to seize a strip of nest material near one end, or mandibulates it along his beak until he has shifted his hold to one end; (2) to double back a strip on itself, especially if the strip is long; (3) to poke one end of the strip with a vibratory motion into the nest mass or alongside some object such

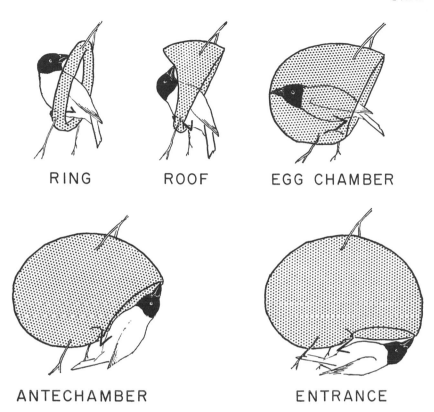

RING   ROOF   EGG CHAMBER

ANTECHAMBER   ENTRANCE

Figure 6.   Normal sequences in nest building by a male Village Weaverbird
include: (1) ring, (2) roof, (3) egg or brood chamber, (4) antechamber,
and (5) entrance.

as a twig; (4) if the strip sticks, to release it, to move his head around
to the other side of the twig or nest mass and again seize the strip; (5) to
bend or wind a strip about objects such as a twig, ring, or about another
piece of nest material; (6) to reverse and alternate the direction in which
he winds a strip; and (7) to poke and pull a strip through holes, normally
the interstices of the nest.

The unit movements used in nest building are quite stereotyped. All of
the weaving is done with the bill, although the bird often uses one or
both feet to help hold a strip. When a piece of nest material fails to stick
on the initial push, the male often shifts to some other spot and tries again,
and this exploratory pattern confers flexibility and adaptability to weaving.
If the movements of No. 4 (see above) are repeated more than once in the
same direction, the strip is coiled about the twig, as seen in Figure 7 (upper
right). Alternately reversed winding (Figure 7, upper), when done repeat-

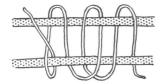

ALTERNATELY
REVERSED WINDING
BETWEEN TWIGS.

COILING ABOUT SINGLE
TWIG, ALTERNATELY REVERSED
WINDING, AND SPLIT STRIP.

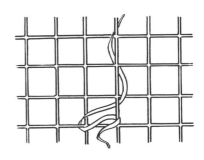

DETAILS OF WEAVING
FROM EGG CHAMBER.

STRIP WOVEN INTO
A WIRE FRAME.

Figure 7. **Details of weaving of the initial strips in a nest (above), and nature of stitches in later stages of weaving (below).**

edly and closely between two twigs, two grasses, or a twig and a grass strip, binds them firmly together and gives a strong suspension for the nest, as well as being important in helping to hold the rest of the nest together. In the egg or brood chamber many of the constituent strips are continued and threaded over and under other grass strips, at times approaching the classic conception of human weaving, as in some simple forms of basketry or cloth (Figure 7). But at times, especially when strips are long, the ends may apparently be lost sight of, and the bird pokes and tucks loops of the strip in as such. Knots of other types than "hitch-knots" are quite rare in the nest of the Village Weaver, but the end of a strip is frequently looped back on itself or on other strips in such a way that pulling on the strip tightens its attachment, *i.e.*, a typical hitch-knot. A hitch is also a type of knot that is characteristically easy to untie or tear loose and conceivably the prevalence of hitchlike stitches in the nest of a weaverbird is related to the habit of the male of tearing down his nest when no female accepts it, and of building a fresh nest in its place (Collias and Collias, 1957a, 1959).

Every nest is unique in the details of its fine pattern. But the repeated use of the same basic mechanisms in weaving by the male results in nests that look extremely similar to each other, whether built by the same or different males, and readily recognizable as belonging to the same species. This is a statistical concept of nest form and structure.

EXPERIMENTAL ANALYSIS OF THE STIMULUS SITUATION AT EACH STAGE OF NEST BUILDING

The different stages of nest building that will be analyzed, omitting the nonwoven lining of the floor put in by the female, refer to the work of the male, and include: (1) gathering of nest materials; (2) the initial attachment; (3) ring; (4) egg or brood chamber; (5) antechamber and entrance; (6) ceiling; and (7) entrance tube. To a large extent these different stages represent a chain of specific stimulus situations and response tendencies. Each stage automatically provides the stimuli for its own termination and for the start of the next stage in construction. This conclusion was reached only after asking and answering, at least tentatively, the question: "What starts and what ends each stage of nest building?" The evidence will next be considered.

Gathering of nest material. For weaving the outer shell the male requires some 300 long strips of grass or palm leaf, and tears and transports each strip individually. For weaving it is important that flexible materials be used. In nature, the most common flexible materials are, of course, the leaves of green plants. That the color green serves as a clue or signal to which the birds are responding can be shown by giving them materials of different colors. Thus, the birds selected green in an overwhelming number of choices when allowed to select between an equal number (10) of toothpicks of each color (Table 1). In fact, green was selected far more frequently than were the other five colors combined. The preference for green was so strong that statistical tests of significance here seem superfluous.

The toothpicks were colored by Tintex Fabric dyes, and the green selected (known as "jungle-green") was one that matched the green of land plants; in other words, it contained some yellow. The stiffness of the toothpicks also indicates the importance of color as a signal, and rules out small differences in flexibility that might have provided cues for the birds had flexible materials of different colors been used.

In this experiment two or three males were kept together in small cages (40 cm wide by 40 cm high by 90 cm long). When colored toothpicks were placed into the cage, an attempt was made to intermix them uniformly,

TABLE 1

TABLE 1

WEAVERBIRDS SELECTED GREEN NEST MATERIALS WHEN PRESENTED
WITH VARIOUS COLORS SIMULTANEOUSLY

(Half-hour test periods of five males for a total of 7½ hours of observation)

	Number of times chosen
Green	686
Black	146
Yellow	105
Red	80
White	70
Blue	33

and toothpicks damaged during the test were replaced by fresh toothpicks for the next test.

An interesting point that developed in the course of the experiment was that young adults showed a gradually increasing preference for green, as if mere selection of "green" had more of a self-reinforcing effect than did selection of any other color. During the tests 1,120 choices were made by the five males, and when these tests are divided into four successive and equal periods, selection of green over all other colors for each period was, respectively: 46 per cent of 178 choices, 60 per cent of 289, 63 per cent of 312, and 69 per cent of 341 choices. Table 1 also shows that yellow was selected three times as often as was blue, and this preference accords with the fact that the greens of natural terrestrial vegetation are generally yellow-greens and not blue-greens.

The possibility that birds were reacting to differences in intensity rather than in color was checked by testing preference for a given color from among various shades of gray (Table 2). Green toothpicks were easily selected when mixed with toothpicks of eight different groups of gray of various intensities. Gray toothpicks were grouped in four sets of 10 toothpicks each because the two intensities within each set were too similar for the human observer to determine accurately in the course of testing. It is assumed therefore that some of these grays must have been very close in intensity to the green color tested. Certainly, the weaverbirds virtually

TABLE 2

WEAVERBIRDS SELECTED GREEN NEST MATERIALS WHEN ALSO PRESENTED
WITH VARIOUS GRAYS SIMULTANEOUSLY

(Half-hour test periods of five males for a total of 2½ hours of observation)

	Number of times chosen
Green	299
White or whitish	45
Pale grays	18
Dark grays	12
Black or blackish	41

ignored the achromatic toothpicks in contrast to their strong predilection for green. It is of interest that, although the toothpicks were unsuited for weaving activities, the birds consistently treated them as nest materials, attempting to poke them into place along the perch or into the wire mesh-work of their cage.

We concluded that green or yellow-green is definitely the preferred color by Village Weaverbirds interested in nest materials, and that this color functions as an index of flexible materials suitable for weaving.

The identification of appropriate nest materials in nature undoubtedly depends on the bird's previous experience with such materials, the example and attractive influence of other birds, and the traditional location of colony trees close to good patches of elephant grass (*Pennisetum pur-pureum*), palm trees, or other good sources of nest material. As we have found out in observations to be published in detail later, a good deal of practice is probably involved before a bird can skillfully bite and tear off long strips of nest material. Once a bird has a piece of nest material in its beak, it seems to be subject to a strong tendency to fly back to its territory and nest site or prospective nest site. Although a weaverbird gathering materials may sometimes tear and drop a few strips before making its final selection, there is probably a self-satiating factor also at work tending to make it return to its nest site. In our tests with colored toothpicks by far most of the active interest of the birds in these sub-stitute "nest materials" took place within the first 15 minutes of the half-hour observation periods.

Initial attachment of the nest. In the initial stages of building the male seems subject to some indecision as to where to attach his nest with-in the bounds of his small (on the order of one-half to one cubic meter) territory in the colony tree. This uncertainty is reflected in the frequency of two or more false starts, in which one or two strips will be attached to a twig and then abandoned, and is particularly evident where a partic-ular substrate pattern is closely replicated several times. One weaver, for example, made five false starts, each along one of five large identically ap-pearing spines along the rachis of a palm frond. Twigs, 0.3–0.6 cm in diameter, seem to be preferred. A series of perches ranging from 0.6 to 1.8 cm in diameter were placed in the aviary, and only rarely did any bird attempt to attach a strip to the thickest perch. Forked twigs are preferred to single twigs for attachment of the nest.

With the attachment of the very first strip of a prospective nest, we at once confront most of the problems involved in analyzing the basic act of weaving itself. The fundamental mechanisms in the stitching or weaving of individual strips have been described above and are illustrated in Figure 7. It remains to give some of the experimental evidence related to these mechanisms. When a male arrives with an initial strip at the twig or fork where he intends to make a nest, he may hold the strip with one or both feet against the twig, seize one end and double back the strip. If a male in a cage with no normal nest materials is given a toothpick, he will often seize it and attempt repeatedly to double back one end until he succeeds in bending back and breaking the tip. In our color preference tests with toothpicks, the usual damage done by the birds was to bite the tips of the toothpicks in trying to bend them back. The tendency of the weavers to poke one end of a strip with a vibratory motion alongside a twig or other strip was observed with especial clarity in the toothpick tests. Repeatedly, the birds would poke and vibrate the tip of a toothpick against the smooth, round, wooden perch in their cage, although there was no chance of fastening this very artificial bit of "nest material" in place. Consequently, the next step in the chain reaction of weaving did not occur, *i.e.*, releasing the material with the bill and moving the head around the perch to the opposite side to seize the material anew.

Analogy with human knots and stitches can be considered part of the experimental approach to the problem of weaving by weaverbirds. In the attachment of initial strips in starting a new nest, the technique of alternately reversed winding, combined with loop-backs (Figure 7), as described above, is frequently used by a male weaver, and it is of interest that "nippering" (a knot used to lash together two parallel ropes—cf. Hasluck, 1942) is a very similar type of fastening.

In fastening the initial strip, not only may the male double back a strip and alternately reverse his winding of one end of the strip between twigs or between a twig and the rest of the strip (Figure 7), but he may also pass the leading end of the strip back through the self-made loops of the strip produced by the act of looping and winding itself. This tendency to poke the strip through any available hole can be demonstrated experimentally by providing the birds in an aviary with smooth, wooden perches through which many small holes have been drilled. Some of our birds then poked and wove strips through these holes. This observation suggests that the bird may respond to any small hole when weaving and not specifically to the normal interstices of the developing nest or to a criss-cross framework as in the weaving that commonly occurs in the sides of the wire cages.

After attaching the initial strip, others are soon added at the same spot, and a small pad of woven material is quickly formed. The reason that strips subsequent to the first one are not diffused over a wide area is probably due to two different response tendencies of the bird: the stronger attraction of the loops and interstices and color of the initial strip as compared with its substrate, and the tendency of the bird to become quickly fixated on one particular site from which to weave. The importance of a fixed point from which to weave was emphasized by the case of a male kept with five other males in a small indoor aviary. This male was strongly subordinate to the others, and he was not allowed to remain long on any one perch, but spread his weaving over an area of some 900 cm² on the wire side of the aviary near one corner. A male prefers to attach his strips in forks or at the site of minor irregularities. We have seen a nest start on a perfectly smooth, round, wooden perch prolonged for 30 cm along the length of the perch as if, in the absence of any irregularities on the perch, the male could not decide where to stop.

Construction of the ring. Building by the male continues, leading to growth of the initial nest mass into the form of a ring that in turn will provide the basic supporting framework for the whole nest. The main force in formation of a ring rather than some other shape is the tendency of the male to keep his feet in much the same place as he weaves in all directions. He also tends to follow the substrate pattern and so weaves first along the twig; or, if the nest is in a palm tree, he tends to follow the support afforded by palm leaflets and spines. Occasionally, slavish adherence to the substrate pattern leads to such abnormalities as incomplete rings, lacking perhaps one entire side (Figure 8).

Saddled rings and nests seem to be built more commonly by young than by adult males, while the latter usually make pensile rings and nests slung within and below a forked twig. Presumably the pensile nest is safer from predation.

Saddled rings are frequently placed in the upright fork of twigs, being built up along the twigs on either side; closing of the upper part of the ring results from the strong tendency of the male weaver to weave over his head. Whether a twig forks up or down, the male weaverbird customarily straddles between the two sides of the fork with one foot grasping a twig on either side. Even in the complete absence of nest materials, Village Weavers will straddle between objects such as the fork of a twig or the side and roof of a small cage with legs and feet spread quite widely.

In the case of pensile rings, closing of the ring beneath the male involves a few special problems. Because of the general tendency to weave over his head and along the substrate, the top and sides of the ring are often completed first. The fixation of the male at a particular site for straddling where the spread required is not too great keeps the male from moving down with his work. A series of dangling ends of strips gradually accumulates on either side. But the tendency of the male to seize the ends of loose strips leads him to reach down and pick up these danglers one by one and then either to weave them up on the same side or to cross them over and attach them to the other side, bridging the gap beneath himself and gradually completing the ring. It is important that the strips be of

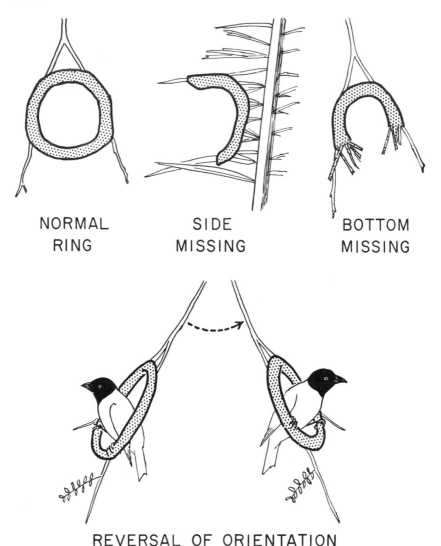

NORMAL   SIDE   BOTTOM
RING   MISSING   MISSING

REVERSAL OF ORIENTATION

Figure 8. Analysis of some factors in the construction of a ring (above), and in the determination of nest polarity (below). Details in text.

sufficient length. When we supplied only strips 12–15 cm long to our colony, only one male succeeded in bridging the bottom of his ring, and the rest of the rings attempted by the birds remained without a bottom (Figure 8), although the top and sides gradually thickened. The one male that succeeded in closing his ring was able to do so because the twigs on either side converged at the point where the bottom of the ring had to be closed.

In thickening the ring (Figure 9) the male enters the ring with one end of a strip in his bill and the rest of the strip crossing his breast and trailing behind. He pokes the strip into the opposite side from the dangling and trailing end, often first inserting one end through one side of the bottom half of the ring. He then reaches around to the other side of the ring and seizes this same end, bending it back and weaving it in. He next shifts to the other end of the strip and weaves in this end. He tends to shift the place in which he weaves successive strips, so the ring becomes thickened rather uniformly.

One of the most striking and significant features of the orientation of the male in his ring is that he almost always enters it from one side and almost always faces the same way, keeping one foot on each side of the bottom of the ring. We made some investigation of the factors determining the orientation of the male. On observing the plane of the ring, we discovered that vertical rings are rare. Generally, the ring tilts backward from a few degrees to a nearly horizontal level, with 10 to 40° being the most frequent tilt. When in his ring the male faces in such a way that invariably the ring tilts toward him. But if a twig containing a ring is swung back so that the tilt is reversed, the male upon his first subsequent visit will immediately reverse the direction in which he faces. Thus the male keeps the same general relationship to the plane of tilt of the ring, *i.e.*, the ring still tilts back toward him (Figure 8). This experiment was done in five cases with similar results.

The significance of this orientation is that it determines the longitudinal polarity of the nest. The male invariably builds the egg or brood chamber out in front of himself from the ring, while he leans over backward to build the antechamber and entrance on the opposite side of the ring, all the while keeping his feet on the threshold of the nest (Figure 6).

Egg chamber and roof. There are several problems in attempting to determine the factors involved in construction of the egg chamber and antechamber: the start, size, shape, details of woven pattern, and ending of these parts of the nest.

What stimuli cause the male to stop building the ring and begin the roof and brood chamber? One possibility is that once the male has done a certain amount of work on the ring, purely internal changes are set up leading him to begin the next stage of building, *i.e.*, the egg chamber, and continuing through each successive stage of the nest building in a gradual unfolding of internal processes. The alternative hypothesis would be that each successive stage of the nest itself provides the necessary stimuli for its own termination and for the beginning of the next stage. An argument against the first supposition is the fact that destruction of one stage of the nest in most instances leads to prompt repair of that defect (Figure 10),

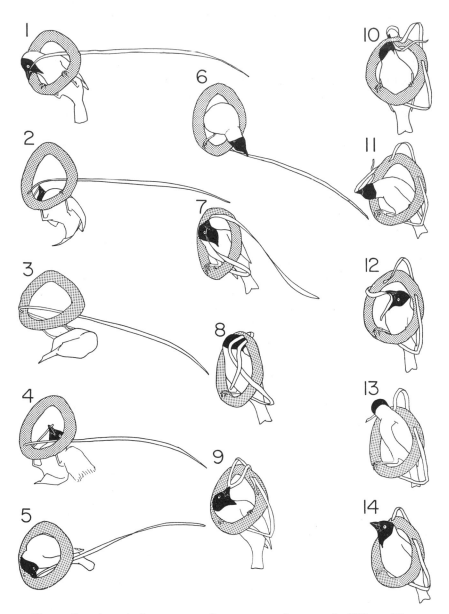

Figure 9. A typical sequence of movements by a male Village Weaver-bird as he weaves a single strip, torn from a leaf blade of elephant grass, into his ring. Selected from motion-picture frames.

regardless of its place in the normal sequence of building. Each type of defect experiment was carried out on at least five fresh nests, with similar results in each case. Thus, if the egg chamber is removed from a fresh nest, the bird does not ignore the defect, as if he had exhausted all of his building potential for that particular stage, nor does he continue merely building at the entrance side. Instead, he promptly builds a new egg chamber on the same nest. The one exception to this procedure is the base of the ring. Whereas all other defects are repaired, a nest with the lower part of the ring cut out (Figure 10) is usually abandoned. When a bird attempts to perch in such a nest, the two sides spread apart to such a degree that the bird finds it difficult to maintain position. The main function of the bottom of the ring seems to be to provide a footrest from which to work. Mr. Richard Burrows and we observed that a nest with the bottom of the ring built around a twig was not abandoned when the strips around the twig were removed. Rarely, we have seen nests built that have only a bare twig as the threshold and footrest.

With each successive strip the ring tends to become thicker and more compact, and it may become more and more difficult for the male to push and pull a strip through the mass of the ring. It would seem that increasing resistance may lead him to weave in less of each end of the strip, resulting in establishment of a system of loose loops extending out from the ring and providing a framework from which to weave out farther. When the male has woven in both ends of a strip and can see only the body of the strip, his pushing-out reaction seems to be activated, *i.e.*, instead of tucking in the loose ends he tends to push the body of the strip away from himself with his beak.

The brood chamber is often started from the roof of the ring, and this tendency is related to the stronger tendency of the male to weave over his head than to one side or beneath himself. A male Village Weaver placed in a small, wire cage often weaves strips of nest material into the roof; less often he weaves into the sides of the cage, and almost never into the wire mesh of the floor.

The male continues to enter from the same side of the ring and to face into the developing egg chamber, perching on the floor of his ring as he works. Each strip he brings is held in his bill, and as he lands, he often pokes that end into the ring to one side. If, as he arrives with a strip, the latter extends backward along his right side, he tends first to weave to the left. But, if the strip extends backward along his left side, he tends first to turn to the right for the initial poke. The far end of a long (30–37 cm) strip is not readily visible to him since it extends across his breast and far behind him. Not being able to see the ends of the strip, he pushes the main body of the strip out in front of him as far as he can,

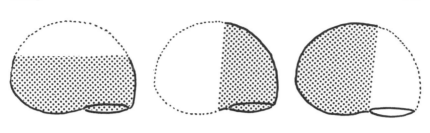

DEFECT REPAIRED

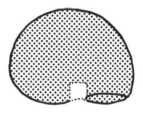

NEST DESTROYED

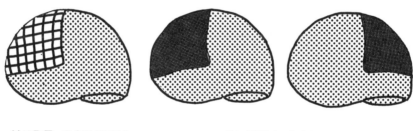

WIRE SCREEN   CLOTH COVER

Figure 10. The male weaver replaces various parts of the nest, with the
exception of the lower half of the ring, when the experimenter cuts away
these parts (above). Nature and location of various artificial covers fastened
over the nest to analyze factors involved in building of the egg chamber (below). Details in text.

the strip crossing the ring from one side to the other. He then leans over
backward and weaves in the second end of the strip. The tendency of the
bird to shift his point of weaving with successive strips, combined with
the pushing-out reaction when he can no longer see the ends of each woven

or tucked-in strip, gradually results in outlining the form of the brood chamber.

The size and globular shape of the egg chamber depend on the size of the bird, combined with his tendency to keep his feet on the bottom of the ring as he pushes out with his beak in all directions as far as he can reach. If a particularly resilient strip springs back to its original position after being pushed out, the male often pushes it out or up repeatedly until it stays in place. We have often seen cases where a male without access to a tree or bush in a cage or aviary becomes fixated on weaving from the perch where the latter adjoins the wire side of the cage or aviary. In such instances, the bird often weaves a round disc of woven materials into the wire frame of the cage, neatly outlining the extent of his reach from the one spot on the perch where he persists in standing as he works. At the same time, his tendency to weave above the level of his belly rather than below it often results in a shallow, bare notch in the bottom part of his disc or mat of nest materials.

The pattern of weaving in the egg chamber depends on the male's tendency to (1) cross preexisting lines of the developing framework at right angles as he weaves in subsequent strips, and (2) to alternate the direction of his winding with each stitch, resulting in an in-and-out threading of the strip as he moves it along by pushing and pulling it with his beak through the loose meshwork of the early stages of the egg chamber.

The tendency to cross strips at right angles is well shown by giving different colored strips of raffia to a male in a cage having parallel vertical wires but not horizontal cross wires on the side. The bird tends to weave each strip in a horizontal plane doubling back and forth and weaving in and out between the vertical wires, the basic pattern being readily revealed by the different colors of each strip. If the bird weaves on a wire meshwork consisting of a pattern of squares or rectangles, there is the same tendency to weave each strip at right angles across preexisting lines. But as more and more strips are added, the pattern soon degenerates into a helter-skelter and kaleidoscopic pattern of variously colored raffia strips interweaving with each other in an apparently irregular and complex pattern.

The tendency of the male weaver to alternate the direction of his winding through the nest interstices is the essence of what we normally think of as "weaving." If the basic framework is very regular, as in a wire meshwork of squares, such threading in and out of strips of nest material may often proceed in perfect alternation (Figure 7). Apparently the weaverbird learns much of this tendency to progressive alternation, the reinforcing stimulus being firmness or resistance to being pulled farther by each successive stitch combined with an early-developed tendency to alternate pushing and pulling or forward and backward movements of the head. Not infrequently, loops are woven in as such in the nest. When palm strips 37–42 cm in length were the only nest materials provided to the birds, loops were more frequent than when only strips of 12–15 cm were given, and the many loose loops projecting from the external surface gave the long-strip nests a rather crude and scraggly appearance.

The stronger tendency of the male to weave above rather than beneath himself is reflected by the much finer meshwork of the roof of the nest compared with the floor of the nest at any given instant (Figure 1).

As the meshwork of the nest becomes filled in, the male tends to do less weaving. This tendency was indicated by an experiment in which a large hole was cut in the roof of the egg chamber and then covered over with a green wire mesh frame (Figure 10). In experiments on different nests, four frames, each with different-sized mesh, were used, and the bird did more weaving in those cases where the meshes were larger. Apparently, the male tended to some degree to accept the meshwork of the wire frame as if it were his own work.

When the bird can no longer see daylight through the egg chamber, he stops building. An initial conjecture that this was the case was based on an abnormal nest in which the egg chamber abutted the rachis of a palm frond that actually furnished the back of the egg chamber. The egg chamber at this point was incomplete, with a large hole in its back.

Ceiling. As the meshes of the nest become smaller and smaller, particularly in the roof, the male shifts quite abruptly from weaving in long, narrow strips to thatching in short, wide strips just under the roof. We believe that this significant shift from weaving to thatching can be explained by the fact that after the meshwork of the roof becomes very fine, the male finds it increasingly difficult to locate or pull through the ends of a strip that he has just poked into place. His tendency to push the main body of the strip away from himself is then automatically stimulated, since he can no longer see the ends or pull them into place. One can actually see this process in operation each time the male puts in a ceiling strip, poking in first one end to one side, then the other end into the roof, followed by pushing the center of the strip well up into the arch of the roof.

A partial experimental test of this theory was made by tying a piece of green fly screen over a hole cut in the roof of the egg chamber. The meshes in the screen were too fine for the weaverbird to thread through any strips readily, and the male put many ceiling strips just inside the screen, but did very little weaving. It seems possible that as a male gathers experience he attains a measure of what we would call "judgment," *i.e.,* he realizes at a glance whether or not he can easily weave another strip through the mesh.

When a male starts to put in a ceiling, a spectacular shift takes place. Instead of gathering long strips adapted to weaving, he gathers short, broad strips needed for thatching. Evidently, the male is able to carry a picture or "set" in his brain of the best type of material for a given stage of the nest. Probably a certain amount of learning is involved in this process, since the ceiling of young males is characteristically sparse, relatively irregular, and composed of strips quite variable in length.

Work on the ceiling is greatly slowed or discontinued by the male when the roof and ceiling become thoroughly opaque. We described above the abnormal nest that was backed against a palm rachis, which replaced the wall of the egg chamber at the place of contact. We made an experimental test by sewing a piece of green cloth over the roof at an early stage of the egg chamber (Figure 10), removing the few ceiling strips present. The male continued to work on the nest at places other than those covered by the cloth, putting in an abundance of ceiling strips under the roof on the entrance side, but almost none beneath the cloth. We also did the reciprocal experiment, hooding the entrance side (Figure 10), and this time the many ceiling strips subsequently put in by the male were almost entirely confined to the egg chamber. The developed ceiling itself provides the stimulus for its own termination, for if one repeatedly removes the ceiling from a reasonably fresh nest, the male repeatedly replaces it.

In conclusion, we believe the ceiling is started at the roof when the roof first reaches the requisite smallness of mesh size, and, secondly, that the work on the ceiling slows or stops as the roof plus ceiling becomes thoroughly opaque.

Antechamber and entrance. In general, little work is done on the antechamber, aside from its roof, before the egg chamber is well-nigh complete. The greater tendency of the male to weave over rather than under himself helps lead to a gradual building down of the roof and sides of the antechamber until it reaches the horizontal at a level approximately with the bottom of the nest (Figure 6). Throughout the process, the male as usual keeps his feet on the bottom of the ring or threshold, leaning over backward as he works. As the antechamber reaches the horizontal, it is gradually narrowed to make the bottom entrance, the narrowing probably being due to the lesser reach of the male as he hangs beneath the threshold. If a nest is rotated backward about its transverse axis, bringing the entrance upward with its margins now vertical instead of horizontal, the male promptly weaves the entrance back down to the horizontal level, and in doing so, builds from top to bottom an extra, abnormal hood on the nest (Figure 11).

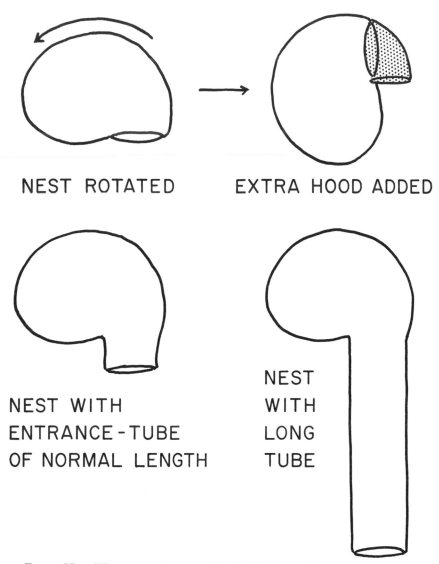

NEST ROTATED   EXTRA HOOD ADDED

NEST WITH
ENTRANCE - TUBE
OF NORMAL LENGTH

NEST
WITH
LONG
TUBE

Figure 11.   When a nest is rotated back 90°, the male weaver again builds the entrance down to the horizontal, adding an extra hood to the antechamber (above). A male weaver was experimentally induced to build a nest with an abnormally long entrance tube (below). See text.

Therefore, work on the antechamber generally can be said to end only when its opening has reached the horizontal. At this stage the male may continue to thicken the nest and ceiling somewhat, but the primary shift in his behavior is toward greatly increased efforts, by means of various displays and vocalizations, to induce visiting females to enter and inspect his nest. If a female accepts the nest and settles down in it, she thenceforth tends to exclude the male from the inside of the nest, although he often looks up into the entrance and sings to her.

The male now generally adds a short tube about the entrance, whereas nests not accepted by a female are likely to be torn down without ever having had an entrance tube (Collias and Collias, 1959). Possibly the stimulation leading to the addition of an entrance tube on brood nests is the added motivation to build afforded the male by having his nest occupied by a female, combined with his exclusion from the interior of the nest that up to that time had furnished the main site for his building activities. If a female disappears or deserts a nest, the male owner soon tears it down.

In weaving the entrance tube, the male for the first time during the construction of his nest begins to work not while his feet rest on the threshold or bottom half of the ring, which provides the supporting framework of the nest, but clings with his feet at almost any place on the rim of the entrance tube as he adds to its length. The entrance tube is not thick— only about one layer of woven material being involved. The abrupt shifting of the male from his normal orientation point of weaving on the bottom part of the ring is perhaps due primarily to the mechanical impossibility of his weaving an entrance tube from that position. The diameter of the entrance tube usually varies little, but occasionally we have noted abnormally narrow or wide tubes, the former being correlated with working positions of the male in which the longitudinal body axis was almost upright, the latter in which it was nearly horizontal.

The male can be caused experimentally to extend the entrance tube to several times the normal length (Figure 11). As he smooths off the rim, tucking loose ends in back from the rim, the male tends normally to terminate work on the entrance tube with attainment of a finished rim. But if the experimenter loosely threads in a strip of nest material by one end into the entrance tube, leaving the main part dangling, the male promptly weaves in the whole strip, instead of merely pulling it out and discarding it. By means of this experimental technique, we have induced a Village Weaver to add a tube 30 cm long to his nest, instead of the normal length of only 5–10 cm.

Finishing of brood nests. While the female incubates, the male quite often works on the outside of the nest. He strengthens the attachment of the nest, weaving in fresh green strips, their bright-green color contrasting with the faded green or brown of the rest of the nest. He also tucks in loose ends here and there over the surface of the nest so that egg or brood nests come to have a smooth, compact, and relatively neat appearance.

Much of the interior work is, of course, done by the female, who adds a thick but nonwoven lining.

Nest destruction. As a rule, the male usually destroys his own nests, including those never accepted by a female and brood nests shortly after the young have departed. In our experiments concerning the repair of various nest mutilations, we found that it was important to use fresh nests. If defects were made in nests that were starting to age, to fade, and to become a little brittle, the male was far more likely to tear these nests down than he was to repair them. But if the nest were fresh, large parts of it could be removed and would be repaired. These results suggest that there is a balance between the tendency to tear a nest down and to build it up. Factors that stimulate building on a nest include not only freshness of materials in the nest, but also position of the male inside the nest in the bottom half of the ring, and a recent visit by inspecting females to the colony tree. Factors that favor destruction of his own nest by a male, on the other hand, include not only the brittleness of aging materials in the nest, but also position of the male outside the nest and failure of a female to accept his nest, or abandonment of the nest by the female, as is normally done when the brood leaves.

DISCUSSION

It is evident in the preceding pages that while we have attempted to analyze the basic mechanisms or causes at each stage of nest building by a Village Weaverbird, we have avoided any detailed discussion of endocrine motivation or of the role of experience. This is because each of these problems is a large and separate topic and will be considered in subsequent articles. We and our colleagues have worked on these problems, and have found that breeding behavior, including nest building by the male of this species, is stimulated by male hormone (Collias, *et al.*, 1961), and that practice plays a not inconsiderable role in nest building by weaverbirds (Collias and Collias, 1962). This article attempts primarily to make a precise and detailed specification and analysis, based on experimental studies, of the significant stimulus situation at each stage of nest building.

Comparable studies in the literature from this point of view are rare and fragmentary. In fact, a detailed, experimental analysis of nest building in any bird seems to have scarcely been attempted before the pioneer studies of Herbert Friedmann on the nest building of the Red-billed Weaverbird (*Quelea quelea*) at the United States Zoological Park in 1922, which appears to have been the first really detailed description of nest building in a bird. But although he carried out some experiments with regard to nest materials, the emphasis of his paper was necessarily on the descriptive aspects of nest building. In India, Ali and Ambedkar (1956) made some interesting observations on the effects of cutting differently shaped holes in completed nests of Baya Weaverbirds, and they found that the birds generally repaired the damage. Crook (1960b) removed different parts of the nest in *Quelea quelea*. But in all but one case this was done subsequent to egg laying, and the nest was soon abandoned and not repaired, except in the case of one male in an aviary who repeatedly replaced the cut-out back of a nest when this was removed.

Two other authors, who have recently analyzed nest building, did not work with weaverbirds. Robert Hinde (1958) has dealt with a much less highly organized nest—the simple cup nest of the canary, which is built in a very different way from the complex woven nest typical of ploceine weaverbirds. Tinbergen (*in* Thorpe, 1956) has suggested a possible causal scheme for the building of the domed and nonwoven nest of the Long-tailed Titmouse in Europe. His scheme is apparently deduced from close observation of the process of nest building in this species, but his causal analysis seems to be entirely theoretical and little or no experimental verification seems to have been published. Our own analysis has proceeded not so much from a theoretical preconception as to the possible mechanisms, but rather by attempting to formulate the problems involved in the construction of each stage of nest building, guided particularly by observation of the building of abnormal nests.

Apart from the developmental aspects, which will be dealt with elsewhere, our conclusions as to some of the more general mechanisms operating in nest construction do not differ greatly from the general conception of instinctive behavior adopted by many authors since the times of Darwin and of Fabre. Stereotyped action patterns are evident in the fixation by the male weaverbird on the lower half of the ring from which to weave, or in such movements as mandibulation and of poking a strip into the nest walls and then moving the head in a deliberate manner to the opposite side before again seizing the end of the strip. Key stimuli guiding the movements of nest building are illustrated by the tendency of the bird to weave or tuck in loose ends of strips, but to push the body of a strip away when its ends are no longer visible. The principle of the chain reaction is illustrated by the regular manner in which each stage of nest building succeeds the other, and by the fact that to a large degree each stage of the developing nest automatically provides the stimuli for its own termination and the starting of the next stage.

Other general mechanisms of species-typical behavior that do not appear to have been so widely recognized include gradients of action tendencies, and exploration tendencies. Thus, the male weaverbird weaves much more readily over his head than beneath himself. If in attempting to place a strip in one place the bird encounters some difficulty, he readily shifts his attempts to some other spot, and this exploratory behavior helps confer some flexibility and adaptability to the weaving of the nest.

The principal contribution of this report is to offer some experimental evidence for the nature of the key external stimuli and conditions that govern the causal sequences of nest building, and an explanation of the basic mechanisms of nest building in a bird as based on this evidence, combined with observation of the details of nest building.

ACKNOWLEDGMENTS

The studies reported here were made during a period of about three years in a large aviary at the University of California at Los Angeles, where the climate is suitable for tropical birds during much of the year. For financial support of the project we are grateful to the National Science Foundation (Grant 9741) and to the Research Committee of the University of California at Los Angeles (Grant 1623). We are indebted to Dr. Jean Delacour, Monsieur F. Fooks, and Monsieur Gérard Morel for their generous aid in obtaining the birds we used. The staff and gardners of our Department of Floriculture and Ornamental Horticulture, on the grounds of which our aviaries are located, helped us in many ways during the course of the investigation. After the first year of the study, various research assistants, including Mr. Richard Burrows and Mr. Brian Kahn, took care of the birds while working essentially on other phases of the weaverbird research program. Dr. Harlan Lewis, Dr. Mildred Mathias, and Mr. Wayne Hansus, all of the University's Botany Department, kindly advised and aided us with regard to plants needed for the birds.

SUMMARY

Study of abnormal nests and experimentally altered nests gives clues to the nature of factors operating in the course of construction of normal nests. Experimental evidence is presented for the following causal factors at each stage of nest building by the male Village Weaverbird, *Textor cucullatus* (Müller).

1. Weaving requires the use of flexible materials. The selection of yellow-green materials, the usual color of fresh, herbaceous vegetation, helps insure this needed flexibility.

2. Initial attachment results from the tendency of the male to poke and vibrate ends of strips into or alongside twigs, to wind such strips about twigs, to reverse his direction of winding between adjacent twigs or strips, and, finally, to poke and pull ends of strips in through holes made by the act of looping back strips.

3. Construction of a ring results from the tendency of the bird to weave in all directions along the immediately adjacent substrata while keeping the

feet in a fixed orientation in the bottom half of the ring. Strips longer than 15 cm are important to the closing of the ring that results from the tendency of the bird to shift its weaving from one end of a strip to the other and from one side of the ring to the other.

4. The polarity of the nest, with the brood chamber on one side and the antechamber on the other, is a consequence of the plane of tilt of the basic ring frame, since the weaver faces so that the ring tilts toward him, and then builds the brood chamber out from the ring before himself. The globular form of the brood chamber results from the fact that the male always stands in much the same place on the ring as he weaves, and then pushes the developing brood chamber out with his bill in all directions as far as he can reach.

5. A ceiling is thatched within the roof when the meshwork of the roof becomes too fine to permit more weaving easily, and the ceiling is terminated as it becomes opaque and blocks off the entrance of light through the roof.

6. The antechamber and entrance are built out from the ring, as the male, keeping his feet in the standard position on the base of the ring, weaves over his head, gradually leaning over backward more and more until he has built the entrance down to a more or less horizontal level.

7. An entrance tube is generally added after a female has accepted the nest. She prevents the male from entering the nest, and he satisfies part of his building drive by adding to the rim of the entrance. One factor tending to terminate growth of the entrance tube is the smoothing of the rim as the male works in loose ends.

8. It is evident that each stage of the developing nest automatically provides the external stimuli for its own termination and for the starting of the next stage.

Literature Cited

Benson, F. M. 1945. Observations on a nesting colony of Mottlebacked Weavers, *Sitagra nigriceps nigriceps* Layard, including attacks by a Harrier Hawk, *Gymnogenys typicus typicus* Smith. Ostrich, **16**: 54–65.

Chapin, J. P. 1954. The birds of the Belgian Congo. Part 4. Bull. Amer. Mus. Nat. Hist., **75B**: 1–846, 27 pls.

Collias, Elsie C., and N. E. Collias. 1962. Development of nest-building behavior in a weaverbird. Amer. Zool. (In press.)

Collias, N. E., and Elsie C. Collias. 1957a. Breeding behavior of the Black-headed Weaverbird, *Textor cucullatus graueri* (Hartert) in the Belgian Congo. Folia Scientifica Africae Centralis, T. III, No. 2, p. 44.

Collias, N. E., and Elsie C. Collias. 1957b. The analysis of nest-building in the Black-headed Weaverbird, *Textor cucullatus* (Hartert). Folia Scientifica Africae Centralis, T. III, No. 2, p. 44.

Collias, N. E., and Elsie C. Collias. 1959. Breeding behavior of the Black-headed

Weaverbird, *Textor cucullatus graueri* (Hartert), in the Belgian Congo. Proc. First Pan-African Ornithological Congress, Suppl. No. 3, pp. 233–241.

COLLIAS, N. E., and ELSIE C. COLLIAS. 1960. Some mechanisms of nest-building by the African Village Weaverbird, *Textor cucullatus*. Bull. Ecol. Soc. Amer., **41:** 53.

COLLIAS, N. E., P. J. FRUMKES, D. S. BROOKS, and R. J. BARFIELD. 1961. Nest-building and breeding behavior by castrated Village Weaverbirds (*Textor cucullatus*). Amer. Zoologist, **1:** Abstract 101.

CROOK, J. H. 1960a. Nest form and construction in certain West African weaver-birds. Ibis, **102:** 1–25.

CROOK, J. H. 1960b. Studies on the social behavior of *Quelea q. quelea* (Linn.) in French West Africa. Behaviour, **16:** 1–55.

FRIEDMANN, H. 1922. The weaving of the Red-billed Weaverbird in captivity. Zoologica, **2:** 355–372.

GRZIMAK, B. VON. 1952. Zum Baltzverhalten des westafrikanischen Textor-Webers, *Hyphantornis (Ploceus) cucullatus*. Zeits. f. Tierpsychol., **9:** 289–294.

HASLUCK, P. N. 1942. Knotting and splicing ropes and cordage. D. McKay Co., Philadelphia, 160 pp.

HINDE, R. A. 1958. The nest-building behavior of domesticated canaries. Proc. Zool. Soc. Lon., **131:** 1–48.

THORPE, W. H. 1956. Learning and instinct in animals. Harvard University Press, Cambridge, Mass. 493 pp.

Departments of Zoology and Entomology, University of California at Los Angeles, and Los Angeles County Museum, Los Angeles, California.

20

Reprinted from *Auk*, **81**(1), 42–52 (1964)

THE DEVELOPMENT OF NEST-BUILDING BEHAVIOR IN
A WEAVERBIRD

Elsie C. Collias and Nicholas E. Collias

Nest building is often considered a prime example of instinctive behavior in birds, but few detailed and systematic studies of the development of this process in young birds seem to have been made. Our object was to make such a study. The observations and experiments to be described were carried out on a colony of captive Village Weaverbirds, *Textor cucullatus*, at Los Angeles, California. These birds belonged to the West African race of the species, *T. c. cucullatus*, and the original stock came from Senegal. The value of this species for the present study is that it makes a complex and highly organized nest. We have elsewhere published an account of the breeding behavior of the race *T. c. graueri* in the wild state in Central Africa (Collias and Collias, 1959), and a detailed analysis of the mechanisms of the nest-building behavior of the adult males, which do all of the weaving of the nests in this species (N. E. Collias and E. C. Collias, 1962). This latter project was extended into the present investigation of the development of the ability of young birds to make a nest, but only a brief abstract of this work (E. C. Collias and N. E. Collias, 1962) has been published. We report herewith some details of our experimental analysis of this problem.

The young birds used were all hatched in a large outdoor aviary on the campus of the University of California at Los Angeles. This aviary measured 16 × 30 × 16 (height) feet and contained a palm tree (*Phoenix canariensis*) and an African acacia tree of unknown species. The birds nested in both trees, but preferred the acacia. The colony was maintained on parakeet seed mixture, lettuce, and mealworms, with grit and cuttlebone continuously present. Mealworms proved inadequate as a diet for the young, but the latter were raised successfully when the mealworms were supplemented with crickets. Mr. Richard Burrows, Mr. Brian Kahn, Mr. Herbert Brown, and Mr. Edward Tarvyd each helped us rear the young brought up in the aviary during one of the three years of the study.

We are grateful to Dr. Jean Delacour and Monsieur Gérard Morel for making it possible for us to obtain the birds for our colony. Dr. W. J. Dixon kindly advised us with regard to statistical treatment of the results. We are indebted to the National Science Foundation (Grants G-9741 and G-22236) and to the University of California at Los Angeles (Grant 1623) for financial support of the program.

Development of Ability to Manipulate Objects

The first use of the bill by young Village Weaverbirds, other than in gaping to receive food from the parent, is to preen disintegrating sheaths off the feathers. This preening action involves biting and nibbling at the feather sheaths, and we have observed it by the time the young bird is two weeks of age. Normally, the young Village Weaverbird spends almost three weeks in the nest, and the eyes open about a week after hatching. The mouthing of the feather sheaths would seem to be the precursor of the ability to mandibulate strips of nest materials, and thus to adjust the position of a strip in the bill.

The development of pecking activities introduces new motor elements needed for nest building. In the third week after hatching, the young bird begins to lunge toward and attempts to seize the food offered by the parent. But it is not before its first week out of the nest that the young bird becomes able to pick up food from the ground for itself. In part, the development of this ability is very likely facilitated by parental example. We have often seen young start to pick up food when next to the parent, shortly after the parent does so.

As it begins to feed itself the young bird develops a strong exploratory urge, i.e., it now picks up and manipulates all sorts of objects with its beak, and spends a good part of its day engaged in such activities. Mrs. Nice (1962) mentions exploratory pecking in young birds of more than a dozen different species, including both altricial and precocial types. Our general impression, after rearing and observing young birds of many species, was that few equalled and none exceeded our young weaverbirds in the frequency with which they manipulated various objects.

SELECTION OF APPROPRIATE NEST MATERIALS

A Village Weaverbird normally uses only fresh, green, flexible materials to weave the nest. The outer shell is woven by the male, of long strips torn from the leaves of elephant grass or palms, and he also adds a non-woven ceiling of short strips of grass or of dicot leaves. An inner lining of soft grass-heads and feathers is put in by the female. The key to the use of appropriate materials suitable for weaving is the green, or rather yellow-green, color of herbaceous flowering plants, and we have found that adults prefer green to other colors of artificial nest materials (N. E. Collias and E. C. Collias, 1962).

An experimental test was made of the color preferences of young weaverbirds upon initial exposure to nest materials. Some 30 young were hatched in outdoor aviaries. Half of them were left in the aviaries to be reared by their parents, and had access to the trees in the aviary and to the giant reed grass (*Arundo donax*) regularly placed in the aviary as a source of nest material. The rest of the young birds were removed from the nest before their eyes opened at the age of five to seven days, and

TABLE 1

SELECTION OF DIFFERENT COLORS OF NEST MATERIALS BY YOUNG WEAVERBIRDS REARED
IN ABSENCE OF NEST MATERIALS*

	Number of birds	Number of times color selected					
		Green	Yellow	Blue	Red	Black	White
Controls	6	311	256	90	76	116	53
Deprived	7	488	79	55	62	58	129

* Based on 14 hours and 40 minutes' observation of controls (37 days) and 8 hours
and 20 minutes' observation of deprived birds (11 days). Birds tested at age of 7 to
10 months.

these young were hand-reared while deprived of any normal nest materials.
However, it seems impossible completely to separate a young bird from
at least some analogue of nest material, for these hand-reared young, in
contrast to the controls, often manipulated their own feathers or those of
cage-mates. It was not uncommon to see such a bird hold a protesting
cage-mate's wing under one foot and attempt to "weave" its wing feathers.
Similarly, at times, one of these birds would reach down and try to weave
its own tail feathers. Since many of the feathers of young weavers are
yellow, yellow-green, or olive-green in color it is evident that we are far
from having a perfect control in this experiment. However, it is true that,
relative to the controls, the hand-reared young had far less experience
with anything that much resembled normal nest materials. Prior to the
experiment now to be described, the aviary-reared young had often manipulated normal nest materials, but had not yet built nests.

For a test of color preferences the birds were exposed to equal numbers
of colored toothpicks, 10 of each color: green, yellow, blue, red, black,
and white. The green was chosen to imitate the color of natural vegetation and contained some yellow, the commercial dye ("Tintex") used
being known as "jungle-green." The toothpicks were mixed completely
at random on presentation to the birds for standard observation periods,
generally half an hour in length. The birds were not given more than two
such observation periods a day. A bird would pick up a toothpick and
treat it as nest material, attempting to poke it into or alongside the perch
with a typical vibratory motion. As the table of results (Table 1) shows,
young male weaverbirds, whether or not they had been reared in the
absence of nest materials, selected green over other colors. The controls,
however, also showed a strong predilection for yellow. Most of the tests
with the aviary-reared young were done under fluorescent lighting; some,
as in the case of all the tests with the hand-reared young, were done under
ordinary incandescent light or else natural daylight.

It is of interest that in the case of the relatively naive, deprived young,

TABLE 2

SELECTION OF DIFFERENT COLORS OF NEST MATERIALS BY FOUR MALE WEAVERBIRDS,
REARED IN ABSENCE OF NEST MATERIALS*

| | *Number of times each color selected* | | | | | | *Per cent green* |
	Green	Yellow	Blue	Red	Black	White	
First two days	51	42	0	11	2	28	38
Second two days	163	16	6	13	1	29	71
Third two days	147	4	6	4	7	18	79

* Based on a total observation time of 5 hours and 50 minutes, over six successive days, when the birds were approximately seven months old.

the degree of preference for green on initial exposure to the artificial nest materials quickly increased with experience (Table 2). Thus, a Chi Square test, comparing the relative preference of green over all other colors combined for the first two days with the second two days of testing showed a significant difference well beyond the 1 per cent level (Chi Square > 50). As seen in the table, only 38 per cent of the selections in the first two days were of the green toothpicks, whereas 71 per cent were selected in the second two days of testing, and 79 per cent in the third two days. Five control weaverbirds, toward the end of their first year, under similar conditions of testing, showed an immediate and sustained preference for green, selecting 54 per cent green toothpicks in the first two days of testing, 48 per cent in the second two days, and 66 per cent in the third two-day period.

As Table 3 shows, the young male birds that were reared in the absence of normal nest materials, also had the normal preference for flexible over rigid nest materials (Collias and Collias, 1959) when tested at approximately eight months of age. These tests were conducted with green vinyl plastic strips of identical shape and weight, but differing in flexibility. Two thickness classes were used; a greater differential between "flexible" and "stiff" in the thicker strips was reflected in a greater difference in the frequency of strips selected (Table 3).

TABLE 3

SELECTION OF NEST MATERIALS DIFFERING IN FLEXIBILITY, BY FOUR YOUNG MALE WEAVERBIRDS REARED WITHOUT NEST MATERIALS*

| Thickness | *Picked up* | | *Wove* | |
	Flexible	Stiff	Flexible	Stiff
0.02 inch	258	160	8	0
0.01 inch	169	127	1	0
Totals	427	287	9	0

*Tested at approximately eight months of age. Observed 13 hours and 20 minutes.

These young males also preferred long over short strips, and when offered equal numbers of one-, two-, and four-inch strips, took the four-inch ones in 44 out of 69 choices. When eight-inch strips were added they chose these in 15 of 17 choices.

Weaverbirds develop an increasing discrimination in the selection and use of nest materials with normal experience, and will come to reject such things as toothpicks, string, and even raffia, when normal materials are available, and sometimes they may reject the artificial materials, even when normal materials are not available. Artificial nest materials are ignored by experienced adults when normal nest materials are available.

GATHERING OF NEST MATERIAL

When a weaverbird obtains a strip from a leaf of elephant grass (*Pennisetum purpureum*) in Africa or, in our aviaries, from 5 to 15-foot tall Mexican reed grass (*Arundo donax*) as a suitable substitute, it perches on the stalk or firm base of the leaf, bites through one edge of the leaf, tearing off part of the strip, and then tears the rest of the strip loose by flying away with it in the general direction of the tip of the leaf. In watching young weaverbirds, whether these birds were reared by hand or in the aviary, it seemed to us that they had to learn many things in carrying out the process of tearing a strip of reed grass properly. While experienced adult males generally fly off, finishing the tearing in one smooth action, the young often made such mistakes as perching in an unstable place, starting the tear too close to the tip of the leaf, or at the very base, or taking too broad or too narrow a bite, or tearing in the wrong direction, or tearing part way and repeatedly starting partly detached strips, or tearing strips that were too short to be woven.

Comparable and systematic observations on other species, in regard to the handling of nest materials, have been made by Dilger (1962), who finds that hybrid parrots of the genus *Agapornis* gradually improve their ability to cut and transport suitable nest materials.

DEVELOPMENT OF THE ACT OF WEAVING

In nature, the first-year males build very crude nests, with relatively sparse ceilings and with many loose loops projecting from the outer surface of the nests, in contrast to the compact, neat nests with thick, well-organized ceilings of the fully adult males (Figure 1). The latter are easily distinguished from the yearling males by their black heads. In India Sálim Ali (1931) had similarly observed, in the case of the Baya Weaver (*Ploceus philippinus*), that the nest of the young male is relatively crude compared with that of the adult male. This age-correlated difference in ability to construct a nest may turn out to be a quite general phenomenon among weaverbirds.

Figure 1. The first nests built by young males (right) are more loosely and crudely constructed than are nests built by adult, experienced males. Photographed by us at Entebbe, Uganda, 1957.

We have gathered experimental evidence that, in the Village Weaverbird, the ability to build improves with practice. We used the same males employed in earlier experiments on color preference with artificial nest materials. When these birds were almost a year old, three control and three deprived males still survived, and these two groups were placed side by side in two identical indoor aviaries, each with a guava bush of the same size. Normal nest materials in the form of palm strips or reed grass were supplied during standard observation periods of one-half hour or one hour. In the interim between this experiment and the earlier one on color preference, only the controls were given any access to normal nest materials. An opaque cloth partition between the two aviaries precluded the possibility of observational learning by the hand-reared young from the more experienced, aviary-reared young.

The initial observations were made in early August, 1961. In the first week the birds were given palm strips, and the hand-reared males wove not a single stitch onto the bush or onto the wire meshwork of the cage, in contrast to all the control males, which wove well in both places. In the second week, with reed grass, the hand-reared young wove a few stitches on the wire only. In the third week, the experiments summarized in Table 4 were started. As shown in the table, the deprived males succeeded only in weaving a very small percentage of strips compared with

TABLE 4

ABILITY TO WEAVE OF YOUNG MALE WEAVERBIRDS REARED IN ABSENCE OF NORMAL
NEST MATERIALS TO ALMOST ONE YEAR OF AGE

	Number of birds	Dates	Hours observed	Strips carried	Per cent woven
Controls	3	11–16 August	3.5	73	62
Deprived	3	10–15 August	3.5	69	26
Controls	3	12–25 November	3.5	95	77
Deprived	3	13–25 November	3.5	158	52

the controls. The difference was significant to the 5 per cent level by the Rank-sum Test (see Dixon and Massey, 1957).

After the first experiment the nest materials were left with the birds, and then a fresh supply of reed grass was maintained constantly in the aviaries. By the middle of November, 1961, after some three months' opportunity to practice, the deprived males had improved considerably and (Table 4) had doubled the percentage of strips woven over their performance in August. The difference in the percentage of strips woven by the controls and the deprived birds was no longer significant by the Rank-sum Test ($P = 0.20$).

In the three-month practice period, two of the males reared in absence of nest materials managed to weave 2 nests. In the same period, the three control males built 11 nests. There was no obvious difference in the quality of nests built by controls and experimentals.

In the case of the simple cup nest of the canary, Hinde (1958) observed that three females reared from eight days after hatching without access to normal nest materials would, when they reached breeding condition, go through all the movements of building in a nest pan. When appropriate materials were presented for the first time these females responded rapidly and built nests that appeared as large and tidy as those built by experienced ones. As in the case of our deprived young weaverbirds, these deprived canaries showed a pronounced tendency to pull at and carry their own feathers or those of cage-mates. Of course, the normal nest materials of canaries, being relatively short pieces of grass as well as feathers, are more like their own body feathers than are the long, flexible strips of grass or palm leaf used by a male weaverbird to weave his nest, and the integration of basic movements involved in building a nest is far less complicated in the canary.

It appears that some practice is important to successful weaving, and the next question that arises is just what the birds learn. It is interesting to watch the initial reactions to nest materials of young male weavers that have been reared without normal nest materials to an age where the control males can weave. These deprived males already possess certain basic motor elements of the act of weaving. They can select suitable materials, hold a strip under one foot or in the beak, mandibulate to one end of the strip, and poke it toward or alongside the perch, or into the interstices of the wire meshwork of the cage. Somewhat later the young birds develop the ability to reach around to the other side of the twig while standing on a strip, seize the strip anew in the beak and wind it around the twig.

In watching the attempts of a young weaver to fasten its initial strips of nest materials and the gradual improvement of weaving, it seemed to us that in general what every young male weaver has to learn is what in subjective terminology we would call "judgement." But we can break down the things apparently learned into more objective descriptions of our impressions after watching a great many instances where a young weaverbird was attempting to "weave." Frequently the young bird attempts to weave with materials that are too short or too stiff. But it is channelized into the right direction by its normal tendency to develop a preference for green, flexible, and longer materials. Before the bird can succeed in firmly fastening a strip in place it must learn to carry out the basic movements in an *effective sequence*. It must learn not to pull a strip out again before it is firmly fastened in place. Often the bird starts to weave one end of its strip and, instead of following through, *shifts* its point of attack in an ineffective way. In contrast, the adult is much more likely to *persist* in its weaving at one end of the strip, or at a given site. Often the young bird will push a strip partly into a wire mesh and just as promptly pull it right out again, making no attempt to let go in order to shift its hold through to the next mesh. Learning *when to let go* of a strip is an important requirement of weaving. But even when let go, the strip frequently may fall out of place, of its own weight or resilience, before the young bird can reach around to the other side of the twig or wire and seize it anew. The bird apparently has to learn to push the strip through *far enough* so that it doesn't fall out or spring back at once when released. Unless the bird has also learned to hold a strip *under foot*, the strip may even fall to the ground when the bird shifts its beak-hold on the strip.

Even when a young weaver can effectively make several stitches, it still fails to make a nest. For one reason or another, the bird generally ends up *removing* each strip before he has thoroughly woven it into place; we have seen this repeatedly. For example, in 1962, a deprived young weaverbird, Male WY, on its first exposure to reed grass, was observed for one hour daily for five days, the reed grass being removed after each hour of observation, and in this time he was seen to weave a total of 36 stitches, gradually increasing the number each day to a total of 21 on the last day. But, in contrast to the usual behavior of controls, not one of these stitches or strips was left in place by the bird by the end of each observation hour, so WY ended his first week of exposure to normal nest materials with not a single strip in place. This tendency to undo one's own weaving we dubbed the "Penelope act" for convenient reference.

INFLUENCE OF OTHER BIRDS ON LEARNING TO WEAVE

An important influence on the development of ability to build would seem to be the restriction of learning opportunities relative to position in the dominance order. Village Weaverbirds have a social ranking system based on habits of aggressive domination and subordination, and, at least in an aviary, this social order is as rigorous and almost as rigid as that seen in the domestic fowl. A much subordinated male has far fewer chances to practice weaving than does a dominant male. No sooner does a male low in dominance gather a strip than a dominant bird is likely to come and take it from him. In each of our groups, the most subordinate male wove a much smaller percentage of strips carried than did other birds, and it was evident that the main, though not the only, reason was his low social position.

The role of tuition by example from other birds, especially older males, is a factor we have not yet quantitatively assessed. It appears likely that some social facilitation exists for nest building, since the birds tend to show interest in nest materials at about the same time, just as in the case of their other activities. Marais (1937), studying the South African "yellow weaverbird," took eggs from the nest and had them hatched by canaries. The nestlings were then reared out of sight of both adult weavers and normal nest materials. These young weaverbirds were able to weave nests, indicating that tuition from older birds was not needed. Unfortunately, he gave no further details, as, for example, concerning possible improvement in skill of the birds in building their nests, nor did he certainly indicate the species concerned.

We do have some information of a qualitative nature bearing on this problem. A male Village Weaverbird was taken from a nest in our colony before his eyes opened and reared in our home in complete visual isolation from all other weaverbirds to an age of more than two years. He was supplied with normal nest materials from an early age, and gradually developed the ability to construct normal nests, the first appearance in development of the different acts involved in nest building being in the following sequence: mandibulation, biting and pulling at various objects with the beak, poking of raffia into any nearby object with the beak, poking raffia into the wire meshwork of the cage, tearing of strips, and standing on strips with the feet. At five and one-half months of age he

Figure 2. A male Village Weaverbird, hand-reared in complete visual isolation from other weaverbirds, but supplied with nest materials from an early age, built the nest illustrated here. The entrance is to the lower left, the brood chamber is at the right.

wove his first stitch into the wire but at once pulled it loose. Then, at six months of age he wove the first strip (with three stitches) that he left in place. At seven months he managed to fasten a strip onto a perch, at eleven months made a ring, and by one year made his first nest (of raffia). He subsequently wove several nests of reed grass or palm strips (Figure 2), which to us appeared no different from those built by other males of comparable age.

We concluded, from the case of this male reared in isolation from other weaverbirds, that tuition by example is not necessary to the development of the ability to build a nest in the Village Weaverbird. However, it is not unlikely that example may have a facilitation role, since two control males were each seen to weave their first strip in the outdoor aviary in which they were hatched by four months and one week of age, more than a month earlier than did the isolated male.

SUMMARY

Fledglings of the African Village Weaverbird, *Textor cucullatus*, begin to manipulate all sorts of materials very frequently soon after they leave the nest, and the males build crude nests long before attaining sexual maturity.

Some 30 young were hatched in outdoor aviaries. Half were reared by their parents. The rest were removed before their eyes opened and were hand-reared in the absence of nest materials. The hand-reared young, unlike the controls, often manipulated or tried to "weave" their own feathers or those of cage-mates. As did controls, when given equal chances for selection, they preferred green nest material to yellow, blue, red, black, or white. The hand-reared young quickly developed a preference for green within the first few days after initial exposure to artificial nest materials. They also had the normal preference for flexible as against stiff nest materials, when tested with vinyl plastic strips.

When about one year old, the survivors were given their first normal nest materials, and the hand-reared males wove a much smaller percentage of strips than did control males. But this discrepancy greatly diminished with about three months' practice in handling strips, and two of three hand-reared males managed to weave 2 nests. Three control males wove 11 nests during the same period.

One male, hand-reared in complete visual isolation from other weaverbirds, but supplied with nest materials from an early age, gradually developed the ability to construct normal nests.

We conclude that practice, channelized by specific response tendencies, but not necessarily tuition by example, is needed for development of the ability to build a normal nest by the male of this species.

LITERATURE CITED

ALI, S. A. 1931. The nesting habits of the Baya (*Ploceus philippinus*). J. Bombay Nat. Hist. Soc., **34**: 947–964.

COLLIAS, E. C., AND N. E. COLLIAS. 1962. Development of nest-building behavior in the Village Weaverbird, *Textor cucullatus* (Müller). Amer. Zool., **2**: 399 (abstract).

COLLIAS, N. E., AND E. C. COLLIAS. 1959. Breeding behavior of the Black-headed Weaverbird, *Textor cucullatus graueri* (Hartert), in the Belgian Congo. Proc. First Pan-African Ornith. Congress, Ostrich, Suppl. No. 3, pp. 233–241.

COLLIAS, N. E., AND E. C. COLLIAS. 1962. An experimental study of the mechanisms of nest building in a weaverbird. Auk, **79**: 568–595.

DILGER, W. C. 1962. The behavior of lovebirds. Sci. American, **206** (1): 88–99.

DIXON, W. J., AND F. J. MASSEY. 1957. Introduction to statistical analysis. 2nd edit. New York, McGraw-Hill.

HINDE, R. A. 1958. The nest-building behavior of domesticated canaries. Proc. Zool. Soc. London, **131** (pt. 1): 1–48.

MARAIS, E. N. 1937. The soul of the white ant. New York, Dodd, Mead and Co.

NICE, MARGARET M. 1962. Development of behavior in precocial birds. Trans. Linnaean Soc. of New York, **8**: v–xii, 1–212.

Department of Zoology, University of California at Los Angeles, and Los Angeles County Museum, Exposition Park, Los Angeles, California.

21

Reprinted from *Viltrevy, Swedish Wildl.* **8**(3), 160–165, 168, 182–203, 254–260 (1971)

OBSERVATIONS AND EXPERIMENTS ON THE ETHOLOGY OF THE EUROPEAN BEAVER (CASTOR FIBER L.)

Lars Wilsson

[*Editors' Note:* In the original, material precedes this excerpt.]

The building of lodges, dams and winter stores

Motor patterns of building behaviour

Transport movements. Long pieces of trunks and branches are grasped near one end with the incisors and are dragged beside the swimming, diving (Fig. 24) or walking animal (*dragging*, Fig. 9).

Fig. 24. Beaver drags a branch, swimming under water towards the dam in the stream tank (dragging). — *Bäver släpar en gren medan den simmar under vattenytan mot dammen i strömakvariet.*

Fig. 25. Beaver swimming under the water in the stream tank with a bundle of short branches in its mouth. — *Bäver simmar under vattenytan i strömakvariet med en bunt korta grenbitar i munnen.*

Short pieces are carried in the mouth (*carrying*, Fig. 25). They are often lifted up with one hand, one by one. Finer material can also be collected and carried in this way, both forepaws being used in rapid, alternating grips to bundle it together.

Most often, however, fine material is collected between the upper sides of the fore legs and the under side of the chin (Fig. 26). As a rule first some bits of branches or small bundles of sticks are collected in the mouth. Then as the palms of the hands are placed on the ground and pushed under the material the chin and the mouthful of sticks are pressed down against it. The hands are then pushed forward along the ground and the material is shoved along (*shoving* Fig. 26 a).

If the surface of the ground is so uneven or steep that the beaver cannot shove the bundle along in front of it, it rises on its hind legs and carries the burden walking bipedally (*carrying a burden*, Fig. 26 b). The arms are stretched forward and the material remains pressed between them and the chin while the animal smoothly raises itself into

a

b

Fig. 26. *a.* The beaver has collected a bundle of branches in its mouth and is shoving an armful of twigs between its hands, which are on the ground, and its chin (shoving). *b.* The beaver has now risen on its hind legs, and is carrying the bundle, walking bipedally (carrying a burden). — a. *En bunt grenar har samlats i munnen och bävern skjuter ett fång kvistar mellan händerna på marken och hakan.* b. *Bävern har rest sig och bär materialet gående på bakbenen.*

a more or less upright position. The tail functions as a balance when the burden is raised or lowered.

The beaver can shove or carry large burdens over long distances, but as a rule only on horizontal surfaces or against the incline of a slope.

On the building site the beaver lifts the burden and puts it on top of the material already collected (*lifting,* Fig. 34 *a*).

When the beaver carries material in its mouth, it often gets up on its hind legs and supports the material with its forepaws or fore legs.

Burdens securely held between the fore legs and the chin in the

Fig. 27. The beaver swims up to the base of the dam with a burden of building material which it then carries, walking bipedally, up the steep upstream side of the dam. — *Bävern simmar under vattenytan med en börda byggnadsmaterial som den sedan bär uppför dammens branta uppströmssida under tåbent gång.*

manner described above are transported most quickly underwater. The beaver can *swim* long distances *with* large *burdens* both on and under the water (Fig. 27). The burdens are most effectively collected on the bottom of the water course.

The beaver can also push material in front of it, applying the method already described in the paragraph on digging (*pushing*, page 155, Fig. 22 f–g).

Building movements. The beaver pins branches and sticks in lodges, dams and winter stores in the following manner (pinning, Fig. 28); the stick or branch is grasped near one end with the incisors and near the other end with one forepaw. The paw is used to guide the stick towards a gap or hole, while the teeth push it forward with jerking movements of the head until it is firmly fixed.

Pinning has also been observed in situations of conflict when it was interpreted as a displacement activity (page 239).

Fig. 28. Female 1A pinning at the age of three months. The stick is directed with one hand towards a crack and firmly fastened with jerking movements of the head. — *Honan 1A, tre månader gammal, fäster en pinne. Pinnen styrs mot springan med ena handen medan den skjuts fast under knyckande rörelser med huvudet.*

Earth or other fine material is tightly packed against the dam or lodge with the palms of the forepaws, the legs moving simultaneously (*packing*). This behaviour is characteristic of many other rodents, although the legs usually move alternately. (Eibl-Eibesfeldt 1951 *b*, Roth-Kolar 1957).

Fine material can also be tightly packed with the chin, nose or with the exposed incisors (cf. digging, page 155).

Peeled sticks are sometimes *stripped* into long and thin strips with the teeth. The resulting material is like wood-wool and is used as bedding in the nest or as filling-in in the walls of the lodge. Rats (Rattus rattus) also manufacture bedding material in this manner (Eibl-Eibesfeldt 1958).

Most of the motor patterns of the building behaviour are briefly described by Shadle (1956).

Lodges

Along the main river there is only one type of beaver nest where the animals spend the winter and give birth to their young. This is the *bank lodge* (Fig. 29) and is also found along the tributaries where the banks are sufficiently high and are composed of material in which the animals can easily dig.

The tunnels from the bottom of the water course often reach or go very near the surface of the ground where they pass the waterline inside the bank. On the ground above the tunnel the beavers always build a "roof" which reaches a short distance out into the water above the tunnel entrance and if possible so far in over land as to cover the feeding chamber (page 155). Sooner or later the earth covering of the waterfilled tunnels and the feeding chamber caves in

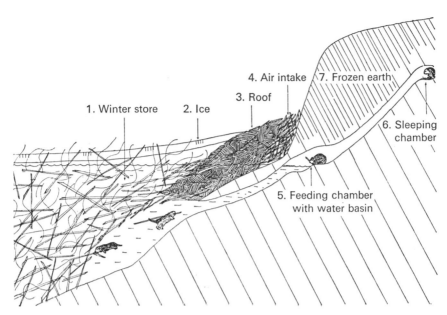

Fig. 29. Bank lodge. *1.* Winter store. *2.* Ice. *3.* Roof. *4.* Air intake. *5.* Feeding chamber with water basin. *6.* Sleeping chamber. *7.* Frozen ground. — *Älvhydda.* 1. *Vinterförråd.* 2. *Is.* 3. *Tak.* 4. *Luftintag.* 5. *Matkammare med vattenbassäng.* 6. *Sovkammare.* 7. *Tjälad mark.*

The ice covers the winter store and the earth is frozen above the tunnels in the bank. The roof on the shore above the waterfilled part of the tunnel system insulates it against the cold and protects the tunnels against erosion by ice and water. The rear part of the roof is not plastered with earth and thus serves as an air intake. Where the tunnels rise above the surface of the water they are widened into feeding chambers, with water basins. The tunnel system branches widely inside the bank, with the sleeping chambers safely above the water. — *Isen täcker vinterförrådet och marken är frusen ovanför gångarna inne i strandbrinken. Taket på stranden ovanför den vattenfyllda delen av gångsystemet isolerar mot kyla och skyddar gångarna mot erosion av vatten och is. Bakre delen av taket har inte tätats med finare material och kan därför fungera som luftintag. Där gångarna når ovan vattnet är de utvidgade till matkammare med vattenbassänger. Gångsystemet är vitt förgrenat inne i strandbrinken där sovkamrar har holkats ut på betryggande höjd ovan vattenytan.*

the beavers remove the fallen earth and the tunnels and the feeding chamber are eventually left covered just by the roof (Fig. 33).

One or several tunnels run obliquely upwards in the river bank from the feeding chamber to one or several *sleeping chambers.*

[*Editors' Note:* Material has been omitted at this point.]

379

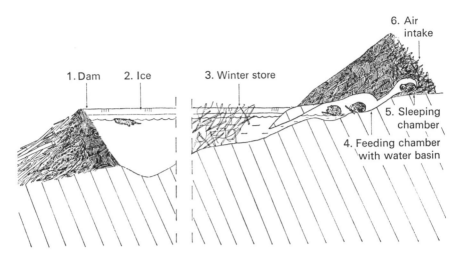

Fig. 31. Brook lodge. *1*. Dam. *2*. Ice. *3*. Winter store. *4*. Feeding chamber with water basin. *5*. Sleeping chamber. *6*. Air intake. — *Bäckhydda*. 1. *Damm*. 2. *Is*. 3. *Vinterförråd*. 4. *Matkammare med vattenbassäng*. 5. *Sovkammare*. 6. *Luftintag*.

Where the banks are low the beavers build lodges of the wellknown conical type. In these the chambers are hollowed out of the material which makes up the lodge. If the beavers build a dam they can regulate the water level in relation to the lodge and in winter they can lower the water level so that an air space is created between the water and the ice. — *På låga stränder bygger bävrarna hyddor av den välkända koniska typen. I dessa är kamrarna urholkade i själva hyddans byggnadsmaterial. Om bävrarna bygger en damm kan de reglera vattenståndet i relation till hyddan och under vintern kan de sänka vattenståndet så att det bildas ett luftrum mellan isen och vattnet.*

If there is nowhere to excavate a tunnel system out from the lodge the chambers are enlarged instead; should the lodge walls then get too thin, they are added to from the outside. Lodges of this type gradually become very large. The largest we have seen rose more than 3 metres above ground level. In one case the lodge contained a single cavity measuring $4 \times 2.5 \times 0.6$ metres. Another lodge had two completely separate tunnel systems.

According to Dugmore (1914) and Bailey (1926) the American beaver lodges have never more than one chamber, which can be from 0.9–3.6 metres in diameter. On the other hand, Pilleri (1960) mentions an American beaver lodge which contained 12 chambers. In our comparative studies in the USA we found that the American beaver lodges are of the same type as the European ones.

Dams

The beavers along the main river and the largest tributaries do not build dams. Most of the Faxälv tributaries are so shallow, however, that the beavers have no safe living space until the water level is raised. A beaver group often has a whole series of dams, usually situated downstream from the lodge.

The raised water level greatly improves the beavers' protection from their enemies. They can quickly take refuge underwater and swim unseen between their different tunnel systems. As a result of dam building the water level remains relatively constant in the water basin of the feeding chamber. The water outside its entrance is deep enough to permit the collection of a winter store, and collecting is facilitated by the fact that the animals can float wood over the whole pool created by the dam.

Dugmore (1914) states that the beavers dig a channel in the crest of the dam during the summer as soon as the water level rises over a certain height. We have observed such openings being made only in winter. Often the flow decreases and water level sinks after the water-course has frozen over. If, however, this does not happen the beavers may, during the late winter, make an opening in the crest of the dam nearest the lodge, thus lowering the water level and creating an air space between the ice and the water. They can then swim on the surface of the water all over the pool: they can find a spot between the ice and the bank where they can gnaw and dig a passage up above the ice, and from there frequently dig a tunnel under the snow far into the woods to suitable food.

We have observed that when the animals can come up on land in this manner in winter, they fetch fresh food as soon as the temperature is above about $-4°C$ and often sit out in the open air and eat. They sometimes fetch food from the land even when the temperature is a few degrees lower, but they then sit in shallow water under the ice, or inside the lodge to eat. When the temperature is below about $-6°C$ they apparently go out only when they are short of food. The beaver's dependence on temperature ought to be more systematically investigated.

Dams upstream of a lodge are relatively unusual, but in places where they are found the animals may make openings in them during the late winter and thus keep open a passage from the lodge to the pool upstream.

Dam building behaviour

We have followed construction work at a number of dams in different types of habitat both out in the forest and in the enclosures where the conditions could be changed experimentally.

When lodge building behaviour is activated, dam building behaviour can also be released.

Bradt (1938), who caught beavers by placing traps in dam openings, noticed that the males are the first to repair a broken dam, but that later all the animals in the group take part in the work. In the two beaver pairs which built dams in the stream tank, the female was the first to hurry to a small leak in the dam, but if the leak was large the males repaired it (page 193).

The beavers use primarily material from the bottom of the brook. Sticks and branches peeled by the beavers sink rapidly to the bottom and large quantities accumulate during the year.

Dam building is carried out from the upstream side (Bailey 1927). According to our observations (Wilsson 1960, 1962) work is always started at a place where the water is rushing noisily between stones or other obstacles (Fig. 37).

If building behaviour is only weakly activated by the acoustic stimuli from the water, the beavers often push mud and other fine material against the source of stimuli. At a stronger activation the beavers build with branches and other coarse material. These they unload higgledy—piggledy over the source of the acoustic stimuli, in depressions in the riverbed or in indentations in the heap of material already collected (Fig. 37 a–d, cf. lodge building page 173).

When the top of this heap is roughly level (Fig. 37 d) the animals react to any kind of indentation between the branches by pinning

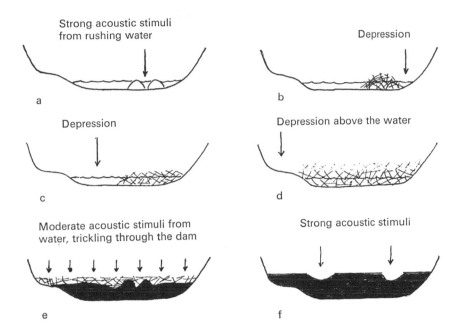

View from the upstream side   View from the upstream side

Strong acoustic stimuli
from rushing water   Depression

a   b

Depression   Depression above the water

c   d

Moderate acoustic stimuli from   Strong acoustic stimuli
water, trickling through the dam

e   f

twigs into the crannies, and pushing and packing fine material against the upstream side of the construction. This activity is stimulated by acoustic stimuli from water rushing through the material (Fig. 37 e). Thus the upstream side is tightened and smoothened out with fine material and the water rises.

Arriving at the dam, swimming with a burden of mud, other fine material or short branches (Fig. 27), the beaver rises on its hind legs and carries the material bipedally up the steep slope of the dam to the top. The material is unloaded on that part of the upstream side of the dam which is above water. Once the dam is tightened the animals pack material against the crest wherever water is pouring over until the crest is horizontal just above the surface of the water (Fig. 37 f, g).

Longer bits of wood are always dragged over the crest by successive grips with its teeth, as the animal either floats in the water in front of the dam or stands on top of it. They are finally pinned firmly against the bottom on the downstream side of the dam (Fig. 38). The stereotyped motor patterns used when fastening the branches (pinning)

Stream direction

←

Shoving and lifting burdens.
Pinning against the bottom
at the downstream side

Moderate acoustic stimuli

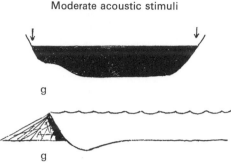

g

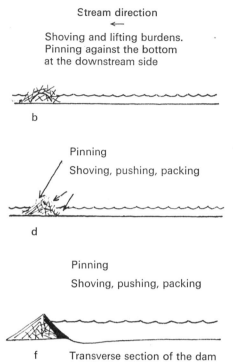

b

Pinning

/ Shoving, pushing, packing

d

g

Pinning

Shoving, pushing, packing

f   Transverse section of the dam

Fig. 37. Stimuli and motor patterns during dam building. *a.* Strong acoustic stimuli from rushing water. *b.* Shoving and lifting burdens. Pinning against the bottom at the down stream side. *c.* Depression. *d.* Pinning. Shoving, pushing, packing. Depression above the water. *e.* Moderate acoustic stimuli from water, trickling through the dam. *f.* Pinning. Shoving, pushing, packing. Strong acoustic stimuli. *g.* Moderate acoustic stimuli. — *Stimuli och rörelsemönster under dammbyggandet.*

result in the latter being placed at about right angles to the crown of the dam to form a support for the entire dam from the bottom of the brook on the downstream side.

Richard (1967) states that the beavers thoroughly select material according to a certain scheme when building dams. We have not been able to confirm this theory. The choice of material seems mainly to be determined by the degree of motivation as described above.

The beavers also excavate the bottom of the brook in front of the dam. This digging together with building with mud and other fine material, make the dam watertight rapidly. Thus the upstream side of a complete dam has an even slope of fine material and earth down to a hollow in the bottom of the brook while the downstream side consists of a conglomeration of branches supported by the bottom and the banks of the brook (Fig. 37 *g*, Fig. 31). Water passes the dam mainly at the ends, where the beaver does not react to a moderate sound of running water.

Fig. 38. Pinning a branch against the bottom of the downstream side of the dam in the stream tank. — *En gren fästes mot botten på dammens nedströmssida i strömakvariet.*

If there are stones on the bottom they are pushed (Fig. 39) or carried to the building site and soon form a wall.

The beavers can also use entire small trees which are placed parallel to the current with the crown pointing upstream. The current presses the twigs and branches of the crowns against the supporting trunks and this contributes effectively to making the dam watertight.

Ontogeny of dam building behaviour

Dam building behaviour could be released in the young 2 D as early as its first winter, at the age of 6–7 months (page 201), but neither 2 D, nor the rest of the young, built complete dams until their second autumn. They then built complete dams at their first encounter with the relevant stimulus situation.

Thus young beavers seem not to build dams until their second autumn. The motor patterns are already fully developed in the first autumn, but the ability to react to the activating stimulus situation is probably not developed until the first winter, when normally building behaviour is not activated.

Fig. 39.. Pushing a stone along the bottom towards the dam in the stream tank.
— *En sten skjutes mot fördämningen i strömakvariet.*

The beavers of group A were kept in still water during their first autumn. During their second autumn the males 4 A and 5 A built dams in the enclosure (no. 1) where they had spent the second summer.

In these males dam building behaviour was observed already in the beginning of August. They began to fix sticks in the part of the wire netting which crossed the brook where it entered the run. When this construction had reduced the sound of running water they changed over to working in a similar manner along the netting at the outlet, which resulted in a rising waterlevel. The animals then only worked sporadically at the inflow and concentrated most of their activity to the outlet, where they built a complete 50 cm high beaver dam.

Later in the autumn we built in the middle of the pond a dam of boards with an opening through which the water rushed strongly. The beavers immediately reacted by building a dam across the opening.

The fact that the young beavers began to build a dam at the inflow of the brook cannot be put down to lack of experience. The two newly captured experienced beavers, which were kept in another enclosure (no. 3) during the same autumn, also began to build in a

Fig. 40. Dam, built by two-year-olds 1A–3A during their second night in the enclosure. — *Fördämning, byggd av tvååringarna 1A–3A under deras andra natt i inhägnaden.*

similar manner at the inflow and only afterwards changed over to
building at the outlet.

1 A and 3 A, which were kept in the stream tank early during their
second autumn began too, to fix sticks to the netting which covered
the inflow to the tank and then changed to building by the netting
across the outlet. When they were moved to enclosure no. 2 at the end
of September they built a dam during their second night there. Unlike
the dams which had been built earlier by the other animals in group A,
this dam was not supported by a net. It was built over the narrowest
part of the brook, between a few stones where the running water
produced the strongest acoustical stimuli within the enclosure (Fig. 40).
Some days later these animals also tried to block the inlet of the
brook.

These observations indicate that dam building behaviour is released
at the point where the acoustic stimulus situation is optimal. The
intensity of building always increased when the animals had heaped
up so much material that the water rushed through it noisily. The
animals pushed earth and mud against the irregularities and openings
in the dam from upstream, and they would maintain this behaviour

even in the absence of stimuli from flowing water, e.g. after the water had been siphoned out through a pipe over the dam. They obviously reacted to the irregularities in the dam exposed when the water level dropped in the same way as they reacted to irregularities in the wall of the lodge (page 174).

When 4 A and 5 A had raised the water level by building a dam inside their run, they began to build a winter store in the water a short distance downstream from the entrance to their lodge. Later we lowered the water level with the help of siphons so the winter store was partially exposed and the water ran audibly through the gaps in it. The animals immediately reacted by fixing peeled sticks in it and by pushing mud against the upstream side. The store was thus quickly transformed to a dam, and the water level rose again outside the lodge.

When after a time the siphons became clogged, the dam at the outlet from the enclosure began to function again. The water level thus rose within the whole enclosure and the winter store was again submerged. The animals then removed the peeled sticks using them for building on the lodge and the dam at the outlet, and again began to fix branches with edible bark at the store.

The animals thus reacted to the sound of the water running noisily through the winter store in the same way as at the dams. When the store was submerged again this stimulus disappeared and there were no irregularities above the water to which the animals could react. They then reacted to the building material fastened earlier in the same way as to building material lying on the bottom of the brook, and to the edible branches as to material for the store.

Peeled sticks and other waste is always taken out of the lodge every morning and, if no building activity is in process, is thrown into the water where it sinks to the bottom. Material deposited in this way during the course of the year constitutes the most important source of building material in the autumn. In places where there are no natural obstacles interfering with the flow of the water the pile of accumulated waste can itself sometimes serve as an obstacle, forming the base on which the beavers begin to build their dam. Such behaviour was observed in animals 4 A − 5 A which, during their second autumn in the enclosure, began to build a dam on the pile of sticks that had accumulated on the bottom during the previous summer.

The enclosures were later equipped with artificial dams at the outlet of the brook. The water therefore ran noisily only at the inlet. All

Table 2. Experiments on dam building in the stream tank (cf. fig. 41)

Interference	Waterlevel (dm)		Day	Time	Dam height in the following morning (dm)	Water level in the following morning (dm)
	Before interference	After interference				
The crest of dam removed	7.0 (fig. 41 a–b)	3.4 (fig. 41 c)	19/9	19.30	5.5 (fig. 41 f)	5.0 (fig. 41 f)
Furrow in the crest of the dam	5.2 (fig. 41 g)	4.4 (fig. 41 g)	20/9	17.00 17.30 21.00		
The crest of the dam removed	5.2 (fig. 41 g)	3.8 (fig. 41 g)		22.00	5.5	4.7
Inflow increased	4.7 (fig. 41 h)	6.5 (fig. 41 h)	21/9	23.00	6.4	6.1
Inflow increased	6.1	6.7	22/9	20.00		
Furrow in the crest of the dam	6.7 (fig. 41 i)	6.1 (fig. 41 i)		23.00		
The crest of the dam removed	6.7	5.0		24.00	8.0 (fig. 41 j)	8.0 (fig. 41 j)
Inflow increased	6.5	7.5	25/9	17.20		
Furrow in the crest of the dam	7.5	6.0		21.55		
Inflow increased	7.5	7.7		22.30		
Inflow decreased	7.7 (fig. 41 k)	6.0 (fig. 41 k)		22.00	7.5	7.5
Inflow increased	7.0	7.5	27/9	17.00		
Furrow in the crest of the dam	7.5	6.5		22.30	7.5	7.5
Inflow decreased	7.5	6.8	28/9	17.00	7.5	7.5
The crest of the dam removed	7.5	6.5	2/10	22.00	7.5	7.5
The crest of the dam removed	7.5	5.8	3/10	17.00	7.0	7.0
The water siphoned past the dam	7.0	5.0	4/10		7.0	7.0

	Waterlevel (dm)				Dam height in the following morning (dm)	Water level in the following morning (dm)
Interference	Before interference	After interference	Day	Time		
The water siphoned past the dam	7.0	4.2	5/10		7.5	7.5
The water siphoned past the dam	7.5 (fig. 41 l)	3.2 (fig. 41 l)	9/10		7.5	7.5
Furrow covered by a glass dome	7.5 (fig. 41 m)	5.5 (fig. 41 m)	11/10		7.5	7.5
Furrow covered by netting	7.5 (fig. 41 n)	5.5 (fig. 41 n)	20/10		7.5	7.5
Furrow covered by a glass dome	7.0	5.0	24/10		7.0	7.0

the beavers, even old experienced ones, built against this rushing water until the noise from it was reduced.

Immediately after the ice forms all dam building always ceases. On all occasions when beavers have spent the winter in enclosures, they have, when the pond has frozen over, made an opening in the top of the dam and thereby made a space between the ice and the water surface. It is possible that in the enclosures the limited space contributed to the early activation of this behaviour. This behaviour was also observed in the handreared beavers during their first winter outdoors.

Experiments on dam building behaviour

Experiments in the stream tank. Although some sporadic dam building activity was observed in 2 B and 3 B when they were placed in the stream tank in their first autumn they did not build a complete dam until their second autumn. To release that activity some stones were placed on the bottom of the tank in late August. Within a week the beavers had completed a dam (Fig. 41 *a–b*) and as soon as the water above the dam began to rise appreciably they started transporting building material to the lodge as well, and soon concentrated entirely on lodge building.

When 4 D and 5 D were released in the stream tank during their

390

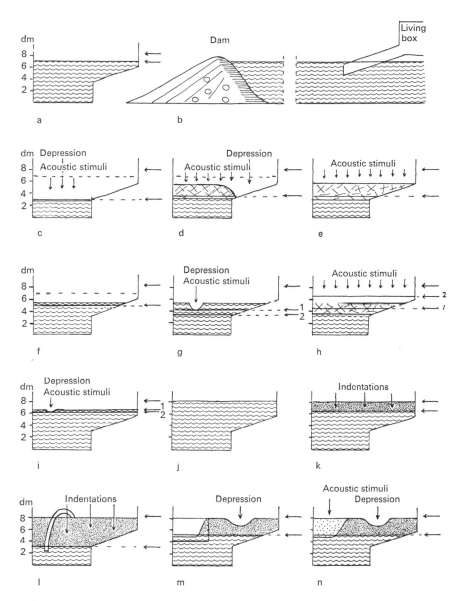

Fig. 41. Experiments on dam building in the stream tank. (Table 2.) Horizontal arrows indicate water levels and the level of the living box respectively. Vertical arrows indicate releasing and directing stimuli *a*. The upstream side of the dam before the experiments. *b*. Longitudinal section of the stream tank before the experiments. *c–e*. The upstream side of the dam during the first evening (19.9). The crest of the dam removed (c), the beavers have built with coarse material across

second autumn they at once built a complete dam, although from the time of capture until they were released in the tank they had spent their lives in still water. Their dam building behaviour tallied with the observations made during the experiments on 2 B and 3 B.

The building material used by the animals in the stream tank consisted of stones of different sizes together with sticks and branches which the animals had peeled the bark off during the year they had been in the aquarium.

After 2 B and 3 B had built a complete beaver dam in the tank they were kept under constant observation between 6 p.m. and midnight for 15 evenings between Sept. 19 and Oct. 24. The conditions in the tank were changed in different ways. The results of activities in the tank before and between the observation periods were also noted (Table 2, Fig. 41).

The following situations were presented to the animals on the different observation evenings:

1. The crest of the dam removed (Fig. 41 c–f); water running over the whole dam with a consequent lowering of the water level (Fig. 41 c).

The male first built with branches at the deepest part of the tank (Fig. 41 d), then continued building until the crest of the dam was level from wall to wall in the tank (Fig. 41 e). He then packed

---

the deep part of the tank (d), the construction is extended across the tank (e). *f.* The upstream side of the dam during the following morning (20.9). The dam is tightended with earth. *g.* In the evening 20.9. A furrow is made in the crest of the dam (water level 1). The crest was later removed (water level 2). *h.* In the evening 21.9. The water is raised from level 1 to level 2. *i.* In the evening 22.9. The water was slightly raised. When the beavers had heightened the dam (water level 1) a furrow was made in the crest (water level 2). The crest was later removed. *j.* In the morning 23.9. The beavers have heightened the dam to the level of the living box. *k.* In the evening 25.9. The water level is lowered. *l.* In the evening 9.10. The water level is lowered with a siphon. *m.* In the evening 11.10. A furrow in the crest is covered with a glass dome. A furrow is made in the crest above the water. *n.* In the evening 20.10. A furrow in the crest is covered with a netting. A furrow is made in the crest above the water. — *Experiment med dammbyggnadsbeteendet i strömakvariet. (Tabell 2.) Dammens uppströmssida (a) och längdsnitt genom strömakvariet (b) före experimenten. Dammens uppströmssida under första kvällen (19.9, c–e), följande morgon (20.9, f), under kvällarna 20.9 (g), 21.9 (h) och 22.9 (i), morgonen 23.9 (j) och under kvällarna 25.9 (k), 9.10 (l), 11.10 (m) och 20.10 (n). Horisontella pilar markerar vattenstånd respektive bolådans nivå. Vertikala pilar markerar utlösande och riktande stimuli.*

fine material along the whole dam and also continued to fasten branches and long sticks in the bottom on the downstream side. The water then passed by the walls of the tank (Fig. 41 *f*).

By morning after the first night the water had almost reached the straight and level top of the dam.

The animals began building as soon as they happened to come close to the dam. If the height of the dam was only moderately reduced by removing the crest they often showed a delayed response. After they had been close to the dam in such situations they would first eat for a long while, then suddenly, stop eating, swim straight to some material in the tank and transport it to the dam. Although the initial response had been weaker, the end result was often the same.

When, on one occasion, the female was sitting on the dam she continually stopped eating, listened and looked as though she was going to build. It seemed as though she was receiving acoustic stimuli from the running water, and this activated building behaviour more and more, balancing the feeding behaviour. Finally the dam building behaviour "exploded" in intense building on the dam.

Leaks in the dam below the surface never released dam building behaviour. If, however, water was spurting from leaks on the downstream side while the animals were there, they sometimes fastened sticks and pushed fine material into the leak. Building on the downstream side, however, was always of short duration and never gave lasting results.

2. Water rushing through a furrow in the top of the dam, with consequent lowering of the water level (Fig. 41 *g*, *i*).

The furrows never resulted in as strong a current over the dam as in experiment (1). The first transport of material towards the dam was released before any of the animals had been in immediate contact with the furrow. Thus on this occasion only acoustic and visual stimuli can have contributed to the directing stimulus situation. Later the animals sometimes investigated the water in the furrows with their noses. The furrows were always tightened during the observation time.

3. The rate of inflow increased, with consequent rise in water level; water overflowing the whole dam (Fig. 41 *h*).

Dam building activity was also released by water overflowing the dam owing to a rise in the water level. Since the dam itself was unchanged in these experiments, stimuli from indentations in the crest cannot have contributed to the releasing of dam building activity.

If a layer of wood-wool was placed on the top of the dam so that the water trickled quietly through it, dam building activity was not released.

The more swiftly the water ran over the dam, the faster dam building activity was released. The flow over the dam was always stopped within the observation hours.

In these situations, once the water had stopped running over the dam the animals began to build on the nesting box. Sometimes they alternated building on the dam and on the nesting box, which on occasion resulted in a repeated carrying of material between the lodge and the dam, which was hardly rational. Presumably this behaviour was caused by the unnaturally short distance between the lodge and the dam. As mentioned earlier (page 182) observations in the enclosures also indicate that lodge building behaviour is activated when the water level rises.

Intense digging activity was released when, at the end of these experiments, the level was raised until water began pouring into the lodge. Under natural circumstances this may have the function of lengthening the tunnels, to bring them again above the surface of the water.

During the night after the water level had been raised as high as possible the animals transported more material to the box than during all the previous evenings together.

4. Inflow decreased and water level consequently lowered (Fig. 41 *k*).

The flow of water through the tank was decreased as much as possible before the animals came out of the nesting box; the water level therefore dropped considerably, and the sound of running water was lessened.

When the female came out she immediately inspected the dam. Then, for a long period, she alternately ate and inspected the dam before she suddenly fetched material and packed it on the upstream side of the dam.

Dam building activity was presumably activated by visual and tactile stimuli from the indentations in the dam which were exposed when the water level was lowered.

5. The water siphoned past the dam and water level thus lowered (Fig. 41 *l*).

In these experiments water passed the dam only through the siphon.

This opened into the drain of the stream tank, so there was no sound of water passing the dam. The beavers pushed mainly fine material on to the upstream side.

Sometimes they showed a delayed response. For example, they went back into the box after they had seen the dam for the first time that evening, but when they came out again after about 30 minutes they swam straight to it, taking material with them. The animals never reacted to the water being sucked into the siphon at the base of the upstream side of the dam.

6. Water rushing through a furrow in the top of the dam; water level consequently lowered. The furrow was covered by a glass dome with a wooden frame level with the top of the dam. In this way the noise of the water flowing over the dam was considerably reduced. A furrow was later made in the top of the dam above the surface of the water, about 0.5 metres from the glass dome (Fig. 41 *m*).

The animals did not build on the dam until the dry furrow was made but this they filled in as soon as they caught sight of it.

While they were working in the dry furrow the animals sometimes suddenly listened, turning their heads towards the downstream side of the glass cap, where the water was gushing out from the furrow beneath it. Both animals then built alternately there and in the dry furrow until the dry furrow was filled in, and the noise of the water spurting from the downstream side of the dam had been reduced. They did not react at all either to the water which was running turbulently under the glass cap on the top of the dam, nor to the water being sucked in below the surface on the upstream side. The building on the downstream side never resulted in a higher water level. As the frame of the glass cap was horizontal and level with the crest of the dam, the beavers could not see any break in the horizontal line of the dam when they faced it from the upstream side. The animals did not build in the dry furrow when the glass cap was placed over it.

The fact that the beavers did not build on the glass dome in the normal way, from the upstream side of the dam, was apparently due to the diminished stimuli from running water and the lack of visual stimuli from a break in the crest of the dam.

7. The glass dome was exchanged for a fine netting. A furrow in the top of the dam above the surface of the water (Fig. 41 *n*).

The human ear distinctly perceived the noise of the water running under the netting, but the water was not visible through the netting

Fig. 42. The beavers have built a "dam" high above the water at the spot where the sound of water running far below under the artificial dam was loudest. — *Bävrarna har byggt en »fördämning» högt ovanför vattnet där ljudet av vattenströmmen djupt under den artificiella fördämningen var starkast.*

and there was no break in the crest of the dam. As soon as the female came near the dam she placed a burden of building material on the netting, apparently hearing the water running beneath. Then she built repeatedly in the dry furrow. The animals did not build in the dry furrow when the netting was placed over it.

This experiment too supports the theory that acoustic stimuli from water running over the dam and visual stimuli from a break in the crest of the dam release and direct dam building activity. Unfortunately the experiments had to be interrupted when the female was accidentally killed and the male then stopped building.

Experiments in the enclosures. In the autumn of 1962 the brook outlet from the enclosure (no. 3) of the wild, experienced beavers was led through a pipe under the dam. The water flow could be heard most clearly immediately above the middle of the dam. The beavers built intensely here during the entire autumn, in spite of the fact that they never came into contact with, or saw the running water or any break in the crest of the dam. A "beaver dam" soon appeared

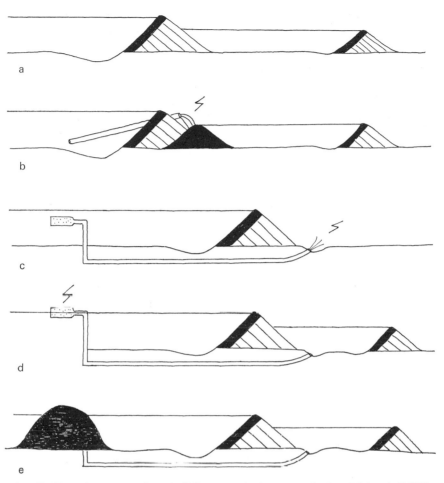

Fig. 43. Experiments on dam building. Freely interpreted after Richard (1967). *a.* The beavers have built two dams, working in the normal way from the upstream side. *b.* When a pipe was inserted in the primary dam the beavers shoved and pushed earth towards the acoustic stimuli from water gushing from the outlet, working from the down stream side. *c.* Water is spurting from the outlet of a pipe, buried under the dam. *d.* The beavers have built a dam, working in the normal way from the upstream side, at the source of the acoustic stimuli (the outlet of the pipe). This dam functions normally and has removed the primary acoustic releaser. The inlet of the pipe, equipped with a strainer, was then moved to the surface; the water is therefore sucked noisily into the inlet. *e.* The beavers shoved and pushed earth towards the spot on the bottom where the acoustic stimuli from the inlet of the pipe were strongest. Thus the inlet was finally plugged. If the inlet of the pipe was placed near the dam, the building behaviour, released by the acoustic stimuli, was directed towards the dam and this too finally resulted in the

well above the water surface and on top of the artificial dam. It was highest in the middle, i.e. above the place where the sound of the rushing water was heard most plainly. The beavers smoothed the side facing the water with mud (Fig. 42). This experiment shows that the beavers react to the sound of rushing water and direct the building towards the source of the sound. It also shows that the profile of the top of the dam is determined mainly by this acoustic stimulus, i.e. the top of the dam is even and level under normal conditions, since the animals build along it until the sound of the water flowing over the dam ceases.

During a series of experiments a pipe, from which water ran into the pond, was placed at different points within two of the enclosures (no. 2 and no. 3). If the pipe was placed on the dam, the animals (3 A and three adult, experienced beavers) always pushed material from the bottom of the pond towards the gushing water during the very first night, but when the pipe was placed at a distance from the dam, the animals did not react. In these cases the dam building behaviour had obviously been already localized to the dam during the earlier activities of the animals in the enclosure.

A loudspeaker was placed at different points on or near the dam in the enclosure, and various sounds reproduced by a tone generator, a tape-recorder or an electric razor. The animals reacted to the recorded sound of running water as they would to the natural sound, but also

---

inlet being plugged. In some other experiments the beavers showed a tendency to continue building at a place where building behaviour earlier had been successful, including when the releasing stimuli were removed a certain distance from the dam.
— *Experiment med dammbyggnadsbeteendet. Fritt tolkat efter Richard (1967). a. Bävrarna har byggt två dammar på normalt sätt, arbetande från uppströmssidan. b. När ett rör stacks in genom den primära dammen sköt bävrarna jord mot ljudet från vatten som strömmade ut ur rörets mynning, arbetande från nedströmssidan. c. Vatten sprutar ut ur ett rör som grävts ner under dammen. d. Bävrarna har byggt en damm på normalt sätt från uppströmssidan vid källan till de akustiska retningarna (rörets utlopp). Denna damm fungerar normalt och har avlägsnat de primära utlösande retningarna. Rörets inlopp, som försetts med en sil, flyttades sedan upp till vattenytan. Vatten sugs därför ljudligt in i inloppet. e. Bävrarna sköt ihop jord mot den plats på botten där de akustiska retningarna från rörets inlopp var starkast. På det sättet täpptes slutligen inloppet till. Om rörets inlopp placerades nära dammen riktades det av de akustiska retningarna utlösta byggnadsbeteendet mot dammen och detta resulterade också till slut i att inloppet täpptes till. I en del andra experiment visade bävrarna en tendens att fortsätta bygga på en plats där byggnadsbeteende tidigare gett positivt resultat även om de utlösande retningarna flyttades inom ett visst avstånd från dammen.*

identically to the sound of the electric razor. On the other hand they did not react at all to constant tones of any frequency.

In 1963 the animals in the enclosures (3 A and a wild female in the one (no. 2), and two beavers captured as adults in the other (no. 3)) had stopped building at their dams as early as the middle of September and then built intensely on their lodges all the autumn. However, they would always build against the sound of running water from a loud-speaker, or water running from a hose on the dam. Owing to low rainfall the inflow into the pond was so slight that it could be led past the dam in such a way that the beavers could not hear it running. These experiments were performed every night during the autumn. The animals pushed one or more bundles of fine material against the relevant stimulus.

Every night during November the animals were presented with the sound of running water from which certain frequencies had been re-moved by means of filters. The frequencies on the tapes that were used are listed below.

Tape	Frequencies (Hz)	Tape	Frequencies (Hz)
1 A	125– 500	1 B	125– 250
2 A	250– 1 000	2 B	250– 500
3 A	500– 2 000	3 B	500– 1 000
4 A	1 000– 4 000	4 B	1 000– 2 000
5 A	2 000– 8 000	5 B	2 000– 4 000
6 A	4 000–15 000	6 B	4 000– 8 000
		7 B	8 000– 16 000

One of the tapes from the A series was played every night from a loudspeaker on the top of the dam from about 7 o'clock in the evening until about 7 o'clock next morning. Each tape was played during three nights. The animals only reacted to the tapes 3 A and 4 A, which both contained the frequencies 1 000–2 000 Hz. When these tapes were played the animals directed all their dam building activity towards the loudspeaker and repeatedly built against it during the night.

During the spring of 1964 there was some building activity in one of the enclosures, where the female was pregnant. When the tapes from the B-series were played on the dam only tape 4 B (1 000–2 000 Hz) released building activity.

The experiments were continued during the autumn of 1964, but then precipitation was so high that we could not prevent the water from rushing noisily through the dam. The animals built against the natural sound, which was stronger than the artificial one. They also built against all the different frequencies from the loudspeaker, but it is possible that this behaviour was released by the natural sound of running water.

An analysis of the releasing acoustic stimulus situation is highly desirable.

Richard (1960, 1961, 1967) studied a pair of adult beavers in a large enclosure. In 1962 and 1964 he observed our experiments and then made a series of experiments in France which according to our view supports the conclusions made in this investigation. Richard's experiments (1967) are freely summarized and interpreted in Fig. 43.

On the basis of his observations on the dam building behaviour Richard stresses the importance of tactile stimulation for the orientation of the building activities. Although experimental evidence is lacking it is highly probable that tactile stimuli can play an essential role but Richard's conclusion that the beaver does not react to visual stimuli while building because building is usually performed in the dark is not valid. Experiments indicate (pages 178, 194) that building can be directed towards visual stimuli from indentations, depressions or holes even if the animal never had the opportunity to receive any tactile information from the source of these stimuli or from running water.

Richard's description of the sequence of dam building activities and the use of different kinds of building material completely differs from our observations.

Dambuilding at first encounter with running water. Animal 2 D was kept in a large wooden tank of still water. When once, during its first winter, it gnawed a hole in the wall of the tank at water level so that the water began pouring out, the beaver started pushing material against one of the corners near the leak. Obviously the young beaver, which had never experienced flowing water, had nonetheless the ability to react to the sound of the rushing water and to the visual stimuli from an indentation near the releasing stimulus, by directing building activity towards it at first encounter.

During its second autumn a loudspeaker was placed behind one of the short walls of the tank. The animal began to fasten sticks and push fine material against this wall as soon as the recording of running water was played. Building activity was often directed towards the corner nearest the source of the sound (Fig. 44). In the beginning a

Fig. 44. Animal 2 D built in still water during its second autumn at the first encounter with the sound of running water from a loudspeaker. *a.* The first night the loudspeaker was placed to the right behind the short wall of the water tank. Building was directed towards the corner nearest the source of the sound. *b.* When the loudspeaker was put behind the middle of the short wall the animals built in both corners. — *Bäver 2D byggde i stillastående vatten så snart den under sin andra höst hörde ljudet av strömmande vatten från en högtalare. a. Under första natten placerades högtalaren bakom högra delen av bassängens kortsida. b. När högtalaren flyttades till kortsidans mitt byggde bävern i båda hörnen.*

piece of dark-coloured paper near the loudspeaker had the same effect as the corner (cf. lodge building, page 179).

2 D reacted in this way every night, although it did not achieve any result in the way of a higher water level. The wall of the tank was covered with netting which enabled the animal to fasten building material to it.

We always changed the water in the tank during the day, when the beaver was asleep. However, when the animal had been building against the artificial sound of water for 15 nights we left a water hose on in the tank, close to the corner nearest the lodge. The following night the beaver pushed material towards the mouth of the hose and it continued building until the whole corner was filled with building material. This building gave a positive result in that it offered an easy passage from the water direct on to the lodge, which the animal then used when it transported lodge building material. At this point it stopped building against the artificial sound and concentrated all building activity in the tank to the corner nearest the lodge.

In the enclosures the animals seemed to learn not to react to the rushing water at the inlet of the brook, since they did not get any positive result when working there. On the other hand, they always built against the sound of running water on or near the artificial dam, even if the work did not result in a higher water level, since here the animals succeeded in fastening the material in the right way.

Observations of free animals also indicate that the beavers continue to build at a place where dam building activity has been released only if they get a result in the form of a lasting construction and a decrease in noise from running water. The observations of the early dam building behaviour in animal 2 D indicate that also other kinds of results, which are of positive value to the animal, can serve as reinforcement.

Hartman and Rice (1963) have investigated the characteristics of one young beaver in instrumental learning and operant conditioning. The results of these experiments show that the beaver does not differ from other rodents except that the beaver works to avoid running water.

Hartman and Orth (1968) also worked with one young beaver, and came to the conclusion that the beaver works to avoid variations in sound, especially the sound of flowing water.

REFERENCES

Bailey, V. 1922. Beaver habits, beaver control and possibilities in beaver farming. U.S. Dept. Agr. Bull. 1078: 1–29.

— 1926. How beavers build their houses. J. Mamm. 7: 41–44.

— 1927. Beaver habits and experiments in beaver culture. U.S. Dept. Agr. Bull. 21: 1–39.

Bradt, G. W. 1938. A study of beaver colonies in Michigan. J. Mamm. 19: 139–162.

Dugmore, A. R. 1914. The romance of the beaver. London.

Eibl-Eibesfeldt, I. 1950. Beiträge zur Biologie der Haus- und der Ährenmaus nebst einigen Beobachtungen an anderen Nagern. Z. f. Tierpsychol. 7: 558–587.

— 1951 b. Gefangenschaftsbeobachtungen an der persischen Wüstenmaus (Meriones persicus). Ein Beitrag zur vergleichenden Ethologie der Nager. Z. f. Tierpsychol. 8: 400–423.

— 1958. Das Verhalten der Nagetiere. Handbuch der Zoologie 8, 12, 10 (13): 1–88.

Hartman, A. M. and Orth, F. 1968. Analysis of conditions leading to the regulation of water flow by a beaver. Idaho State Univ. (In press.)

Hartman, A. M. and Rice, H. K. 1963. Observations of a beaver, Care, 24-hour activity records and performance on simple learning tasks. The Psychol. Rec. 13: 317–328.

Hinze, G. 1950. Der Biber. Berlin.

Pilleri, G. 1960. Biber. Die Umschau in wissenschaft und Technik 14: 420–424.

Roth-Kolar, H. 1957. Beiträge zu einem Aktionssystem des Aguti. Z. f. Tierpsychol. 14: 361–375.

Richard, P. B. 1958. Les Pattes du Bièvre. Mammalia 22: 566–574.

— 1960. Essai préliminaire sur l'adaption à des problèmes simples chez le Castor (Castor fiber). J. de Psychologie normale et pathologique 4: 421–430.

— 1961. Le déblaiment chez le Castor: rapport entre le déblaiment et la réalisation des canaux et des barrages. Vie et Milieu XII: 507–515.

— 1967. Le determinisme de la construction des barrages chex le castor du Rhóne. La Terre et la Vie 4: 339–470.

Shadle, A. R. 1930. An unusual case of parturition in a Beaver. J. Mamm. 11: 483–485.

— 1956. The American Beaver. Animal Kingdom. 59, 60, 61.

Wilsson, L. 1959. Inventering av bäverstammen vid Faxälven med biflöden inom Ångermanland. (An inventory of the beaver population in Ångermanland, Sweden. In Swedish.) Svensk Jakt 6.

— 1960. Observations on inborn and acquired behaviour patterns in the European beaver. (Castor fiber L.) EOARDC Contr. No. Af 61 (052)–195. Technical Summary Report Nol. Appendix III.

— 1962. Inborn and acquired behaviour patterns in the beaver. (In Swedish with an English summary.) Svensk Naturvetenskap, pp. 305–319.

AUTHOR CITATION INDEX

Addicott, A., 147
Aitken, E. H., 80
Akerman, C., 166
Ali, S. A., 373
Allee, W. C., 19, 233
Altmann, S. A., 222
Amadon, D., 149
Ambrose, E. J., 316
Andrew, R. J., 222
Andrews, E. A., 129
Appleman, P., 19
Apstein, C., 166, 316
Archer, A. F., 166
Arey, L. B., 129
Armstrong, E. A., 147, 278
Asdell, S. A., 222
Ashby, E., 270
Austin, O. L., 147
Avis, V., 222
Azuma, S., 222

Baerends, G. P., 279
Bailey, R. E., 279
Bailey, V., 403
Baldwin, S., 278
Ball, W. P., 129
Bally, G., 222
Balogh, I. J., 166
Bamford, C. H., 316
Barfield, R. J., 148, 362
Barnett, L. B., 289
Bartholomew, G. A., 148, 233, 251, 278, 279, 289
Bartlett, A. D., 222

Bates, R. W., 334
Batra, S. W. T., 203
Baum, R., 166, 168
Beatty, H., 222
Beg, M. A., 224
Bell, A. L., 316
Benson, F. M., 361
Bent, A. C., 148, 279
Bentley, J. R., 251
Benton, J. R., 166
Berger, A. J., 149, 278
Berghe, L. van den, 222
Berkson, G., 222, 223
Bernstein, I. S., 222
Betten, C., 184
Bingham, H. C., 222
Birch, H. G., 222
Blackwell, J., 316
Bohart, G. E., 203
Bolwig, N., 222
Bond, J., 149
Botosaneanu, L., 184
Bouillon, A., 94, 133
Bourgin, P., 167
Bradt, G. W., 403
Brant, D. H., 251
Braunitzer, G., 166, 316
Bristowe, W. S., 166
Brody, P. N., 149, 333
Brody, S. B., 251
Brooke, J., 222
Brooks, D. S., 148, 362
Brown, H. T., 189
Budd, A., 222

Bugnion, E., 129
Butler, R. A., 222

Cade, T. J., 148, 251
Carpenter, C. R., 222, 333
Chapin, J. P., 94, 148, 361
Chapin, J. T., 129
Cheng, M-F., 302
Chiang, M., 222
Christiansen, A., 166
Chun, C., 80
Cleveland, L. R., 129
Cloudsley-Thompson, J. L., 166
Collias, E. C., 94, 148, 233, 279, 289, 302, 361, 362, 373
Collias, N. E., 19, 94, 148, 222, 233, 279, 289, 302, 361, 362, 373
Comstock, J. H., 166, 316
Copeland, M., 184
Corley Smith, G. T., 279
Couper, W., 80
Cowles, R. B., 129
Craig, W., 333
Crome, W., 166
Crook, J. H., 148, 362
Cross, E. A., 203
Cross, H. F., 222
Crowell, S., 184
Cullen, E., 148
Cummins, K. W., 289
Cunningham, J. T., 129

Dabelow, S., 167
Darby, C. L., 222
Darchen, R., 19
Darlington, P. J., Jr., 129
Darwin, C., 5, 19, 129, 233
Davies, S. J. J. F., 148
Davis, E. A., Jr., 251
Davis, M. B., 184
Dawson, W. R., 148, 251
De Bont, A.-F., 251
de Bournonville, D., 222
Desneux, J., 129, 133, 189
DeVore, I., 222, 224
DeWilde, J., 167
Dilger, W. C., 148, 373
Dixon, C., 148
Dixon, W. J., 373
Dodd, F. P., 80
Doflein, F., 80
Downing, R. L., 148
Drent, R. H., 279
Dreyer, W. A., 129
Dudley, P. H., 129
Dugmore, A. R., 403

Eberhard, M. J. W., 6, 19
Ehrhorn, E. M., 129
Eibl-Eibesfeldt, I., 403
Eklund, C. R., 278
Elder, J. H., 224
Elliott, A., 316
El-Wailly, A. J., 279
Emerson, A. E., 6, 19, 94, 129, 130, 133, 148, 189
Emlen, J. T., Jr., 148
Enders, R. K., 130
Escherich, K., 80
Escombe, F., 189
Evans, F. E., 148
Evans, H. E., 6, 19, 94
Evans, M., 19, 94
Exline, H., 167

Fabre, J. H., 19, 48, 53
Farner, D., 333
Fielde, A. M., 80
Forel, A., 80, 130
Frank, L. K., 222
Freisling, J., 167
Friedmann, H., 148, 289, 362
Frisch, K. von, 6
Frith, H. J., 148, 233
Froggatt, W. W., 130
Frumkes, P. J., 148, 362
Fuller, C., 130

Gartlan, J. S., 222
Gemsell, C. A., 333
Gerhardt, U., 167
Gertsch, W. J., 167
Gilbert, C., 222
Gilbert, L. I, 184
Gilliard, E. T., 148
Gillman, J., 222
Goodall, J., 222
Grandi, G., 203
Grandjean, F., 167
Grassé, P. P., 130, 133, 189, 302
Green, E. E., 80
Grzimak, B. von, 362

Haartman, L. von, 148
Haddow, A. J., 222
Hall, K. R. L., 222
Hanby, W. E., 316
Hancocks, D., 6
Hanks, J. B., 303
Hansen, E. W., 223
Hardy, J. W., 148
Hare, L., 130
Harlow, H. F., 222, 223
Harlow, M. K., 223

Harper, E. H., 333
Harris, G. W., 333
Harrisson, B., 223
Hart, J. S., 251
Hartman, A. M., 403
Harvey, P. A., 130
Hasluck, P. N., 362
Haverschmidt, F., 148
Hayes, C., 223
Hayes, K. J., 223
Hegh, E., 130
Heijkenskjöld, F., 333
Heroux, O., 251
Herrick, F. H., 148
Herscher, L, 223
Hill, G. F., 130
Hinde, R. A., 148, 223, 224, 302, 303, 333, 362, 373
Hindwood, V. A., 148
Hingston, R. W. G., 19, 130, 303
Hinze, G., 403
Holdaway, F. G., 130
Hollis, J. H., 224
Holm, A., 167
Holmes, S. J., 130
Holmgren, N., 130
Hoof, J. A. R. A. M. van, 223
Hooper, P. D., 270
Hopfmann, W., 167
Horváth, O., 148
Howell, T. R., 148, 233, 278, 279, 289
Howse, P. E., 6
Huber, F., 19
Huber, P., 80
Hudson, J. W., 289
Hudson, W. H., 148
Huggins, R. A., 278
Hutchinson, C. E., 167
Huxley, J., 19

Imanishi, K., 223
Immelmann, K., 148
Imms, A. D., 130
Irving, L., 148, 278
Itani, J., 223
Izawa, K., 223

Jacobson, E., 80
Jahn, L. R., 148
Jay, P., 223
Joustra, T., 279

Kaestner, A., 167
Kahler, S. G., 303
Karawaiew, W., 80
Kaston, B. J., 94

Kaufmann, J. H., 223
Kawamura, S., 223
Kemner, N. A., 130
Kendeigh, S. C., 148, 149, 278, 279, 289
Kerfoot, W. B., 203
Khaskin, V. V., 279
Kiel, W. H., Jr., 149
Kimzey, S. L., 289
Kinder, E. F., 233
King, J. R., 279
Kinsey, A. C., 130
Kluijver, H. N., 279
Klüver, H., 223
Koenig, M., 316
Kofoid, C. A., 130
Koford, C. B., 223
Köhler, W., 223
Kohts, N., 223
Kooij, M., 223
Kortlandt, A., 223
Kossack, C. W., 278
Krog, J., 148, 278
Kullman, E., 167
Kummer, H., 223

Lack, D., 149, 278
Lahr, E. L., 303, 334
Lange, R. B., 203
Lau, D., 19
Lawick-Goodall, J. van, 223
Ledoux, A., 19
Lehmensick, R., 167
Lehrman, D. S., 149, 333
Lenormant, H., 316
Lewis, F., 270
Light, S. F., 131
Lindauer, M., 19, 20, 234
Linsdale, J. M., 149
Lisk, R. D., 303
Liversidge, R., 149
Loewenstein, W. R., 316
Lofts, B., 333
Loizos, C., 223
Longman, H. A., 167
Lorenz, K. Z., 223
Lundy, H., 278
Lüscher, M., 234

McCook, H. C., 167
MacGinitie, G. E., 6
MacGinitie, N., 6
McKeown, K. C., 167
MacKinnon, J. R., 94
Maclean, G. L., 289
MacNeill, C. D., 203
Maeterlinck, M., 131

Malyshev, S. I., 203
Marais, E. N., 373
Marler, P., 223
Marples, B. J., 167
Marples, M. J., 167
Marshall, A. J., 149, 333
Marson, J. E., 167
Martin, H., 19, 20
Martinez-Vargas, M. C., 303
Mason, W. A., 222, 223, 224
Massey, F. J., 373
Matthews, L. H., 333
Mattingley, A. H. E., 270
Mayer, G., 308
Mayr, E., 20, 149
Medway, Lord, 149
Meier, F., 308
Menzel, E. W., 224
Merfield, F. G., 224
Michener, C. D., 6, 20, 94, 189, 203, 303
Miller, H., 224
Miller, J. W., 303
Millot, J., 167
Milne, L. J., 184
Milne, M. J., 184
Mjöberg, E., 131
Monterosso, B., 167
Montgomery, T. H., 167
Moore, A. U., 223
Moore, R. T., 279
Moreau, R. E., 149
Morgan, L. H., 303
Morrison, P. R., 251

Napier, J. R., 224
Needham, J., 131
Nice, M. M., 149, 373
Nickell, W. P., 149
Nielsen, A., 184
Nielson, E., 167
Nissen, H. W., 224
Noirot, C., 94, 133, 189
Norgaard, E., 167
Norris, A. T., 149

Ordway, E., 203
Orth, F., 403
Osborn, R., 224
Oshima, M., 131
Osten-Sacken, R. von, 80

Palmer, R. S., 149
Park, T., 129
Patel, M. D., 333
Paynter, R. A., 234
Peakall, D. B., 316, 317

Pearson, O. P., 149, 251, 279
Peters, H. M., 167, 303, 317
Petrunkevitch, A., 167
Petrusewiczowa, E., 308
Pilleri, G., 403
Pitman, C. R. S., 149
Pocock, R. I., 167
Postuma, K., 279
Preston, F. W., 149
Prychodko, W., 251

Reed, C. F., 317
Reitchensperger, A., 131
Rettenmeyer, C. W., 203
Reynolds, F., 224
Reynolds, V., 224
Ribbands, C. R., 20, 234, 303
Rice, H. K., 403
Richard, P. B., 403
Richmond, J. B., 223
Richter, G., 167
Riddle, O., 303, 334
Ridley, C. V., 80
Riopelle, A. J., 222, 224
Rivolier, J., 149
Roberts, N. L., 167
Rogers, C. M., 224
Rohlf, F. J., 279
Romanes, G. J., 94
Romanoff, A. J., 278
Romanoff, A. L., 278
Rösch, G. A., 303
Ross, H. H., 184
Ross, J. A., 270
Roth-Kolar, H., 403
Rowell, T. E., 223, 224
Rozen, J. G., 203
Rueping, R. R., 223
Ruschi, A., 149
Russell, J. A., 303
Ryser, F. A., 251

Sade, D. S., 224
Saeki, Y., 334
Sakagami, S. F., 203
Savage, T. S., 224
Saville-Kent, W., 80, 131
Savory, T. H., 167, 303
Saxon, S. V., 222
Schaller, G. B., 94, 224
Schiller, P. H., 224
Schmalhausen, I. I., 184
Schmidt, R. S., 133, 189
Schmidt-Nielson, B., 234
Schmidt-Nielson, K., 234
Schneiderman, H. A., 184

Schooley, J. P., 334
Schusterman, R. J., 222
Scott, J. P., 224
Sekiguchi, K., 167, 317
Serventy, D. L., 149, 270
Shadle, A. R., 403
Sharpe, L. G., 224
Shelley, F. W., 222
Sibley, C. G., 278
Siddiqi, M. R., 224
Sielmann, H., 149
Silver, R., 302
Silvestri, F., 131
Simonds, P. E., 224
Sjöstedt, Y., 131
Skaife, S. H., 303
Skead, C. J., 149
Skutch, A. F., 149, 278
Sladen, W. J. L., 149
Smith, L. G., 222
Smith, W. J., 149
Smythies, B. E., 149
Snyder, T. E., 131
Sokal, R. E., 279
Southwick, C. H., 224
Sparks, J., 224
Spencer, H., 131
Spencer-Booth, Y., 223
Spradbery, J. P., 6
Stein, R. C., 149
Stockhammer, K. A., 203
Stresemann, E., 149
Sudd, J. H., 6, 20
Swales, B. H., 149
Szlep, R., 167

Talbot, M. W., 251
Tanabe, Y., 334
Taylor, J. K., 270
Teale, E. W., 20
Theuer, B., 168
Thorpe, W. H., 224, 362
Tilquin, A., 168
Tinbergen, N., 223
Titus, E. S. G., 80
Toyoshima, A., 222
Trelease, W., 80

Uyemura, T., 168

Van der Kloot, W. G., 6, 184
Van Dobben, W. H., 149
Van Lawick-Goodall, J., 94
Van Someren, V. G. L., 149
Van Tyne, J., 149, 278
Victoria, J. K., 303

Wagner, H. O., 149
Walkinshaw, L. H., 149
Wallace, A., 224
Warburton, C., 317
Warren, E. R., 303
Warren, R. P., 333
Warwicker, J. O., 317
Washburn, S. L., 224
Wasmann, E., 80
Weismann, A., 131
Wellington, W. G., 234
Welty, J. C., 149
Wetmore, A., 149
Wheeler, W. M., 20, 80, 131
White, F. N., 233, 279, 209
Whitehouse, H. L. K., 278
Whiting, P. W., 131
Whitman, C. O., 149
Whittell, H. M., 149, 270
Wiehle, H., 168
Williams, C. M., 6, 184
Williams, G. C., 20
Williams, O. L., 131
Wilson, E. O., 6, 20, 94, 234, 303
Wilson, R. S., 317
Wilsson, L., 303, 403
Witschi, E., 149
Witt, P. N., 166, 168, 317
Wolff, D., 166, 316
Wortis, R. P., 149, 333
Wright, S., 20, 131
Wuycheck, J. C., 289
Wyman, J., 224

Yaginuma, T., 168
Yamada, M., 224
Yerkes, A. W., 224
Yerkes, R. M., 224

Zetek, J., 131
Zimmermann, R. R., 223
Zuckerman, S., 224

SUBJECT INDEX

Ant nests and accessory structures, 17–18,
  64–79
  Camponotus, 71, 76
  classification of nests, 198–199
  Formica, 69, 77
  loss of nest building ability, 75
  monodomous, 70
  Oecophylla, 18, 71–76, 78
  paths (trails), 77–78
  polydomous, 70
  Polyrachis, 71–72, 77–78
  tents (sheds) for aphids and coccids, 78–79
Apicotermes, 3, 87, 110, 116, 133 (see also
  Evolution, termite nests)
Apis mellifera (see Honeybee)
Araneus diadematus, 309–317
Araneus nauticus, 293–294
Avian incubation, 231, 271–279
  attentiveness
    and environmental temperature, 273–
      274, 276–277
    modulation of, 277–278
    and nest insulation, 274
    and sensory receptors, 277
  in different species, 271–272
  egg temperature and embryonic heat
    production, 275–276

Beaver
  building, 374–402
    dam, 298–299, 381–382
    dam-building behavior, 382–385
    experiments on dam building, 390–402
    lodges, 378–380
    motor patterns, 374–378
    ontogeny of dam-building behavior,
      385–390
  canals, 300
  meadows, 299–300
Bees, 191–202 (see also Honeybee)
  bumble bee (humble-bee), 37, 199
  mason bee, 15–17, 49–63
  Melipona (stingless bee), 37, 199–200
  social bees, 16–17, 198
  solitary bees, 15–17, 198
Building, mechanisms of, 301–302
  by *Araneus* spider, 294
  by beaver, 374–402
  effect of crowding, 293
  key stimuli, 301
  by ring dove, 318–332
  sequences, 292–294, 297, 301–302
  stigmergy, 292–294
  in termites, 293
  by weaverbird, 298–299, 355–361
Building behavior, development of, 4, 302
  endocrine glands, 296
  in honeybee, 295–296
  learning, 298, 302
  maturation, 295–296, 301, 304–308
  in mice, 297
  practice, 298, 302
  wax glands, 294
  in weaverbird, 297–298, 363–373
  in *Zygiella* spider, 295, 304–308
Building materials, 2–3, 152–154, 309–317,
  364–367
Burrows, 3, 88–89, 128, 196–198, 202, 227

Caddisworm, cases, 86–87, 91–92
Camponotus, 71, 76
Castor spp. (*see* Beaver)
Caterpillar, 23, 234–242
Chalicodoma muraria, 15–17, 49–63
Chimpanzee (*see also* Tool-using)
  nests, 204–212
    development of nest-making ability, 211
    nest-making behavior, 204–211
    sleeping and resting in, 211–212

Endocrine effects on building, 296–297
  birds, 296–297
  in mice, 297
Energetics
  and building behavior, 4, 36
  and nests (*see also* Homeostasis; Temperature, regulation)
    avian energetics, 227
    behavioral energetics, 228
Evolution, 1, 5, 17, 21 80, 82–202
  ant nests, 18
  Apicotermes nests, 185–190
    evolution and phylogeny of, 186–187
    functions, 187–188
  bee nests, 37–38, 87–90, 92, 191–203
    cell lining, 191–193
    cell shape, 195
    cocoons, 193–194
    food masses, 194–195
    homologies and constructs, 201–202
    nests, 196–200
    phylogeny of bees, 192–202
    summary, 202–203
  bird nests, 85–86, 92, 134–149
    classification of, 134
    compound nest, 144–145
    figures of weaverbird nests, 150–151
    megapodes, 135–136
    nesting in cavities, 136–138
    open nest, 138–141
    origin in evolution, 135–136
    roofed, 141–144
    and speciation, 145–146
    summary, 147
  caddisworm cases and nests, 86–87, 91–92, 173–184
    evolution of behavior patterns, 176–181
    in geological time, 183–184
    homology in behavioral steps, 181–182
    hormonal influence on behavior, 182–183
    morphological evolution, 174–176
    phylogeny, 178
  methods of study, 84

principles illustrated by architecture, 88–92
regressive (*see* Loss of ability to build)
spider webs, 86, 90–91, 152–172
  cocoon (egg sac), 158
  construction of orb web, 154–155
  irregular and sheet webs, 161–163
  loss of web, 162, 166
  orb webs, 163–166
  origin of silk, 153–154
  phylogeny of spiders, 155–156
  silk glands, 153
  silk-lined burrows, 159–160
  uses of silk by spiders, 152–153
  of young spiders, 158
termite nests, 82, 84, 87, 90–91, 95–133, 189–190 (*see also Apicotermes*)
  and behavior evolution, 97–98
  caste that constructs nest, 112–113
  divergence within genus, 119–124
  function of nests, 113–118
  Kalotermitidae, 99 101
  Mastotermitidae and Hodotermitidae, 101–103
  nomenclature changes, 132–133
  pre-isopteran nesting behavior, 98–99
  Rhinotermitidae, 103–104
  selection of nesting site, 118, 119
  summarized, 128–129
  Termitidae, 105–112
External construction
  classification of, 2–4
    ant nests, 70–71
    bird nests, 134
    by building materials, 2–3
    by function, 3–4
    spider silk, uses, 152–153
  function of, 227–233
  historical background, 1, 8–19

Formica (*see* Ant nests)

Group selection, 11–14, 39–41

Habitat control, 3–4, 227–233
Harvest mouse, metabolism, 243–251
  and activity, 248
  and body size, 243–244, 245
  and huddling, 247
  and insulation of fur, 247
  and insulation of nest, 247
  method of measurement, 243
  summary of bioenergetics, 250
  and temperature, 244–247

Homeostasis, and nests, 4 (*see also* Energetics, and nests; Temperature regulation)
  avian incubation, 231, 271–279
  harvest mouse, 243–251
  honeybee, 227
  kangaroo rat, 227
  mallee fowl, 252–270
  sociable weaver, 280–289
  tent caterpillar, 234–242
  termites, 230
Honeybee, 9–11, 27–38, 90, 200, 227, 229–230, 296

Instinct, 8–11, 22–42, 43–63
  Darwin's summary, 41–42
  defined, 15, 16, 22
  modern view, 16
  nest building as example, 335

Learning, and building, 298, 302
Leiopoa ocellata (*see* Mallee fowl)
Loss of ability to build, 42, 75, 79, 92, 111, 126, 134, 162, 166, 177–178, 188

Macrotermes, 3, 111–112, 292–293
Malacosoma spp. (*see* Tent caterpillars)
Mallee fowl, mound, 135–136, 252–270
  construction, 252–253
  nests of megapodes, 135–136, 252
  photographs, 267–269
  temperature
    experiments in control of, 256–260
    factors affecting, 254–256
    maintenance of, 260–267
Maturation, 295–296, 301, 304–308
Melipona spp., 37, 199–200

Nasutitermes exitiosus, 3, 110
Natural selection, 8–13, 41–42, 125–128
Nest photographs
  ants, 65, 66, 68, 69
  Apicotermes, 189–190
  bees, 197, 200
  chimpanzees, 207
  mallee fowl, 268–269
  sociable weaver, 150, 282
  tent caterpillars, 235–237, 240–241
  termites, 100, 105, 107, 109, 110, 112–113, 115, 116–117, 120, 122, 124, 189–190
  village weaverbird, 337–339, 372
Nests, 2
  ants, 95–133, 185–190
  apes, 92–93
  Apicotermes, 185–190

bees, 27–38, 191–203
birds, 42, 85–86, 92, 134–147, 231–232, 271–279, 280–289, 318–325
chimpanzee, 92–93, 203–212
defined, 134
harvest mouse, 243–251
social insects, reviews, 2
termites, 3, 95–133, 185–190
weaverbirds, 88–89

Oecophylla, 18, 71–76, 78

Pan troglodytes (*see* Chimpanzee)
Philetairus socius (*see* Sociable weaver)
Physiological effects of nest, 296–297
Polyrachis, 71–72, 77–78
Procubitermes spp., 3, 122–123

Reithrodontomys megalotis (*see* Harvest mouse)
Reprints, selection of, 5
Ring dove, reproductive cycle, 296, 318–332
  description of cycle, 318–321
  effects of prolactin, 331–332
  external stimuli to hormones, 328–331
  external stimuli to incubation, 321–325
  hormonal stimuli to incubation, 326–327

Sociable weaver, nest, 145, 147, 280
  effects of nest size, 284–285
  evolution, 145, 147
  function, 288–289
  measurement of temperature, 280–281
  and metabolism, 286–288
  photograph, 151, 282
  and temperature control, 232–233, 233, 283–285
Spider
  Araneus diadematus, 309–317
  Araneus nauticus, 293–294
  Zygiella x-notata, 304–308
Spider silk, snythesis, 309–317
  in *Araneus diadematus*, 309–317
  secretory stage, 311, 312
  silk glands, 309–311
  silk sensory receptor, 313, 315–316
  stimuli to synthesis, 311
  synthesis of new protein, 311–313
Spider webs (*see also* Evolution, spider webs)
  photographs, 168–172, 305–307
Stigmergy
  spider, 293–294
  termites, 292–293
  weaverbird, 297
Streptopelia risoria (*see* Ring dove)

Temperature regulation, 226, *(see also* Energetics, and nests; Homeostasis)
  honeybee, 229–230
  leading investigators, 227
  mallee fowl, 231
  rat, 230
  sociable weaver, 232–233
  tent caterpillar, 234–239
  village weaverbird, 231–232, 271–279
Tent caterpillars, 234–242
  leadership, 241–242
  orientation, 239–240
  temperature regulation, and tent, 234–239
Textor cucullatus (see Village weaverbird)
Tool-using, by chimpanzees, 93–94, 213–221
  aimed throwing, 213–214
  for ant and termite feeding, 215–217
  development and evolution, 219–221
  hitting, 215
  leaves, body wiping, 219
  leaves, drinking, 218–219
  as olfactory aids, 217–218
  photographs, 207
  throwing behavior, 214–215

Trichoptera (see Caddisworm; Evolution, caddisworm cases and nests)

Ventilation of nests, 187–188, 229–230
Village weaverbird
  building mechanisms, 335–361
    description, 338–342
    as example of instincts, 335
    experimental analysis, 343–358
    summarized, 360–361
  energetics, 228, 231, 273–278
  ontogeny of building, 363–373
    influence of other birds, 371–372
    manipulation, 363–364
    ontogeny of weaving, 367–371
    selection of materials, 364–367
    summarized, 372–373

Wasps, 16, 46–48
Web-building ability, maturation of, in *Zygiella x-notata*, 304–308
Webs, 2, 88–89 (see also Evolution, spider webs; Spider webs; Stigmergy)

Zygiella x-notata, 304–308

About the Editors

NICHOLAS E. COLLIAS is Professor of Zoology in the Department of Biology at the University of California, Los Angeles, where he teaches and directs research on the subject of animal behavior. He graduated Phi Beta Kappa in 1937 from the University of Chicago, where he received a Ph.D. in Zoology in 1942. He has served on the faculties of Amherst College, the University of Wisconsin, Illinois College, and since 1958 at UCLA.

Professor Collias was formerly Chairman of the Division of Animal Behavior of the American Society of Zoologists and Chairman of the Section of Animal Behavior and Sociobiology of the Ecological Society of America. He is a Fellow of the Animal Behavior Society, which he helped to found. He is also a Fellow of the American Ornithologists Union, and his principal research interests and publications deal with the behavior of birds. He and his wife have written a book entitled *Evolution of Nest-Building Behavior in the Weaverbirds (Ploceidae),* published by the University of California Press and based on field work in Africa and Asia.

ELSIE C. COLLIAS is Research Associate in Ornithology of the Los Angeles County Museum of Natural History and also Research Associate in Zoology with the University of California at Los Angeles. She received her B.A. in 1942 from Heidelberg College where she was valedictorian, and her Ph.D. in Zoology from the University of Wisconsin in 1948. She has been an entomologist for the U.S. Public Health Service, and has taught at Heidelberg College, Illinois College and the University of Wisconsin. She is a charter member of the Animal Behavior Society. Most of her publications are reports of research on the behavior of weaverbirds and other birds done over the last 25 years.